HUMANS

— *in the* —

LANDSCAPE

An Introduction to Environmental Studies

Kai N. Lee
WILLIAMS COLLEGE AND THE DAVID AND LUCILE PACKARD FOUNDATION

William R. Freudenburg
LATE OF THE UNIVERSITY OF CALIFORNIA, SANTA BARBARA

Richard B. Howarth
DARTMOUTH COLLEGE

W. W. NORTON & COMPANY

NEW YORK • LONDON

W. W. Norton & Company has been independent since its founding in 1923, when William Warder Norton and Mary D. Herter Norton first published lectures delivered at the People's Institute, the adult education division of New York City's Cooper Union. The firm soon expanded its program beyond the Institute, publishing books by celebrated academics from America and abroad. By midcentury, the two major pillars of Norton's publishing program—trade books and college texts—were firmly established. In the 1950s, the Norton family transferred control of the company to its employees, and today—with a staff of four hundred and a comparable number of trade, college, and professional titles published each year—W. W. Norton & Company stands as the largest and oldest publishing house owned wholly by its employees.

Editor: Aaron Javsicas
Managing Editor, College: Marian Johnson
Associate Managing Editor, College: Kim Yi
Copy Editor: JoAnn Simony
Editorial Assistant: Catherine Rice
Senior Production Manager: Benjamin Reynolds
Photo Editor: Michael Fodera
Permissions Manager: Megan Jackson
Permissions Clearing: Bethany Salminen
Text Design: Jillian Burr
Composition: Jouve International—Brattleboro, VT
Manufacturing: Quad/Graphics

The text of this book is composed in DIN with the display set in Bembo Standard. This book is printed on permanent paper containing 10 percent post-consumer waste recycled fiber.

Library of Congress Cataloging-in-Publication Data

Lee, Kai N.
Humans in the landscape : an introduction to environmental studies / by Kai N. Lee, William R. Freudenburg, and Richard B. Howarth.—1st ed.
p. cm.
Includes bibliographical references and index.
ISBN 978-0-393-93072-6 (pbk.)
1. Environmental sciences—Study and teaching.   I. Freudenburg, William R.
II. Howarth, Richard B.   III. Title.
GE70.L43 2012
304.2—dc23   2012028652

W. W. Norton & Company, Inc., 500 Fifth Avenue, New York, NY 10110-0017
wwnorton.com
W. W. Norton & Company Ltd., 15 Carlisle Street, London W1D 3BS
4 5 6 7 8 9 10

To the students of ENVI 101 (Williams College),
ES 1 (University of California, Santa Barbara), and
ENVS 3 and 55 (Dartmouth College),
who helped us to learn and to teach better.
—Kai Lee, William Freudenburg, and Richard Howarth

For Dana, Katherine, Dylan, Julia, Anna, John, and Eleanor,
with love for my family and hope for the
future they are building.
—Kai Lee

Brief Contents

Contents

PART III STRATEGIES

Boxes

About the Authors

Kai N. Lee is Rosenburg Professor of Environmental Studies, *emeritus*, at Williams College, and program officer for science at the David and Lucile Packard Foundation. Kai was trained as an experimental physicist at Columbia and Princeton Universities. He taught environmental studies and political science at the University of Washington before going to Williams to direct its Center for Environmental Studies. In both institutions, he taught the introductory course in environmental studies that forms the basis of this book. Kai is the author of *Compass and Gyroscope* (Island, 1993), a book on adaptive management that has been widely used in graduate courses in environmental science. In his grant making at the Packard Foundation, Kai is developing related means of linking knowledge with action. He has served on more than a dozen committees of the National Research Council, advising government agencies on a range of policies where scientific issues play a critical role. He is currently vice-chair of the NRC's committee to advise the U.S. Global Change Research Program. Earlier in his career Kai was a White House Fellow and he represented Washington state on the Northwest Power Planning Council.

William R. Freudenburg was Dehlsen Professor of Environmental Studies at the University of California, Santa Barbara (UCSB), at the time of his death in 2010. He studied at the University of Nebraska, his native state, and at Yale University, earning a Ph.D. there in 1979. He held professorships at Washington State University and the University of Wisconsin before arriving in UCSB in 2002. Bill pursued scholarly interests in risk analysis and rural sociology, and he served as president of the Rural Sociology Society. A popular and devoted teacher, Bill was co-founder of the Association for Environmental Studies and Sciences. He was a prolific author, most recently as coauthor of two studies of the Gulf of Mexico coast: *Catastrophe in the Making:*

The Engineering of Katrina and the Disasters of Tomorrow (Island Press, 2009) and *Blow-out in the Gulf: The BP Oil Spill Disaster and the Future of Energy in America* (MIT Press, 2010).

Richard B. Howarth is Professor of Environmental Studies at Dartmouth College and the Editor-in-Chief of *Ecological Economics*. After receiving an A.B. in Biology and Society from Cornell University in 1985, he pursued an M.S. in Land Resources at the University of Wisconsin, Madison (1987) and a Ph.D. from the Energy and Resources Program at the University of California, Berkeley (1990), where he specialized in the economics of natural resources and sustainable development. Prior to his arrival at Dartmouth in 1998, he held appointments with the Environmental Energy Technologies Division of the Lawrence Berkeley National Laboratory and with the Environmental Studies Department at the University of California, Santa Cruz. He has published widely on topics that include theories of intergenerational fairness; the economics of energy efficiency; climate stabilization policy; the valuation and governance of ecosystem services; and the links between economic growth, environmental degradation, and human well-being.

Preface

This is a book that our students taught us to write. The three authors each taught introductory courses in environmental studies programs in our three rather different institutions. Each of us is, by current disciplinary identification, a social scientist—political scientist, sociologist, and economist—though we are also, in other professional contexts, a lapsed physicist, a rural sociologist who contributed to risk analysis, and an ecological economist who is as much a scientist and philosopher as anything else. Throughout our careers, we have been, by temperament, scholars more interested in solving problems than in disciplinary boundaries.

As we worked to articulate an integrated yet accessible introduction to the sprawling study of the environment, each of us felt uneasy about the prevailing way in which environmental studies was taught. This is an approach, still common, in which the biophysical sciences describe and define a "natural" world being transformed by human activity. Based on that definition of a natural order, human activities are to be restrained or redirected, mainly through public policies. But a new generation of teachers, many of them from interdisciplinary backgrounds, have arrived in the classroom, and their approach is richer by far than this older way of thinking about the field. However, textbooks have not kept up with these important changes. Teachers are supplementing textbooks more and more, and some have abandoned textbooks altogether. The traditional approach isn't wrong, exactly, but it is seriously incomplete, for it leaves out the central actors in the story: the people. It is the people who, after all, drive the environmental problems of most concern: loss of biodiversity, climate change, and a perilously unsustainable, increasingly global economy. What of them—or, rather, us? The courses we devised struggled to impart a fuller picture of the structure and function of human institutions and the way that humans make their homes in landscapes—often by transforming them. Ours is a world in which the biophysical and the social now intertwine; where humans, for better or worse, are now responsible for the biosphere that we and our fellow forms of life depend on. This book is our attempt to introduce readers to that strange and familiar planet, the one that contains and has the potential to sustain intelligent life.

ACKNOWLEDGMENTS

Textbooks are not original scholarship, and we necessarily stand upon the shoulders of others. Our intellectual debts are broad in a book that attempts to open a window on inhabited landscapes—so much that we cannot effectively detail them even to ourselves. Some of the broad outlines are hinted at in the passages we quote in the text, in the suggestions for further reading, and in the many images and diagrams we have appropriated.

It is a pleasure to acknowledge more specific contributions. We thank the autumn 2006 honors section of Environmental Studies 1 at the University of California, Santa Barbara (UCSB), who read the first draft of Chapter 2 and gave us sound advice on how to approach it and the other chapters; Tony Gengarelly of the Massachusetts College of Liberal Arts, whose guest lecture on the Hudson River School artists laid the groundwork for Chapter 2; Karen R. Merrill of Williams College, who enlivened the discussion of American environmental history in Chapter 3 and other parts of this book; Antonia Foias of the Department of Sociology and Anthropology at Williams College, whose clear and thought-provoking lecture on the fall of the Maya is excerpted in Chapter 4; William T. Fox, Professor Emeritus of Geosciences at Williams, who read and improved Chapter 5; James T. Carlton of the Williams-Mystic Maritime Studies Program, whose compelling lecture on introduced species informed Chapter 9; and Charles Benjamin of the Near East Foundation, whose discussion of community governance of resources is summarized in Chapter 10. The students in the spring 2007 section of Environmental Studies 101 at Williams read most of the chapters and provided detailed suggestions that were provocative and helpful; we tried to heed them as much as we could.

For the past five years, the Packard Foundation's conservation and science program, led by Walter V. Reid, has been a remarkable place for Kai Lee to pursue a second career; he thanks the Foundation for its encouragement and support for his work on this book and for providing access to the Stanford University libraries. Richard Howarth acknowledges the generous support provided by the Pat and John Rosenwald Professorship during the period when this book was written. We are grateful for help in the late stages from Robert Gramling of Louisiana State University; Lisa Berry of UCSB; Raymond Huey, John M. Wallace, and Yen-Ting Hwang of the University of Washington; and David Peart of Dartmouth College. All of these benefactors shared their thoughts, time, and knowledge with an open-hearted generosity that lies at the heart of our hopes for the sustainability of a world inhabited by humans.

Aaron Javsicas, our editor at W. W. Norton, and his talented production staff have turned rough lecture notes into a handsome book. And we are deeply grateful to Leo Wiegman, who brought this project to Norton and in fact drew the three authors together as collaborators. Leo was entrepreneur, coach, and therapist; this book exists because Leo believed in it. We are grateful to the reviewers that Norton brought in, to test our words against their understandings and their needs as teachers. We thank them for their candor and rigor, both of which have made a difference:

Margarita Alario, University of Wisconsin, Whitewater

Mark Anderson, University of Maine

Matthew Auer, Indiana University

Valerie Banschbach, Saint Michael's College

April Baptiste, Colgate University

Robert Chapman, Pace University

Alan Diduck, University of Winnipeg

Randolph Haluza-DeLay, The King's University College

Stephanie Kaza, University of Vermont

Michael E. Kraft, University of Wisconsin, Green Bay

Heather Leslie, Brown University

Eric Maurer, University of Cincinnati

Mark Mysak, doctoral student, University of North Texas

Laurel Phoenix, University of Wisconsin, Green Bay

James Proctor, Lewis and Clark College

Anne Rademacher, New York University

Theodore L. Steck, University of Chicago

Trileigh Tucker, Seattle University

Chris Wells, Macalester College

Tom Wilson, University of Arizona

Our coauthor, fellow teacher, and friend, Bill Freudenburg, died in December 2010 as the first full draft of this book neared completion. That autumn, Bill taught Environmental Studies 1 a final time. He finished all the lectures. At a memorial service for Bill, his sister, Patti Freudenburg, offered some thoughts about her brother. He was a good listener, she said, and he had known when to let his younger sister make her own mistakes so that she could learn from them. He also knew to be persistent, to plant seeds—seeds of doubt among the ignorant and recalcitrant, seeds of enlightenment and hope for everyone—knowing that some would grow in time. This book is one of Bill Freudenburg's seeds sprung to life. We hope that it is a book that speaks clearly while also offering a sense of how all three of us have tried to listen. We miss Bill. His ideas are here, and that is good.

Kai N. Lee
Los Altos, California

Richard B. Howarth
Etna, New Hampshire

August 2012

HUMANS

— in the —

LANDSCAPE

PART I

Forces

Chapter One

HUMANS IN THE LANDSCAPE: ENVIRONMENTAL STUDIES AND ENVIRONMENTALISM

FORCES, GRAND CHALLENGES, AND HOPE

In this chapter we meet humans in the landscape—ourselves, as we live in the environment. Human societies have always existed within natural systems that support them. Yet over the past two centuries, people have become an environmental force of global proportions. In 1800, nearly all people lived in one place for their whole lives. They ate food from gardens and fields nearby; they wore clothes made from fibers and skins grown in the valleys and villages around them; and they worked in ways that directly contributed to (and competed with) their neighbors. If the village tannery poisoned the local stream, the people and animals who drank from it were the ones that got sick and died.

By the year 2000, economic development and political reorganizations had swept across the globe, transforming **ecosystems** and societies in their wake. People's activities and their material livelihoods are no longer local. Food from another continent is so common in our supermarkets that it is rarely labeled as imported. A shirt from Nigeria may be worn with jeans sewn in Turkey and sneakers made in China; we might notice the look, but we rarely think about where the clothes came from. We receive unsolicited e-mail from many countries, and the money we pay for gasoline might end up in Venezuela, Scotland, or Iran. All these distant transactions have environmental implications, but these are hard to see now that the impacts might affect another continent. The same is true of impacts that might emerge decades after the damage is done, as is the case with greenhouse gas emissions that alter climate.

Unlike our ancestors, we all live on multiple scales of space and time now. Our actions affect people and ecosystems near and far, now and in the future. The impacts are too numerous to notice or to understand in detail. Was the banana I had at breakfast grown with pesticides? Did the chemicals affect the health of the farmer and her family? What about the greenhouse gases emitted by the trucks and ships that brought the fruit to my kitchen? There are too many questions, if you think about them, so it's easier not to think.

Although citizens recognize that the human footprint on nature has often become so large as to be destructive, governments and social institutions as presently constituted have made little progress toward building a sustainable economy. With our power to reshape nature, we have become a dominant species in the biosphere, wiping out plant and animal species, changing weather patterns, and transforming landscapes and even the sea. Despite all of our power, humans remain an undomesticated species in the environment, even though billions of people and growing numbers of businesses are trying to live responsibly.

And yet we rely on the natural world for everything—from the air we breathe, to the water that grows our crops and slakes our thirst, to the minerals we use to produce our goods, to the fossil fuels we burn to power our industrial economy. Although we do not often see our connections to nature as directly as our ances-

---

Learning Objectives
When you have finished studying this chapter, you should be able to

- ⬊ ask questions about how humans interact differently with their environments at different times and in different places, and to recognize the large inequalities found in human populations;

- ⬊ think about long-term trends in population, energy use, and the environmental modifications that people make, and to recognize that those trends will alter the world in which your career and life will unfold;

- ⬊ recognize the sometimes complicated linkages among people and ecosys-

tems that produce environmental problems;

- ⬊ describe some of the ideas associated with the words "sustainability" and "environment," as well as "institution," "classical environmentalism," "sustainable development," and "integrative problem solving";

- ⬊ give examples of environmentalism and of environmental studies, showing the difference between the two.

tors did, we need only look at the suffering caused by crop failure in Africa or air pollution in China to see that our ways of life and standards of living rely on the functioning and integrity of the environment, both near and far.

A large question faces students now in college. *Can those environmental systems be sustained over the decades and centuries ahead?*

During the working lives of today's students, humanity will make major choices.

- In the twenty-first century, human numbers are projected to level off at roughly 9 billion, approximately one-third larger than today's world population.

- The planet's weather will continue to change in unprecedented ways, bringing surprises ranging from more frequent wildfires to earlier spring blossoms to a melting polar ice cap. Humans are now beginning to focus on the climates we have unwittingly been changing, and a long, likely painful process of adjustment is just starting.

- The habitats of species on land and sea are shifting, too, as humans alter ecosystems either for human benefit or as an accidental consequence of their actions. Fish populations have dwindled in the oceans, particularly those species that people like to eat. Tropical forests continue to be cleared, destroying ways of life and driving species extinct before they have even been identified by scientists. In many parts of the world, including the United States, expanding demand for water is depleting rivers and undermining the animals and plants that depend on the rivers' waters and wetlands.

- The world's urban population is likely to double from 3 billion to more than 6 billion in the coming half-century, with a large majority of these additional people living in countries that are poor today. This urbanization will generate pollution and threats to public health of a magnitude unknown in human history. Yet even in poor cities, people are demanding decent living conditions as they strive for prosperity.

These issues—**climate**, **biodiversity**, and **urbanization**—arise within a dynamic global economy. Each issue constitutes a **grand challenge** of sustainability. We call these *grand* challenges because each involves a set of problems and opportunities for which there is no clear path from the unsustainable present to a durable future. The grand challenges are entangled with the profound and widening inequality of human communities. Within the United States, large differences are found between rich and poor, and as in many nations across the globe, large disparities exist in the ability of communities to gain the food, water, energy, and knowledge that underlie prosperity. We will investigate these inequalities as a central component of **sustainable development**, the fourth grand challenge—a challenge that includes the other three, as you can see. Collectively, these grand challenges threaten the continuation of material civilization as we know it today.

Things will change. As we describe in this book, enough is known, even about the grand challenges, to take action, even though not enough is known to assure success in every case. In short, there is hope for humans, if we can learn and are willing to act. But it is essential that we learn, and that we learn to act.

The challenges will not be easy ones to solve. It can be hard to grasp the implications of humans in landscapes, and harder still to anticipate those implications before they emerge as crises. This is the usual situation for environmental problems. Environmental problem solving must accordingly begin by understanding this difficulty. Because the character of environmental problems may be seen more clearly at a local scale, we begin to learn about global problems by first turning to a remote corner of the United States, the Channel Islands off the coast of Southern California. There, in a landscape as nearly wild as one can find in North America, we can begin to see the ways in which local events are connected to the global forces of human society.

THE CHANNEL ISLAND FOX

The Channel Islands make up a handsome, little-known archipelago about 30 miles off the coast of Southern California. One island contains a retreat popular with Hollywood's elite, but for the most part, the Channel Islands have no human inhabitants and look pretty wild. The islands are managed by the National Park Service and the Nature Conservancy, a major nonprofit organization, and they are surrounded by underwater kelp forests teeming with marine life to delight divers and sport fishers. Living in the beautiful, rugged hills is a species of animal found only on these islands—the Channel Island fox, known to scientists as *Urocyon littoralis* (Fig. 1.1).

For more than ten thousand years, this fox population remained stable, falling modestly at times but rising again. Then, in the six years between 1994 and 2000, the foxes nearly disappeared—dropping from nearly four thousand on three of the large Channel Islands to fewer than two hundred.

The story of the Channel Island fox is valuable because it illustrates the complex interconnections between the natural world and human activity. This is a story of unintended consequences, of the way that nature draws together the threads of humans and landscapes, often weaving patterns that no person expects to see. As is often the case, unintended consequences are not simple.

TRIPLE THREATS CONVERGE ON THE CHANNEL ISLAND FOX

The near-extinction of the Channel Island fox surprised people, because the fox did not become endangered as the direct result of one event or human action. It was only because of what happened in three intersecting stories, playing out over

FIGURE 1.1
The Channel Island
fox.

three different scales of space and time, that the fox experienced a sudden and unexpected misfortune. In none of these stories was the fox a visible player. Put another way, the decline of the Channel Island fox at the end of the twentieth century reflected changes in three overlapping ecological systems. Human actions had altered the patterns and rhythms of nature in each of them. The fox belonged to all three ecosystems, and when the altered patterns came together in the 1990s, *Urocyon littoralis* was caught in a perfect storm.

The first story is about a cheap, effective, and apparently safe way for humans to exterminate insect pests. Dichloro-diphenyl-trichloroethane (better known as DDT) first came into use against insects during World War II. DDT's rapid suppression of disease-carrying insects revolutionized public health and won its developer a Nobel Prize. Because the pesticide did not seem to have toxic effects on humans, it was routinely sprayed on both children and adults who were exposed to lice and other insects. Although it has not been tested on them, DDT probably has no detectable effect on Channel Island foxes either. But DDT did have an unanticipated effect on fish and birds. In particular, it builds up in the fatty tissues as one organism eats another. As it moves up the food chain, DDT can become so concentrated that it impairs the ability of some bird species to metabolize calcium. As a result, their eggshells become so thin that the eggs break before the young can survive. The insects were killed swiftly, but bird populations were dwindling, too, out of sight and slowly enough that this decline was not noticed until DDT had been widely distributed in the environment.

One person who noticed was a biologist and writer named Rachel Carson. Carson was a respected writer, but in a field that had few women scientists, she was not regarded as a "real" scientist by some (male) biologists. In 1962, she published a book, *Silent Spring*, warning that widespread release of pesticides might reduce bird populations so much that we would no longer hear them singing in the springtime. The response to the book was enormous, and *Silent Spring* helped to launch the modern environmental movement. Sweden and Norway banned DDT in 1970, and the United States banned it in 1972. DDT stays in the environment for a long time, but about ten to twenty years after it was banned, bird populations began to recover. Among the species whose numbers rose again was the national symbol of the United States, the bald eagle.

A second story gathered force in the dry hills of Southern California. This one also did not involve the Channel Islands or their foxes, but took place in the rapidly growing metropolitan area of Los Angeles. Before the city's rapid development in the twentieth century, the Los Angeles basin had provided prime habitat for the golden eagle, a majestic bird of prey that rarely shares habitat with the bald eagle. The golden eagle had never spent much time in the Channel Islands, a terrain dominated by the American bald eagle for thousands of years before it was brought low by pesticides. Among the differences between the two, bald eagles eat fish and the somewhat larger golden eagles eat mammals. Bald eagles feasted on the fish around the islands, showing no interest in the Channel Island fox. Although DDT cut into the bald eagle populations, the fox was unaffected.

Had it not been for this second story, the golden eagles would most likely have stayed in the Los Angeles basin. But as the populations of both eagle species began to recover, the golden eagles found that their former habitat on the mainland had become the world's iconic example of suburban sprawl. By 1990, Los Angeles had a population half the size of New York City, but the Angelenos took up twice as much land as the New Yorkers. In the years before World War II, Southern California had been a dry, open landscape of grasslands and shrubs—good habitat for hunting birds such as golden eagles. Roughly half a century later, Los Angeles had become an urbanized landscape of parking lots, housing subdivisions, freeways, and shopping centers—prime habitat for automobiles. Making good use of their strong wings, the golden eagles looked for other places to hunt, places that still looked like the old Southern California. They found the Channel Islands.

The third story began about a hundred years before the other two, and it was the most localized, taking place on the Channel Islands themselves. As California was settled in the wake of the mid-nineteenth-century gold rush, farmers and ranchers colonized the Channel Islands, introducing livestock such as pigs. Although farming and ranching proved to be profitable for a time, agricultural operations on the islands ultimately came to an end, with descendants of the original farmers either giving or selling their lands to the Nature Conservancy or the National Park Service. In the interim, however, some of the pigs had escaped, and they and their descendants survived as feral (wild) animals. Yet, the foxes and the feral pigs appear to have had little effect on one another. The Channel Islands ecosystem was one in which the foxes lived, as environmentalists would now say, sustainably.

As the golden eagle population climbed in the late twentieth century, the three stories converged. Before World War II, the islands had provided far better habitat for fish-eating bald eagles than for mammal-eating golden eagles. Fifty years later, the feral pigs turned out to provide enough of an additional food source that, for the first time ever, the golden eagle was able to outcompete the bald eagle, preventing the bald eagles from returning to the islands. Partly because both species feed heavily on carrion (the remains of dead animals), one species winds up chasing out the other. With ready access to a much larger food source thanks to the presence of the feral pigs, the golden eagles were now the ones that rose to the top of the food chain on most of the islands.

Like many species accustomed to living in island habitats, the foxes had had few predators for perhaps a thousand generations. This may be why foxes living on the islands at the end of the twentieth century moved about during the day, providing easy targets for the golden eagles. This seems to be the reason why the population fell more than 90 percent in only six years. The decline was detected by a graduate student who arrived in 1993 to study the foxes. When he began to see evidence that the foxes were being killed by a new kind of predator in 1994, he sounded the alarm and humans intervened once again in the Channel Islands ecosystem.

Initially, big organizations were slow to believe the story that he put together, but in 1999—late in the day for the fox (Fig. 1.2)—the National Park Service trapped many of the remaining foxes and began breeding them in captivity. Next, they removed the golden eagles, and after several years of effort, they ultimately exterminated the feral pigs. Then, starting in 2002, bald eagle chicks were released on the islands. The goal of these direct manipulations of the ecosystem was to recreate a landscape in which bald eagles and foxes could coexist again, while saving enough Channel Island foxes so that they retained sufficient genetic diversity to adapt to the landscape once humans released them from captivity. Together, these actions may have saved the foxes from extinction on their native islands. The captive breeding program was ended in 2008 as fox populations rebounded. In 2010, the citizens' group Friends of the Island Fox reported that more than five hundred foxes were living on San Miguel Island.[1]

To people in nearby Los Angeles, the Channel Islands appear to be wild and uninhabited. As these three stories make clear, however, extensive human effort may be the only reason that the "wild" populations of the native species on the island survived into the twenty-first century. Ironically, the fox's ecosystem is now returning to something that looks reasonably similar to its original state, but it is

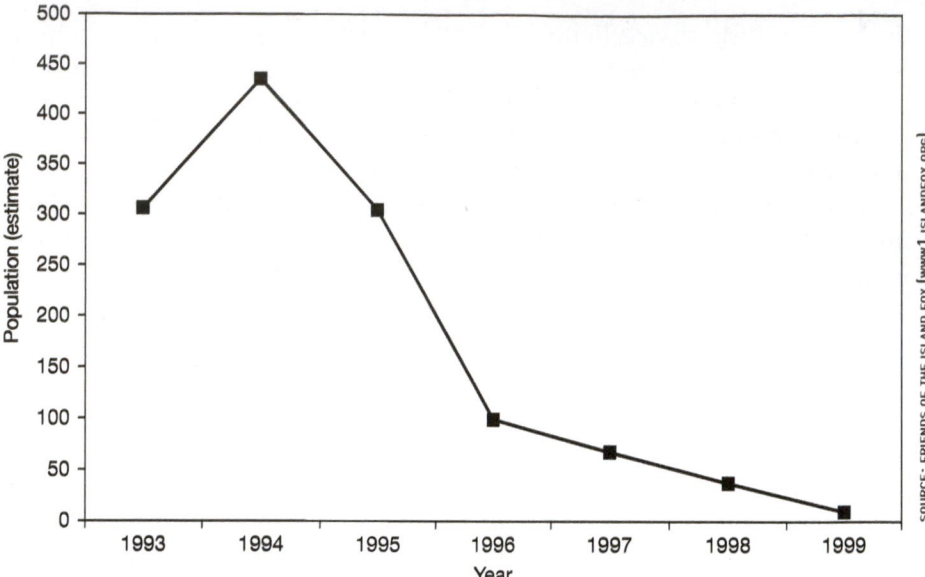

SOURCE: FRIENDS OF THE ISLAND FOX [WWW1.ISLANDFOX.ORG]

FIGURE 1.2
The estimated number of Channel Island foxes on San Miguel Island, 1993–99.

doing so only because of extensive engineering by humans trying to counterbalance earlier changes of human origin.

The Channel Island fox reminds us that our knowledge of the natural world is limited. No one could have foreseen the convergent ecological patterns that nearly drove the fox to extinction. The introduction of nonnative species such as the feral pig can be a scourge in many ecosystems, but that wasn't the fox's problem. The decline of valuable plants and wildlife from indiscriminate release of pesticides is a serious problem in many ecosystems, but that wasn't the fox's problem either. The conversion of large fractions of land and coastal waters to human uses such as agriculture and urbanization has eliminated the habitats occupied by many ecosystems, but not even that was the fox's problem. The fox's problem was the accumulation of multiple ecosystem imbalances, each caused by humans, that came together in the drastic change of an ecosystem that, on the surface, seemed to be free from human influences—a landscape of wild creatures that included eagles, foxes, and feral pigs.

Yet when scientists can warn of disaster in time, and if governments and private organizations are able to respond, there is reason for hope. It was only through the efforts of a graduate student that humans recognized the plight of the fox, and it was only through the efforts of science-based conservation that the fox was saved. There is hope for humans and the environment—if we will recognize our imprint on nature and society and then take corrective action before irreversible damage is done.

ENVIRONMENT AND SUSTAINABILITY

This book is about a world that is being utterly transformed by one species—humans—not just in places such as Los Angeles, where human influences are obvious, but even in places that still appear to be wild. To cope with this complexity, we have organized this book around two ideas, environment and sustainability. Each of these words evokes ideas and values about humans and the landscapes we inhabit.

The word "environment" is deeply influenced by the natural sciences of biology and geology. Biologists (notably those who call themselves ecologists) and earth scientists studying land, water, and atmosphere have tremendously expanded our understanding of the world—the world we still call "nature." In studying the natural world, scientists have found the imprint of humans everywhere: pollution in the air, water, and soil; a climate pushed by forces of human origin; species dwindling and sometimes dying out altogether under the pressure of hunting or destruction of their habitats; transformations of natural systems by farming, urbanization, fishing, and logging—often made worse than intended through ignorance or carelessness. Even remote and seemingly pristine areas such as the Arctic and Africa's Congo basin show the imprint of human activity. Polar bears that live in some of the most remote parts of the world now have high levels of persistent chemicals in their bodies because of emissions from factories that are thousands of miles away. In addition, the future survival of polar bears may be threatened by global climate change, as higher summer temperatures melt the ice that provides critical habitat for this species.

Over the past half-century, "environment" has become a political word, too, in the United States and most of the rest of the world. Major public policies designed to protect the environment and reverse environmental damage have redirected many economic activities. A few polluting factories have been closed, many others have changed their ways, and quite a few business people have found new opportunities, such as providing outdoor recreation services and equipment, constructing energy-efficient buildings, or developing hybrid vehicles. Many of those economic changes have been resisted, at times fiercely. As a result, much remains to be done.

The word "sustainability" has also become political, although it has not been nearly so controversial. The idea, articulated by a United Nations (UN) commission in advance of the 1992 Earth Summit, a gathering of the world's nations in Rio de Janeiro, Brazil, is a simple one. It holds that humans can and should make use of nature to meet their needs, but that we should do so only in ways that respect the needs of future generations.

In ordinary language, a "need" is not the same thing as a desire or a want or an economic demand, but in life the differences are sometimes harder to define. How many passengers boarding an airplane pause to consider the implications of their flight on the climate their grandchildren may face? Figuring out whether flying is meeting a need is not a simple matter.

Sustainability is a moral claim: it is about what people should or should not do. Few would dispute that taking future generations and the natural world into account is a duty we all must shoulder. In some cases, such as the recovery of endangered bird populations after DDT was banned in most developed countries in the 1970s, or for that matter in the apparent success of saving the Channel Island fox from extinction, humans have produced notable success stories. In many other cases, humans have continued activities that are unsustainable, partly because we have had a hard time developing the rules, organizations, and habits that could enable us to do otherwise.

With few exceptions, environmental education has combined natural science with discussion of environmental policies and the political conflicts that have surrounded them. College courses on sustainability focus on promising innovations such as green design or renewable energy, with discussion of the literature that grew up around the Earth Summit's highlighting of sustainable development as a global goal.

Those approaches can be valuable, but in this book we try something different. We put humans at the center of both environment and sustainability. In 1800, environmental problems were things that nature did to people, causing floods, disease, wildfires, earthquakes, and other catastrophes. Today, most environmental problems—from toxins in the groundwater below us to holes in the ozone above—are the result of what humans have been doing to nature. Other environmental problems, such as hurricane damage or water shortages, have larger consequences because humans have defied nature by locating activities where the risks were already well known. It is human activity that needs to be made sustainable—an idea that is widely thought to be obvious, though hard to achieve. At the same time, we seek to persuade you of something that is not at all obvious—that changes in human *institutions* are basic to both environmental protection and sustainable development, and that bringing about those changes requires a grasp of the complicated world illuminated by social as well as environmental science.

In the chapters that follow, we explain in detail what we mean by the term "human institution" and explain why it is much broader than government regulation. For now, it is enough to say that institutions are the rules that define how members of society interact with one another, shaping people's actions and giving rise to patterns of behavior. One good example involves the complex institutions that have arisen around the automobile, such as suburbs, freeways, body shops, speeding tickets, car payments, and much else. These play a prominent role in American life and in our contribution to climate change. Changing these and other institutions isn't easy, as we can see in the intense struggle to limit greenhouse gas emissions from cars as part of our effort to stop global warming.

In the chapters to come, we will also demonstrate how an *integrative* approach to environmental problems—combining natural science, social science, and the humanities—can suggest new ways of solving those problems. Notice that we are saying that a focus on a single discipline is not sufficient, even though mastery of specific disciplines such as mechanical engineering or moral philosophy may be necessary. In short, this is not a book about easy things you can do to save the planet. It is about how hard things can be understood so that each of us can contribute to building a durable presence for humans in the landscapes we have come, for better or worse, to dominate.

CLASSICAL ENVIRONMENTALISM

Humans have solved some vexing and seemingly intractable environmental problems. The ozone layer in the upper atmosphere shields life on Earth from solar ultraviolet radiation. In the middle of the twentieth century, widely used industrial chemicals began to erode the ozone layer. This was essentially a planet-wide DDT problem: we were losing the protective shell that nurtured humans and ecosystems. Thanks to hard work and a commitment to sensible policies on a worldwide basis, production of these chemicals is being phased out. Those chemicals also last a long time, but as they have begun to break down in the atmosphere, the ozone layer has begun to heal.

Protecting the ozone layer is an example of the leading approach to solving environmental problems, an approach we will call **classical environmentalism**. To see how it works, consider this simplified outline: (1) Scientists discover threats to human health or ecosystem stability. These threats are traced to human interventions in the natural world, which arise from economic activities. (2) The threats galvanize citizen concern, leading to politically significant demonstrations of concern and the formation of new citizen groups focused on the threats. (3) In response, politicians and government organizations adopt laws, regulations, and budgets, and create new administrative agencies, to regulate the human activities that create the threats. In the face of regulation, engineers and business firms seek out and adopt technological solutions and mitigations that permit the economic activities to continue while avoiding or limiting the damage to humans and ecosystems (Fig. 1.3).

Antipollution and wilderness-protection policies in the United States fit this historical pattern. It is the story of the modern environmental movement, and it has

FIGURE 1.3
Classical environmentalism.

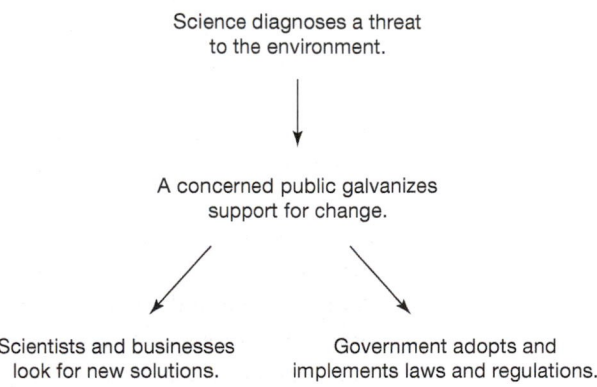

Science diagnoses a threat to the environment.

A concerned public galvanizes support for change.

Scientists and businesses look for new solutions.

Government adopts and implements laws and regulations.

helped to bring about significant gains. The rescue of the Channel Island fox, for example, was facilitated by its status as a species listed under the Endangered Species Act, a federal law created through classical environmentalism in 1973.

We call the approach in Figure 1.3 *classical* environmentalism because this pattern—scientific discovery prompting a regulatory response—has come to define environmental problem solving. Over the past generation, environmental quality has become recognized as a legitimate component of social welfare in almost all nations. The key word is "legitimate": people have come to accept that environmental quality is important and that societies and their governments should protect it. One sign of this consensus is the degree to which people increasingly accept that the global climate is changing and that humans are responsible for this change.

That the quality of the environment has become a legitimate concern does not mean that conflict over environmental issues has ceased. Indeed, the reverse is true. Environmental disputes are common and often intense. But even those bitterly opposed to a proposed solution to an environmental problem do not suggest that environmental problems, in general, are unimportant—only that their opponents are misguided or being deceptive or dishonest.

Classical environmentalism has relied on governments to adopt and implement regulations. This model of lobbying and law making has enjoyed important successes, but it is not succeeding at the global scales of climate change or biodiversity conservation, in part because there are no "higher levels" of government, in the UN or elsewhere, that are powerful enough to impose binding rules on sovereign governments. The United States faced no sanctions, for example, when the American government spurned the Kyoto climate treaty shortly after President Bush assumed office in 2001, although this decision proved to be deeply unpopular in other nations. These frustrations suggest that existing institutions may be inadequate to address some of the challenges that we face. They also suggest that classical environmentalism itself cannot be the whole answer.

SUSTAINABLE DEVELOPMENT AND THE RIDDLE OF WIDENING INEQUALITY

Is there something that can replace classical environmentalism? One possibility is the idea of sustainable development. In 1987, the World Commission on Environment and Development defined sustainable development as "development that meets the needs of the present, without compromising the ability of future generations to meet their own needs."[2] This idea was carried from the international committee that wrote that definition to the Earth Summit in 1992, a UN gathering of all the world's nations, where it was warmly endorsed. Although a large number of specific proposals to implement this idea were embraced there, strikingly little progress has been made in carrying them out in the years since.

Why the lack of progress? Because people everywhere have been drawn into a global system of economic development that takes little account of sustainability. The guiding vision of this system is that free markets, international trade, and a primary focus on promoting economic growth offers the best path forward for global society. In support of this vision is the fact that, over the past three hundred years, human population and economic welfare have risen in a fashion unknown in the earlier history of our species, thanks in part to technology and fossil energy, as we will discuss in detail in Chapters 6 and 7. But this vision misses a critical part of the story: economic development has been strikingly uneven, and billions of people are still mired in deep poverty. In addition, some of the environmental impacts of economic growth are deeply troubling.

Three-quarters of humanity has yet to prosper in the way Americans take for granted. Roughly a third of the human race lives in what the UN calls "absolute poverty," with an income of less than $2 per day—a level few Americans can imagine. Yet if one looks at averages in economic statistics, virtually all the nations of the world have made significant gains in material welfare over the past two hundred years. In the twentieth century, our ability to transform the resources of the world into material goods and services benefiting humans became global. Throughout this time, the rich have gotten richer, decade by decade, and the large majority that remained poor has seen their economic fortunes improve much more slowly, so that the gap between rich and poor has steadily widened. The incomes of the people in the richest nations, such as Sweden and the United States, are now about 100 times as high as those in the poorest.

Here is the hitch with sustainable development. Actually, there are two hitches. First, the resources of the world are finite. In many instances, such as the world's commercial fisheries, these resources are dwindling. We now know, too, that burning fossil fuels is altering the climate in a way that threatens many ecosystems and puts pressure on people and their livelihoods in most parts of the world. These facts mean that, contrary to the tenet of sustainable development, we *are* compromising the ability of future generations to meet their needs. Second, billions of people cannot meet their needs by any reasonable definition. The World Commission on Environment and Development concluded that "the gap between rich and poor nations is widening—not shrinking—and there is little prospect, given present trends and institutional arrangements, that this process will be reversed"[3] That is, we are not even meeting the needs of the present.

In short, current practices fail both of the tenets of sustainable development. At the same time, these practices also constitute a trajectory to which the world's peoples are currently committed. This is a moral problem, as the language of sustainable development might suggest, but it is also a practical problem. It is extremely unlikely that the more than 7 billion people now living on the planet could tolerate the costs associated with a rapid elimination of fossil fuels, for example. And even if we did, that would not bring back the fisheries depleted by overfishing or even bring to a halt the climate change now under way. The human race is stuck in a paradox: what we are doing isn't working; we aren't yet moving toward an economy that looks like it will be sustainable; and we are having trouble even making the improvements that seem sensible and inevitable from the perspective of the natural sciences.

This paradox spells out why there is no simple path to sustainable development. We do not know how to construct such a path with today's knowledge, today's institutions, and today's beliefs in what humans should do. We are committed to a way of life that cannot continue in its present form; yet we do not have a full picture of the changes we need to make, and we lack a way to persuade or compel the people of the world to make even the changes that seem sensible, such as fishing sustainably. What we do have at present, on the other hand, is a growing understanding of the overall contours of the changes we need to make. And that may be the first thing we need before we build a complete and concrete plan for constructing such a path, and perhaps even for developing the political will to take more than a few hesitant steps toward goals that seem reasonable.

INTEGRATIVE PROBLEM SOLVING

What both classical environmentalism and sustainable development emphasize is the significance of the connections that people are unaware of. This is the message of the Channel Island fox and also of many other environmental issues. How many Americans consider the impact on global warming when they buy a new car? If the answer is only a tiny minority, are the rest being reckless? If not, why do they make choices in a way that an environmentalist would consider reckless?

Note that we are giving you more *questions* to think about than answers to learn. Note also that even questions involving the potential for moral judgments can be approached in ways that are logical, not merely judgmental. Why *do* different groups of people, and the institutions they form, reach such different conclusions and act in such different ways? These questions are difficult but they are also important.

Learning to think carefully and deliberately about questions, as well as thinking through potential answers, are some of the central skills of environmental studies. Unlike most academic pursuits, this is a field full of unanswered questions, such as how to salvage endangered species or manage urbanization. Moreover, the questions to which we don't have answers are not merely academic. Instead, they are questions of momentous importance tied to some of the most politically contro-versial, socially wrenching issues that you and your fellow students will face over the decades to come.

A major concern of this book, in short, is to come to terms with the open-endedness of the questions of environmental studies. As we will demonstrate in later chapters, a strategy is needed to grapple with these questions in an integrative, problem-solving way, drawing together the knowledge of different analytical disci-plines, understanding that knowledge is incomplete, and weighing the uncertain-ties and risks in order to inform social choices.

Environmental studies can be thought of as the *search for a responsible role for humans in the landscape.* This search comes at a time when technology allows the "natural" world and the human one to intersect with increasing frequency, creat-ing significant peril for both.

OUR APPROACH

The search for an environmentally responsible and sustainable future is an intel-lectual pursuit separate from environmental activism. As a citizen, you may choose to transfer what you learn in your studies into activism, or you may not. The academic field of environmental studies is rooted in study, not activism. There are two other, practical reasons to separate activism from scholarship. The first is that solutions are likely to work only if they reflect a clear-headed understanding of the underlying problems. The other is that scholars and scientists lack power and are influential only to the extent that they are credible, whereas advocates look to power as well as influence in pursuing their goals in the midst of intense conflict.

This book, accordingly, is focused on learning and analysis, not on activism. At the same time, we recognize that the search for environmental solutions includes finding opportunities to combine activism with analysis—whether we are talk-ing about Rachel Carson's work in the last century, about the scientists who are working to save the Channel Island fox today, or about people tackling the grand challenges of sustainability.

This book does not advance solutions to the grand challenges, but does something more basic. We explain (1) how environmental challenges emerge, (2) what the current challenges are, and (3) what strategies people and institutions are using to address them. We believe this book offers a sound foundation for students who are preparing to meet these grand challenges of sustainability.

The book is divided into three interlocking parts:

↘ Part I, "Forces," examines how people's relationships to landscapes have reshaped human understanding and the world we inhabit. In doing so, we investigate the question of what is natural, and we examine how environmental problems emerge from human institutions.

↘ Part II, "Grand Challenges," describes the challenges that threaten the sustainability of the planet as we know it: global population trends and urbanization, energy consumption and impacts on climate, the fragility of biodiversity, and the search for sustainable development.

↘ Part III, "Strategies," introduces strategies that are now being pursued to address the grand challenges and alter the forces that propel them: environmentalism and public policy, adapting markets to reconnect humans with nature, rethinking consumer behavior, and reducing harmful activities, as well as green design, philanthropy, religion, and the role of adaptive management and institutional learning.

The story of the decline of the Channel Island fox draws together many threads: the ecology of predators and prey, the social values reflected in the banning of DDT and in setting aside the Channel Islands as nature preserves, the wealth of a society that could afford those choices, and the ability of scientists and managers to implement them (rather than being stymied either by a lack of information or an inability to act in the face of opposition). All of these themes illustrate why tackling environmental problems requires the breadth of vision found in liberal education and the skills of people who have encountered the subject matter of the liberal arts and sciences.

In this book, we are saying something radical: environmental problem solving requires a broad, *liberal* education—not just the technical skills of engineering, law, or political advocacy, but all of these and more. This is something that isn't at all obvious at first, and we don't expect you to agree with us today. But read on.

KEY TERMS

biodiversity	climate	sustainable
classical environmentalism	ecosystem	development
	grand challenge	urbanization

Chapter Two

WHAT IS NATURAL?

LANDSCAPES AND THE IDEA OF PLACE

Are humans *natural*? Is our species' extensive and expanding modification of ecosystems a natural process, like the spread of a virus or a forest fire? Or is it instead a series of choices grounded only in our culture and material interests, completely detached from the environment? Such questions signal semantic puzzles; for example, what does the word "natural" mean? In the eyes of many scientists and scholars, "nature" is a term for the all-encompassing reality in which people and society are constituent parts. Environmentalists, however, sometimes view "nature" as a domain that is separate or apart from society, and often consider humans to be violating the natural order. In this chapter we will begin our journey into environmental studies by opening up three perspectives on this indictment:

1. The idea of pristine nature—indeed, divine nature—has a history.
2. Our ability to identify changes in the natural world has been tremendously expanded by environmental science, but many environmental changes remain invisible in everyday life.
3. A sense of naturalness can be designed into landscapes.

Developing these perspectives also demonstrates the wide range of disciplines and ways of knowing that make up environmental studies. All this complicates the question of what is natural, and as we will see in later chapters, these different ways of seeing things help us to grasp more clearly the tremendous impact of humans on the rest of the natural world.

THE DISCOVERY OF LANDSCAPE

Although we think of environmental awareness as a recent social movement arising over the past half-century, it has much deeper roots in American culture. We first examine those roots in American art, through the painters of the Hudson River School, a group of artists who worked in the middle decades of the nineteenth century. Their paintings reflected the larger Romantic movement that produced William Wordsworth and John Keats in British poetry, Franz Schubert and Johannes Brahms in German music, and Ralph Waldo Emerson, Emily Dickinson, and Henry Thoreau in American literature. Romanticism was, in short, an international cultural movement that foreshadowed the way rock music and movies today can resonate in Tokyo and Cape Town as well as in Atlanta and Chicago. There was a popular culture in the nineteenth century similar in some ways to what see now, and these paintings, symphonies, and poems were among the "greatest hits" of the time.

The nineteenth century was also a period of major transformation in the nations that now comprise the developed world. In western Europe and North America, industrialization and urbanization radically reshaped societies, economies, and the natural world. The transformation emerged from the **Industrial Revolution**, which had begun in the later years of the eighteenth century. What emerged is the developed world—a set of middle-class societies in which few children die in infancy, a small fraction of the population feeds everyone through mechanized farming, and massive use of fossil fuels propels manufacturing, automobiles, and

Learning Objectives
When you have finished studying this chapter, you should be able to

- ➘ identify images in contemporary culture that evoke the idealized vision of nature expressed by the Hudson River School artists and the Transcendentalist writers of the nineteenth century;

- ➘ see in many elements of culture, such as these images and the way some parks are designed, that the ideal of pristine nature has a history that can be seen in the human environment;

- ➘ describe how the question "What is natural?" can be analyzed from multiple

perspectives, including those of other courses you are taking;

- ➘ spot examples of the "invisible present" in everyday life;

- ➘ describe how your college has become a place in your own experience, and how its buildings, landscape, and setting have acquired meanings that are shared among the students who attend the school, as well as the faculty and others who belong to the community.

other pollution-generating activities that impose large-scale impacts on nature and its inhabitants. Out of this transition to modernity came environmental concerns: worries about the costs to landscapes and humans of industrial technologies, and a sense of loss for the countrysides and rural ways of life that were left behind in the rush to prosperity. Romanticism in art and literature was a cultural movement that contributed in important ways to the subsequent rise of environmentalism. In this chapter, we focus on the nostalgia for a rural past, and in Chapter 6 we will return to the ecological transformations of modernization, a topic that will concern us in the rest of this book.

There is quite a lot of art to look at in the pages that follow, but we are not asking you to learn details such as the name of an artist or painting or the date it was painted. These are less important for our purpose than understanding the historical context and the statements that can be seen in the artworks. We will discuss paintings made in the middle of the nineteenth century, a time of urbanization, industrialization, the American Civil War, and the westward expansion of settlement that made the United States a continental nation. Although human activities had always impacted the environment, the Industrial Revolution substantially increased the scale and pace of social and environmental change, setting the stage for the major challenges we face today. The artists did not say in words what they meant to convey in their images, but viewing those paintings now, we can observe both what they portrayed and how they did it. This kind of interpretation is what art historians do. It is also what every viewer does, and we ask you to look at the images in this chapter and consider what you see in them.

Frederick Church painted Mount Katahdin in Maine in 1853 (Fig. 2.1). This was the year before Henry David Thoreau published *Walden*, a book whose perspectives on nature and humans still resonate in environmentalism. Today, the mountain's peak stands less than 30 miles west of Interstate 95 in Baxter State Park in north-central Maine. The Appalachian Trail, a famous hiking route that traverses the eastern United States, ends at Mount Katahdin. Church was a leading figure in the Hudson River School, a group of artists active between 1825 and about 1870—the two generations that lived in the years leading up to and through the Civil War.

The Hudson River School painters were the first American artists to focus on landscapes. *Mt. Ktaadn* was completed in 1853, more than two hundred years after the Puritans landed and more than a lifetime after the United States became an independent nation. Church's painting is much closer to us in time than it is to the Puritans or the French voyageurs or Spanish conquistadors. What does it mean that it took two centuries for Americans to develop an aesthetic appreciation of nature?

In their paintings, the Hudson River School made the claim that America has grand landscapes equal in their magnificence to the cathedrals of Europe or the

ruins of Greece and Rome. This notion that our landscape is a key part of our national heritage led to the creation of national parks. The first of those, Yosemite, was set aside in 1864, at the same time the second generation of Hudson River School painters was showing their canvases.

The idea that rural landscapes and farm life, such as that shown in the foreground of *Mt. Ktaadn*, could be beautiful is still reflected on calendars and postcards today, and in the number of tourists who flock to the northern states each autumn to enjoy the colorful leaves. These landscapes inspired much of what we now call environmental awareness in American culture, and they still inspire feelings of awe and attachment to nature. We care about nature and what is "natural" in part because of this legacy. Though environmental issues have other dimensions that do not seem connected to aesthetics—for example, struggles over releases of toxic chemicals from factories in urban neighborhoods—here, too, our understanding of "nature" colors the way we think about policy decisions.

THE HUDSON RIVER SCHOOL

The Hudson River School was founded by Thomas Cole, a young Englishman who moved to America with his family as a teenager not long after the War of 1812 confirmed the independence of the United States. Frederick Church was one of Cole's most prominent students.

Cole found many dramatic landscapes in his adopted land, a wide-open, little-known country. Cole and his followers painted large canvases, such as *The Oxbow*

(Fig. 2.2), for display in large halls (more on this painting below). One might think of these paintings as an early version of movie theaters. Urban crowds would line up to pay admission to see them, in the same way people do now when museums hold blockbuster exhibitions. Cole, and the Hudson River School he founded, became a mass media sensation among the middle classes of America's growing cities—the beginning of a line that led over time to Hollywood, Nashville, and YouTube. Hudson River School paintings remained popular for much of the rest of the nineteenth century.

The nineteenth century was a time of rapid change in America. Apart from Maine, New England's forests were extensively cleared by the 1830s. The opening of the Erie Canal in 1825 dramatically lowered the cost of transporting goods and crops. This led to the growth of Chicago and created stiff competition for northeastern farmers, including the ones who worked the fields in *The Oxbow*. In the second half of the nineteenth century, the gold rush spurred the settlement of San Francisco and the beginnings of today's West Coast. And from 1860 to 1865, the Civil War was fought to determine whether the United States would remain a single country.

All of these major changes are reflected in the Hudson River School's canvases and the idealistic portrait of America they presented. By the early nineteenth century, nature was no longer seen as adversarial, as it had been by Puritans

FIGURE 2.2
Thomas Cole, *The Oxbow*, 1836.

and other early colonists. Barbara Novak, an art historian, wrote of this cultural shift: "Only when the colonist had cleared the forest and made it fit for 'man's abode' could he approach the luxury of loving it."[1] As Romanticism gained widespread influence in Europe and America, nature came to be considered a source of beauty rather than the wild, threatening place that the initial settlers had to tame in order to survive.

A message prominent in many Hudson River School paintings is that beautiful landscapes are filled with *religious* significance and that nature is divine (see Box 2.1: Nature and Belief on page 27). Consider *Early Morning at Cold Spring*, painted by Thomas Cole's friend Asher Durand (Fig. 2.3). There is a church spire at the center of the scene, and the arching trees remind the viewer of a Gothic cathedral framing a meditative figure. The bright sky is reflected in water—the light of

FIGURE 2.3
Asher Durand, *Early Morning at Cold Spring*, 1850.

God. Like the shepherd in *Mt. Ktaadn*, the human figure in Durand's painting contemplates grand nature. What is natural? Everything; and humans are a small part of God's grandeur, or for the mystic, a component part of the divine order.

This school of art parallels the rise of a school of writers, including William Cullen Bryant, Ralph Waldo Emerson, Henry David Thoreau, and Emily Dickinson, who developed a religious philosophy called Transcendentalism. Transcendentalists saw nature as a cleansing, spiritual power. The landscape, as a manifestation of sublime goodness, was to be exalted as well as observed. Here is Emily Dickinson's poetic statement of the Transcendental belief, written around 1863:

> Nature is what we see,
> The Hill, the Afternoon—
> Squirrel, Eclipse, the Bumble-bee,
> Nay—Nature is Heaven.

Nature is what we hear,
The Bobolink, the Sea—
Thunder, the Cricket—
Nay,—Nature is Harmony.

Nature is what we know
But have no art to say,
So impotent our wisdom is
To Her simplicity.[2]

Ralph Waldo Emerson put a similar thought in his book *Nature*, published in 1836:

In the woods, we return to reason and faith.... Standing on the bare ground,—my head bathed by the blithe air, and uplifted into infinite space,—all mean egotism vanishes. I become a transparent eye-ball. I am nothing. I see all.... I am part or particle of God.... I am the lover of uncontained and immortal beauty. In the wilderness, I find something more dear ... than in streets or villages.[3]

Today, the idea of seeing God in nature has become controversial in a surprising way. Some fundamentalist Christians champion the idea of intelligent design, the claim that the awesome intricacy of the natural world must have a supernatural cause. This idea rejects Darwinian evolution and its assertion that random mutations give rise to the variety of living things. Intelligent design proposes that God does not play with dice at all; rather, His perfection is embodied in the complex beauty of creation. This is an idea that conflicts with the current consensus in biological science and medicine. The Transcendentalist worldview was formulated before the appearance of Darwin's *Origin of Species* in 1859, the book that announced the theory of evolution. One can see that intelligent design in the present day shares the reverence expressed in Durand's *Early Morning at Cold Spring* or in Dickinson's mystical admission "So impotent our wisdom is" (see Box 2.1, page 27).

The Hudson River School painters became famous for beautiful landscape paintings like *The Oxbow* and *Sunday Morning*, which were created deliberately as artifacts. These were artistic images, statements not of literal reality but selective portrayals influenced by ideas and social conventions, sometimes more so than by the landscapes themselves. Thomas Cole's paintings were created in his studios in the town of Catskill, New York, and in New York City. He worked from sketches done in the field, but he did not make paintings directly outdoors, as the French Impressionists did later in the nineteenth century. Nature for the Hudson River School painters was an ideal, a perfect world that was disappearing under ax and plow as the country moved from agriculture toward urban settlements and industry.

FIGURE 2.4
Asher Durand,
Kindred Spirits,
1849.

When Thomas Cole died in 1848, his colleague Asher Durand painted *Kindred Spirits* as a memorial to his friend (Fig. 2.4). Cole is shown standing with the Transcendentalist poet William Cullen Bryant, one of the most famous literary figures of the time. In the foreground is a dead tree, a symbol of the transience of life. The two men are standing in a dramatic landscape, but notice that they seem to be dressed for the city rather than roughing it. Moreover, the painting reflects Durand's choices rather than a faithful representation of nature. This scene does not really exist, but is a composite of two nearby sites in the Catskill Mountains. The lower falls is Kaaterskill Falls, which is not fed by the stream tumbling over a rock wall in the central part of the painting. Nature has been selected and enhanced to make an artistic and emotional statement, and to celebrate the artist who founded the Hudson River School. What looks natural may not be literal.

Every autumn, visitors to New England marvel at the beautiful foliage as the seasons turn. How beautiful nature seems on a bright, sunny day in early October. Very few of these leaf peepers, as the locals call the tourists, know that they are looking at second-growth forest. The hills of New England were, for the most part, pastures and farm fields until the 1930s, when farms failed during the Great Depression and were abandoned. Is today's landscape natural? How should we compare it to Asher Durand's enhanced vision of the Catskills in *Kindred Spirits*? And what should we say about images manipulated using computers? Before we judge Durand's artifice, consider that the European colonists who settled America also enhanced nature, as they saw it, though they did so by removing trees, building walls, and planting crops. The rearrangement of ecosystems happens on real landscapes, too, as well as in the virtual reality of a painter's canvas.

Cole's landscape *The Oxbow* (Fig. 2.2), painted in 1836, shows the Connecticut River from Mount Holyoke, south of Northampton, Massachusetts. One can drive up to the Summit House now, where Cole made the sketches for this painting, and it's an enjoyable outing on a fine day. The summit of Mount Holyoke was a spot often chosen by artists because of its sweeping vistas overlooking a wide river valley.

BOX 2.1

NATURE AND BELIEF

In the United States, religion is considered a private matter, one that should have little bearing on legal rights or one's ability to participate in society. We imagine religious duty as being mixed, usually indistinctly, with morality, and we sometimes think of both moral and religious practice in negative terms, as restraints on individual behavior. Or perhaps we think of religion as a family tradition, one that provides solace in times of grief, succor in times of trouble, and a community to celebrate happiness. Some think of prayer as magic, a way to compel recalcitrant fortune, but the hope that prayers will be answered is faint nowadays among many of those who consider themselves educated.

When we contemplate religion, what often comes to mind are church buildings and ceremonies. Many of us do not associate nature or the environment with religious belief. But the border between environmental values, on one side, and intellect and reason, on the other, is not so clearly marked as one might think at first.

Consider this Emily Dickinson poem, composed around 1863. Her language is characteristically antiquated to our ears:

My Faith is larger than the Hills—
So when the Hills decay—
My Faith must take the Purple Wheel
To show the Sun the way—

'Tis first He steps upon the Vane—
And then—upon the Hill—
And then abroad the World He go
To do His Golden Will—

And if His Yellow feet should miss—
The Bird would not arise—
The Flowers would slumber on their Stems—
No Bells have Paradise—

How dare I, therefore, stint a faith
On which so vast depends—
Lest Firmament should fail for me—
The Rivet in the Bands[1]

The first line, "My Faith is larger than the Hills," said something about her own modernizing, industrializing world that we now accept as an anthropological fact: *religion embodies a culture's assumptions about the nature of reality.* The oddity of her poetic diction ("How dare I, therefore, stint a faith / On which so vast depends—") tempts us to dismiss Dickinson as a mystic. Yet her point is one we now take for granted: that the natural world is fundamental, something in which to have faith. (More than a century later, the ecologist Paul Ehrlich compared the loss of species in ecosystems to rivets coming loose in an aircraft in flight. For a long time, nothing much happens, but then the plane can no longer stay airborne. Similarly, when a species goes extinct, the bands of the world lose a crucial rivet.)

Dickinson and the Transcendentalists did more than celebrate nature. Through paintings and poems like "My Faith is larger than the Hills," they taught society to appreciate nature. The success of their teaching has been so sweeping that it is hard for most of us to see how it could be otherwise. But it could. Many cultures have seen the natural world as cruel and arbitrary, ruled by gods as capricious as they are powerful. We now think of nature simply as reality—a sign that this belief is so deeply entrenched in our own culture that it does not seem to be a belief at all.

1. "My Faith Is Larger than the Hills," by Emily Dickinson. Reprinted by permission of the publishers and the Trustees of Amherst College from *The Poems of Emily Dickinson*, Thomas H. Johnson, ed., Cambridge, Mass.: The Belknap Press of Harvard University Press. Copyright © 1951, 1955, 1979, 1983 by the president and fellows of Harvard College.

The major difference between the view that Cole offered and what one sees today is that the trees have regrown in the foreground by the river, and that these trees hide the suburban housing in much of the scene. What we see in small-town New England is true of North America as a whole: forested land has increased, not decreased, in recent decades.

The Oxbow is a celebration of American democracy, a vision of a countryside populated by self-sufficient farmers. Cole painted this landscape when Andrew Jackson was president, and Jackson was known for regularly inviting ordinary citizens to the White House. Cole used a conventional composition: dead trees to the left and, in the distance, eternal mountains and the grandeur of storms. But in the middle distance farms can be seen, a peaceable presence on the land. The view is toward the southeast. To the east lies a landscape obscured by storm. Some art historians see in the sunlit west a promise of harmonious development on the frontier, in contrast to the clouded skies to the east and toward the Old World.

The Hudson River School and the Transcendentalists saw nature as sublime, a vision of a benign world shaped by God, whose presence may be sensed in the awe inspired by a grand landscape. These Romantics projected this vision just as the Industrial Revolution was sweeping across America, and the urban crowds drawn to see the paintings of Church and Cole were touched by nostalgia as well as worship.

PROGRESS, AND A SURPRISE

Even as the Hudson River painters were celebrating tranquil nature, the American landscape was being transformed by large-scale, industrial technologies.

John Gast painted *American Progress* (Fig. 2.5) in 1872 in anticipation of the U.S. centennial in 1876. Although the landscape is visible in this image, Gast was far from the Hudson River School artistically. His fantastic symbolism is as explicit as the symbolism found in medieval art. In this painting, Gast portrayed the idea of Manifest Destiny. The female figure is Columbia, a symbol of the United States. She carries a coil of telegraph wire on her arm as she wafts westward, leading the railroads, stagecoaches, and Conestoga wagons of the pioneers. American progress was to have a significant impact on the landscape and its inhabitants (note the bison and American Indian peoples being driven into the shadows on the left). Yellowstone

FIGURE 2.5
John Gast, *American Progress*, 1872.

FIGURE 2.6
Marc Riboud,
Wuhan,1971.

National Park was created in the same year this painting was made. Gast's canvas shows that the tension between developing landscapes and admiring eternal nature was already apparent in American culture more than a century before the environmental movement.

Progress is not merely an American obsession, of course, nor was it a concept confined to the nineteenth century. Figure 2.6 is an evocative image captured by the French photographer Marc Riboud in Wuhan, China, in 1971. The communist smoke billows to the left, obedient, it seems, to the guidance of Mao Zedong, the "Great Helmsman." This photograph suggests how pollution can be a source of pride as well as shame, a marker of prosperity as well as destruction. Many of the world's poor today put up with illness and pollution from industrialization in the pursuit of growth and development. The values embodied in this image pose important challenges to environmentalists' hopes of conserving biodiversity in the tropics or of tempering China's rapidly expanding use of coal, which has major implications for global climate change.

Figure 2.7 is an image you are likely to have seen. It was taken in 1972, a year after Riboud visited China. This photograph was made just a century after *American Progress*, and it could easily bear the same title as Gast's painting. This, too, is a celebration of the technological prowess of a colonizing society.

But what is surprising, indeed amazing, is the cultural sensibility that this image symbolized and helped to spread around the world. For Earth from space is something we see not in terms of progress at all, through Gast's eyes or those of Mao's masses. We see it, instead, through the eyes of Frederick Church and Thomas Cole. This is an image of sublime nature: the fragile blue planet that is our home, the unspoiled place that we are paving over and polluting.

FIGURE 2.7
NASA Apollo 17 Mission, *Earth from Space*, 1972.

The environmental awareness stimulated by this photograph parallels the Hudson River School's natural theology of more than a century earlier. Both arose after major despoliation of the environment for economic purposes was well under way. Both articulate a nostalgic longing for a disappearing past, both natural and social. And both endured, becoming strands of the economy as well as cultural and educational themes of lasting value. In 1972, when *Earth from Space* was taken, the United Nations (UN) convened the first international conference on the environment, a meeting that led to the founding of the UN Environment Programme, which was established to act as the voice of the world's governments in the protection and management of the environment.

Much of our visual aesthetic of nature is clearly traceable to the Hudson River School, just as the literature of environmentalism has its roots in Thoreau, Dickinson, and other Transcendentalists. These idealized landscapes, as journalist Bill Moyers said, portray "a world that Americans would prefer to remember."[4]

In the nineteenth century—as is still true in the twenty-first—the spiritual elements of unspoiled nature are often more apparent to urban and suburban people than to those living in rural settings who win a livelihood from the landscape. As Durand's *Kindred Spirits* suggested, art is an urban phenomenon, even art celebrating wilderness. In the Hudson River School paintings, nature was seen as a source of spiritual repose, not something to be subdued or exploited. This was a view held by middle-class city dwellers, much as environmental activism today finds a ready audience in the middle-class suburbs.

Indeed, suburban landscapes, also influenced by the aesthetic of the Hudson River School, evoke an American pastoral vision. The house, set in a garden screened by trees, is a private Eden, a secure place of domestic tranquility. So it is not only environmentalism that comes from the natural theology of the Hudson River painters. So do suburbanization and sprawl and, as discussed above, intelligent design. These are descendants of American Romanticism that environmentalists might wonder about.

As for the question of what is natural, we can see now that our definitions of nature have a history. There were no landscape paintings in the New World until the Hudson River School. The culture that admired these artists' canvases embraced the idea that nature is beautiful and touched by the divine. Earlier generations of European colonists saw the landscape as something to overcome rather than admire. What is natural? This is a human question, it seems, answerable only by humans.

TWO APPROACHES TO SCIENCE

In this book, we often treat science as an offshoot of the humanities—a source of insight into vitally important matters that we may not ever understand completely and that we cannot change or may even want to preserve. This is an unusual way to think about science, which we normally think of as inquiry that leads to the development of technology in the same way that physics underlies computers or that social psychology has informed advertising. Much environmental destruction has been described as reflecting "rational" applications of "science," and environmentalists have frequently highlighted the conflict between technology and human values. It is, of course, also the case that many solutions to environmental problems involve rational applications of science. How we think about science is important at the outset. This concern leads to another perspective on what is natural: not an aesthetic one but a scientific one, concerned with different ways of approaching nature.

One of the distinctive features of world culture over the past three centuries has been the flowering of science, and the systematic study of humans and nature, in search of knowledge that is backed by empirical evidence. Science has developed in many different directions, and here we want to highlight a major division among modes of scientific inquiry—between approaches that rely mainly on experiments and those that rely mainly on observations. Real scientists use both approaches in their work, so we are oversimplifying to make the point that understanding the environment relies on careful, systematic observation of both landscapes and humans.

Since the eighteenth century, one model of science has enabled rapid gains in knowledge. In the process, this model has gained enormous prestige, so much so that the word "science" is now usually equated with this approach. We call it **experimental** or **theory-driven science** in this book. As these names indicate, the method of scientific inquiry is one that often depends on experimentation. Experiments probe the predictions of theories (which scientists call "hypotheses"), and successful experiments connect the concrete world of the laboratory and test tube to the abstract understanding and equations of theory.

The sciences of chemistry and physics have become the definitive examples of theory-driven science. Their elegant equations form the foundation of the engineering sciences, the knowledge that informs engineering and technology. On that foundation rest the structures—industrial processes and materials and information processing—of contemporary society. In the twentieth century, the branches of biology that we now call molecular biology adopted the experimental paradigm of chemistry, spawning genomics, proteomics, and synthetic biology. Some of the social sciences, notably psychology, have also embraced the experimental paradigm, again with great success. And over the past generation, ecologists, too, have used experimental techniques to study patches of the natural world and have gained new insights.

Yet there is another kind of science. It is exemplified by astronomy or political science, two arenas of human experience that do not lend themselves to the laboratory-scale manipulations that characterize most experiments. We call these **observational** or **field sciences**.

Collecting things and observing them closely grew out of the tradition of natural history, along with a respect for the complexity of the world as we find it. Systematic observing shapes the sensibilities of many museums and other places where people collect things, such as libraries or junkyards.

Systematic observation is more than collecting, however; it can spawn theory. It is from careful observation that Charles Darwin discovered the principles of evolution in the nineteenth century; that Johannes Kepler inferred the orbits of the planets in the seventeenth century; and that inventors spurred the Industrial Revolution of the eighteenth century.

These examples all demonstrate that from the complexity of the world there sometimes emerges a deep simplicity—the elegant simplicity of theory. Studying the observational patterns found by Kepler, Isaac Newton realized that the force that causes an apple to fall from the tree is the same as the force that propels the Moon around Earth. That is, the Moon is continually falling toward Earth as the Moon traces out its orbit. This meant that the *same* equations could be used to describe both—an astounding claim, if you think about it. We now call these equations the Newtonian theory of gravitation and the laws of motion of classical mechanics.

Darwin knew that, given enough time and careful selection by human breeders, Chihuahuas and Great Danes could be bred from a single species of dog. And he saw that the natural world might do the same, so that fish and birds and trees might all stem from a common form of life. This is called the "theory of natural selection."

The French scientist Sadi Carnot studied steam engines, which inventors had been developing by trial and error for more than a century. In 1824 he published a paper describing these engines in terms of an idealized machine that we now call a heat engine. Carnot's heat engine provided the first clear way to explain the theoretical concepts of thermodynamics, the science of heat. Those ideas gave crucial guidance to the engineers who harnessed fossil fuels in the Industrial Revolution, which would eventually power machines from automobiles to power plants. These and other theories have suggested many different lines of investigation.

Scientists have invented experiments to probe the implications of these and many other theories. The power of experiment stems from a surprising fact. Often it is possible to cut nature into little pieces, isolating phenomena so that they can be observed and measured in a laboratory. Then (this is the part that many find surprising) one can often reassemble, conceptually and sometimes by fabrication, a functioning process from an understanding of the parts. This is a machine-like model of natural phenomena, very different from the organic picture one might develop from watching a seed turn into a flowering plant. If you think about it, it is astounding that the machine model works at all, let alone as well as it often does. Why would we expect, for example, that inserting a gene from a fish into a strawberry plant would give the plant some of the fish's ability to resist freezing?

The name given to this approach of cutting a complex phenomenon into parts is "reductionism." To an extent that few would have thought possible in the eighteenth century, reductionist science, driven by theory and based on experiments, has succeeded brilliantly. By laying the groundwork for engineering everything from bridges to the Internet to heart valves, theory-driven technology has reshaped human life and economies so much that we have emerged as a force on the planet of geological proportions. The decoding of the human genome is a recent milestone in this program.

Yet many important scientific questions continue to resist an experimental approach, many of them in the study of the environment. The behavior of large ecosystems, such as an ocean or one of the Great Lakes, is an obvious case. Most of what we care about in the environment happens too slowly, and on too large a scale, to be reproducible in a laboratory. Think of global climate change or the extinction of species in a rain forest. In other fields, such as anthropology or oceanography, we still have not found effective ways to capture enough of what we think is significant in a test tube. In such fields, systematic observation remains indispensable. Careful observation, in turn, breeds awareness of how complicated the behavior of natural systems can be.

Contemporary environmental science draws upon *both* of these scientific traditions (Table 2.1). For example, the scientific understanding of climate change is grounded on well more than a century of careful observations of air temperatures, wind speeds and directions, and ocean currents and temperatures from all over the planet. These observations are combined with detailed models of the motions of the atmosphere and oceans into computer models formulated with the simplest laws of physics. These large computer models, many of which run on supercomputers, are elaborate theories: statements of how the world's weather, given a specific starting point, would evolve. The observations supply that starting point. The models are then tested against history. For example, how well do they reproduce the climate patterns of the past century? By improving and tuning their theories, climate scientists have gradually evolved more than a dozen models, each with slight differences in their

TABLE 2.1 TWO APPROACHES TO SCIENCE

	Experimental or theory-driven science	Observational or field science
HOW UNDERSTANDING EMERGES	Experimentation to test and build theory	Systematic observations to infer processes from surviving structures
EXAMPLES	Chemistry Physics Molecular biology Social psychology	Anthropology Astronomy Political science History

internal assumptions. These models are then used to project future climates. Those projections show what citizens all over the world now call global warming.

Although environmental sciences from climate to ecology all make use of experimental science and the theories built from experimentation, *observation remains crucial* to environmental studies. This is true for a simple reason. The world is a complicated place, and its behavior is often surprising to humans, even when we are the cause of environmental change. Or perhaps the right thing to say is, especially when we are the cause of environmental change.

Note that both "social" and "natural" sciences are listed in Table 2.1. Environmental studies require the study of both human and nonhuman nature. The contrast between experimental science and observational science is more useful than separating natural from social science disciplines for our purposes. In that respect, the question of what is natural may not matter as much as the question of where in the disciplinary organization of knowledge we might find better ways of understanding how humans respond to and shape landscapes.

THE INVISIBLE PRESENT

As our powers of observation have expanded through the invention of more powerful and more wide-ranging tools, we have discovered something important. The scales of space and time to which humans are accustomed are not the only ones that matter in the natural world. Consider this statement from ecologist John Magnuson:

All of us can sense change—the reddening sky with dawn's new light ... and the changing seasons.... But it is the unusual person who senses with any precision changes occurring over decades. At this time scale, we are inclined to think of the world as static, and we typically underestimate the degree of change that does occur. . . . Processes acting over decades are hidden . . . in . . . "the invisible present."

The invisible present is the time scale within which our responsibilities for planet earth are most evident. Within this time scale, ecosystems change. . . . This is the time scale of acid deposition, the invasion of non-native plants and animals, . . . CO_2 [carbon dioxide]-induced climate warming, and deforestation. . . . Although serious accidents in an instant of human misjudgment can be envisioned . . . destruction is even more likely to occur at a ponderous pace in the secrecy of the invisible present.[5]

The **invisible present**—in plain sight but changing so slowly that humans and our institutions don't notice—implies that observations at scales that lie outside ordinary human experience can produce surprising insights. For example, decades before scientists could see the retreat of glaciers, they knew the chemical composition of the atmosphere was changing, so they knew to be alert for the signs of global warming.

Sometimes, the invisible present emerges from the shadows, if one knows what one is seeing. There is a striking example on a hillside above North Adams, a small town in western Massachusetts. Look northward from the main highway going through North Adams in late summer, and you will see a green forest on the slope. Drive through town again in winter, and the slope is a scene of bare trees. But from mid-September to early October, the hillside provides a fine New England display of autumn leaves in full color. Partway up the hill is a patch of red maples, scarlet in the morning sun. But the patch is a curious one: it has straight edges, which can clearly be seen from the road below.

On this patch of land was an open pasture surrounded by woods. When the pasture was abandoned in the 1930s, the first trees to take root were ones that grow well in open sunlight. These included many red maples. In the woods surrounding the pasture, where the soils have been shaded for many decades longer, other species flourish. The red maple is called a "pioneer species" by naturalists, because it is one that grows well when cleared land is left to regrow without human intervention. The farm field that is now a stand of red maples is thus a visible marker of the history of land use, a sliver of the invisible present that one can see if one looks up from the road on a bright autumn day.

We will return to the idea of the invisible present in later chapters (see Box 2.2: The Invisible Present of New England Winters, page 37). For now, think about what you can see changing on a scale of decades, such as the aging of grandparents or the spread of suburban development. We are used to thinking of the natural world as unchanging, the way the Hudson River School painters did. But with

BOX 2.2

THE INVISIBLE PRESENT OF NEW ENGLAND WINTERS

Williams College in Williamstown, Massachusetts, has long snowy winters, and the memories of the college's alumni often feature skiing or snowball fights. But how much snow was there? The figure shows the yearly snowfall in Williamstown during the twentieth century.

The first thing to notice is that the annual snowfall varies a lot, from a high of more than 110 inches down to less than 30 inches, a factor of about 4. The line connecting the dots jumps around a lot. When we average the snowfall in four-year intervals, however, the variations settle down, as one would expect. Years of high snowfall do not cluster together, and the average over four years is not nearly so variable as the annual snowfall levels.

Four years, of course, is the length of a college education. The class of 1969 graduated after seeing an average of more than 80 inches of snow in their winters, whereas the class of 1992 experienced an average of about half as much snow. The snowfalls in these graduates' memories were quite different. The larger point is the one made by John Magnuson in the passage quoted in the text—that human experience often unfolds on a different scale of time or space than the natural world.

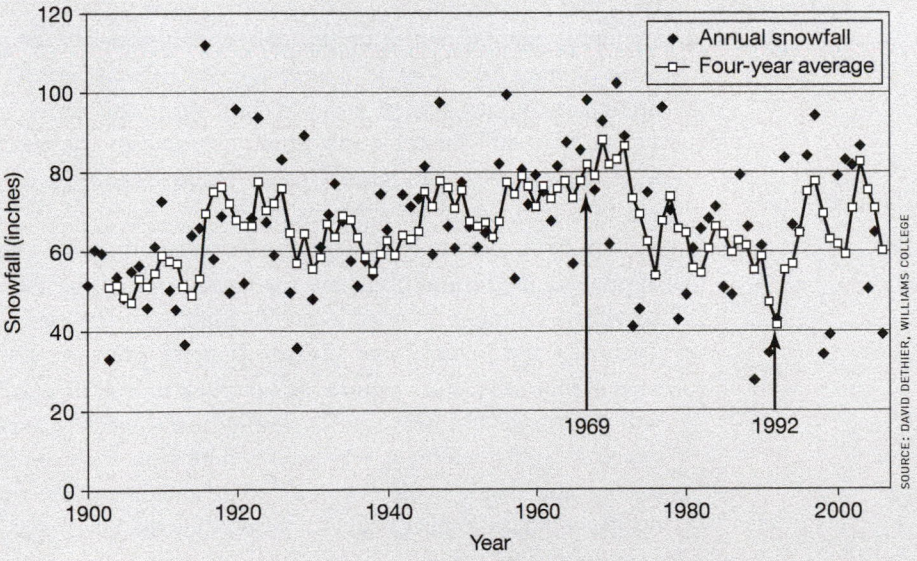

Snowfall in Williamstown, Massachusetts, 1900–2002.

SOURCE: DAVID DETHIER, WILLIAMS COLLEGE

global warming or the dwindling of fisheries or the erosion of farmland soils, we can see that much of the nonhuman world can be changed dramatically—albeit slowly—by human action. Those changes are hidden from view, right under our noses, in the invisible present.

A SENSE OF PLACE

What the Hudson River School painters brought to their audiences was the experience of place, of being in a dramatic landscape and seeing its grandeur in a single glance. What is it about some locales that leads people to feel that a place is special in this way? Why is a beach at sunset more beautiful than a freeway at rush hour?

Human experience has been, until very recently, primarily local, tied to particular locales. The memories that college alumni recall when they return to visit campus are often tied to specific locations—dining halls, courtyards, sports facilities, perhaps a library study space—that have particular meaning.

Sailors have navigated the oceans for centuries, but it is only within the past generation that images of the planet from orbit, or the paths of storms on a regional weather map, have become part of everyday experience. This has profoundly changed our perspective. It helps us, for example, to imagine global warming.

Yet to understand the human presence on the planet, we must grasp the human scales on which choices have been made and continue to be made, in which traditions and ways of life have been etched. Places invested with meaning constitute a tangible past and visible present, places that people want to preserve or change. Most humans have now become aware that national, continental, and global phenomena exist, but human scales in both space and time—place and community and family—still matter a lot. Environmentalism has taken root at the human scale of places, even though environmental science often tells us of phenomena or warns us of threats from different scales.

This book seeks to bridge the gap between human and environmental scales by developing the idea of place and the cluster of ideas called "sense of place." These are elusive notions grounded in people's subjective experiences and reactions, their memories and loyalties and dislikes. It is tempting to dismiss sense of place as merely subjective. But consider any popular movie. In a darkened theater, people will see images and hear sounds that induce subjective experiences, too. But the expectation that audience reactions will be shared among the majority of people is so reliable that a movie critic can write a review of the film knowing that most readers will have similar reactions when they see the movie. So "subjective" is not a synonym for "unreal."

Places are spatial locations, of course. But if your violin teacher told you that long years of practice might bring you to N 40°45'54", W 73°58'49", would you know he was inspiring you to play a concert at Carnegie Hall in New York City? Places are at once geographical coordinates and clusters of meaning. People often care about places, sometimes passionately—as they do in Jerusalem, for example. Students often trace their own environmental consciousness to a time when a favorite grove of trees or an open field was suddenly transformed into houses or a shopping center. Some people now oppose electricity-generating wind turbines because they will spoil a familiar landscape, even though the windmills avoid the generation of greenhouse gases.

The naturalist and poet Gary Snyder wrote that "the world is places."[6] By this he meant that people experience the world, not as a whole, but as localized places. For this reason, people's experience of place is a major arena where environment and human feelings and knowledge come together. Snyder proposed to divide geography into three parts: home, commons, and wilderness.[7] One might call this a developmental concept of place; each of us learns where we are by experience as we grow up. Home is the place we know from earliest consciousness. Wild places are those we may sometimes visit but which are not inhabited. Their character comes as much from the legends that grow up around them as from direct experience. In this sense, places such as the UN or the Amazon rain forest are wild for most of us. We know them by their legends, rather than by direct experience. After Hurricane Katrina in 2005, many Americans were surprised to discover that New Orleans, which they thought of solely in terms of the French Quarter, also had a large population of poor people, who were previously on the fringes of the New Orleans legend. Snyder then focuses on those places that lie between home and wilderness, which he calls "commons." This is a very important idea for environmental studies, to which we'll return in the next chapter. But for now, it's worth noticing that the commons are lands that are used (not wild) but are not owned by individuals, as a farm would be.

Snyder's use of the word "home" suggests something warm and cozy. Yet as we know from many biographies, it is possible to feel that a place is familiar without feeling especially comfortable in it. Here is the way the writer Joan Didion described her home:

There is something uneasy in the Los Angeles air this afternoon, some unnatural stillness, some tension. What it means is that tonight a Santa Ana will begin to blow, a hot wind from the northeast whining down through the Cajon and San Gorgonio Passes, blowing up sandstorms out along Route 66, drying the hills and the nerves to the flash point. For a few days now we will see smoke back in the canyons, and hear sirens in the night. I have neither heard nor read that a Santa Ana is due, but I know it, and almost everyone I have seen today knows it too. We know it because we feel it. The baby frets.

. . . Los Angeles weather is the weather of catastrophe, of apocalypse, and, just as the reliably long and bitter winters of New England determine the way life is lived there, so the violence and the unpredictability of the Santa Ana affect the entire quality of life in Los Angeles, accentuate its impermanence, its unreliability. The wind shows us how close to the edge we are.[8]

This ominous sense of place is surely subjective, and there are many other views of Los Angeles that people might say is the *real* Los Angeles, whether it is the *barrio* or Hollywood or the unstable geology of the San Gabriel Mountains. And yet, even a casual visitor to Southern California is likely to acknowledge that Didion has evoked something recognizable about that place.

Figure 2.8 shows a wild place as depicted by Edward Abbey in his famous book *Desert Solitaire*. Working as a ranger in the Arches National Park in the Utah desert, he came across a pair of snakes engaged in what looked like a dance. Quietly, he crawled closer for a look:

The two gopher snakes are nearly identical in length and coloring. . . . I cannot even be sure that they are male and female, though their performance

FIGURE 2.8
Arches National Park, Utah.

resembles so strongly a *pas de deux* by formal lovers. They intertwine and separate, glide side by side in perfect congruence, turn like mirror images of each other and glide back again, wind and unwind again. . . .

Suddenly . . . they discover me. . . . The dance stops. After a moment's pause the two snakes come straight toward my face, the forked tongues flickering, their intense wild yellow eyes staring directly into my eyes. For an instant I am paralyzed by wonder; then, stung by a fear too ancient and powerful to overcome I scramble back, rising to my knees. The snakes veer and turn and race away.[9]

This is a vision of wild nature, of beings, as Abbey wrote, driven by "beautifully selfish reasons of their own."[10] And in his primeval panic as the snakes came at him, he also was wild, part of the legend that he would tell.

DESIGNING EXPERIENCE

Now consider an example of what Snyder would call commons—the entrance to Prospect Park in Brooklyn, New York (Fig. 2.9). The footpath leads from the bustling Grand Army Plaza, a major intersection where several streets converge. The path passes a low hill with some trees growing on it and heads toward a pedestrian tunnel called the Endale Arch. The hill screens the noise of the traffic, and the long,

FIGURE 2.9
The Endale Arch in Prospect Park, Brooklyn, New York.

wide tunnel of the arch opens out. Here is how urban environmentalist Tony Hiss described the scene:

> If you were to come with me along the entrance path, through the Endale Arch, and a dozen steps beyond it, you would find the surroundings changing dramatically. . . . We would have just reached the north end of the beautiful undulating Long Meadow . . . the longest continuous open space in any urban park in the United States. . . . Someone can look straight ahead for almost a mile and all there is to see is grass and the sky above it. . . .
> . . . On almost any day of the year the whole area is flooded with light. . . . A stroller may feel that a much vaster scene perhaps repeats itself many times over, distance upon distance.[11]

Is the handsome vista down the Long Meadow a natural scene? Not at all. The park was designed by American landscape architects Frederick Law Olmsted and Calvert Vaux just after the Civil War. All of its features are manmade. The hill at the entrance was put there by municipal workers. When the park opened in 1873, one of its streams flowed with water driven by a steam-powered pump.

Artificial as the Long Meadow may be, when people walk into the park their experience is tranquil, like a Hudson River School painting. What is going on? The landscape design evokes the serenity of Durand's *Early Morning at Cold Spring* (Fig. 2.3). Prospect Park, too, was meant to create a sublime experience, like the "transparent eyeball" evoked by a Transcendentalist landscape painting. Here is what Olmsted and Vaux wrote in their proposal for Prospect Park:

> A scene in nature is made up of various parts; each part has an individual character and its possible ideal. It is unlikely that accident should bring together the best possible ideals of each separate part . . . and it is still more unlikely that accident should group a number of these possible ideals in such a way that not only one or two but that all should be harmoniously related one to the other. It is evident, however, that an attempt to accomplish this artificially is not impossible. . . . The result would be a work of art [that] we denominate landscape architecture.[12]

Remember that Asher Durand also rearranged nature in *Kindred Spirits* (Fig. 2.4) to portray "the best possible ideals" of Thomas Cole. By designing a landscape that is *not* "a scene in nature," Olmsted and Vaux could summon up the feeling of being in nature. It is a feeling that Tony Hiss described more than a century later.

This is a startling idea at first. How can subjective experience be designed and continue to be effective generations later? Recall the movie producer who creates an experience on film. Of course, a theater is a controlled environment, like a test tube, where it might be possible to manipulate the audience's responses. A park in the middle of a large city is not so readily controlled, though the hill at the entrance does separate the walker from the noise of the traffic outside the park. And think about how you dressed "appropriately" for going to a party or a job interview. Were you designing others' experience of you? Were you successful in inducing the perceptions you wished? The next time you are in a supermarket, look around and see if you can perceive how the store's layout and lighting might be intended to create perceptions on the part of shoppers. Also, think about experiences that are not designed, such as accidents, construction sites, weather events such as storms, or Joan Didion's brooding Los Angeles. These scenes also contribute to one's sense of a place—a dangerous intersection, or an unruly sea of mud and girders.

When the design of a place succeeds, the people who inhabit and use it create a history there and a larger meaning. Think about the places the al-Qaeda terrorists chose to attack on September 11, 2001. The World Trade Center and the Pentagon are not just office buildings but symbols of a larger order. The attack killed many innocent people, but it was the meaning of the places that was the target.

By looking at art, science, and design in this chapter, we have laid groundwork for the ways in which we will discuss environmental studies in the chapters to come. The model of classical environmentalism described in Chapter 1 pointed to the importance of science and public policy as ways to respond to environmental problems. But what do we count as environmental problems? In this chapter, we have seen that the value placed on the natural world emerged from a historical process, in response to the rapid industrialization and urbanization that transformed agricultural landscapes and the humans who inhabited them. That transformation has produced major problems, such as the loss of biological diversity. But it has also inspired passions—the passion for nature that one sees in the Hudson River School paintings, and the passion to solve problems that may have brought you to this book.

The knowledge to inform those passions is broad ranging, and part of the work of this chapter was also to surprise you (*Art? Subjective experience that is designed? Astronomy as a branch of natural history?*). The Hudson River School painters did not imagine global warming, but they helped to create a cultural setting in which environmental problems could be recognized. And they helped to make the question "What is natural?" both important and complicated.

FURTHER READING

Gaddis, John Lewis. Chapters 3 and 4 in *The Landscape of History: How Historians Map the Past*. New York: Oxford University Press, 2008.

Hiss, Tony. *The Experience of Place*. New York: Vintage, 1991.

Novak, Barbara. Chapter 8 in *Nature and Culture: American Landscape and Painting, 1825–1875*. New York: Oxford University Press, 1980.

Snyder, Gary. "The Place, the Region, and the Commons." In *The Practice of the Wild*, 25–47. San Francisco: North Point Press, 1990.

KEY TERMS

experimental (theory-driven) science	Industrial Revolution	invisible present	observational (field) science

Chapter Three

COMMONS

THE ORIGINS OF ENVIRONMENTAL PROBLEMS

In this chapter, we explore how humans cause environmental problems. Nearly all of us have felt that we cannot do much, as individuals, to deal with these problems; we have no choice, somehow, even though we feel guilty. We will probe two troublesome situations that arise as humans access nature: the commons and inequality. Both leave individuals without good choices, and the two together account for a large share of environmental problems.

We are saying something that may surprise you: even though we first see an environmental problem as an impairment of nature, such as a declining fish catch or polluted air, environmental problems arise from the way that human behavior is organized and guided. As we come to understand the social roots of environmental problems, we will also begin to see ways to solve them through social change.

How does human behavior produce problems in nature? Consider Figure 3.1. Humans live among the ecosystems that make up Planet Earth. Our livelihoods depend on the resources and services we gain from those ecosystems—crops grown in the soil and water for our farmlands; energy from the sun and fossil fuels; air we can breathe and water that is safe to drink; the aesthetic pleasures of scenery; and much more (top arrow). We notice environmental problems as they affect the quality and quantity of the resources and services we obtain from ecosystems (bottom arrow). Two hundred years ago, most people would have understood "environmental problems" to mean the things that nature did to people, such as floods, droughts, and earthquakes. Today, the vast majority of the things we see as "environmental problems," from holes in the ozone layer above us to contaminants in the groundwater below, result from things that humans have done to nature. Today, even the weather bears the imprint of human activity.

One of the achievements of environmentalism as a social movement has been to raise awareness of human impacts on the environment. This impact has often been indirect, as we saw in the story of the Channel Island fox in Chapter 1. As scientific understanding of the natural world has grown, we have greatly increased our ability to diagnose the environmental impacts that we experience. In some circumstances, we have been able to manage those impacts technologically, as we have done in controlling sewage and in developing medicines to combat infectious diseases. In other circumstances, however, a scientific understanding of how a problem arises does not lead to a straightforward solution because the causes are embedded in stubborn patterns of human behavior. The right arrow in Figure 3.1 indicates that the way humans connect to nature matters, and that these connections have two important dimensions: technology and institutions.

Learning Objectives

When you have finished studying this chapter, you should be able to

---

↘ diagnose commons—situations in which the incentives facing individuals lead them to exploit a resource in a way that is harmful to the community that shares the resource;

↘ recognize important social institutions, including property, regulation, and markets, and to give concrete examples of contemporary institutions;

↘ analyze different ways in which a community attempts to correct the shortcomings of commons by imposing "mutual coercion mutually agreed to";

↘ inquire into who loses and who wins in the governance of commons;

↘ see the relationship between environmental problems and the problems of governing community resources of all kinds;

↘ discuss the principles of community governance in analyzing an environmental problem;

↘ analyze mismatches between the rhythms and cycles of ecosystems and the responsibilities shouldered by the humans who affect those ecosystems;

↘ explain how the relationship between the environment and humans is one that involves ethical, political, economic, and social institutions.

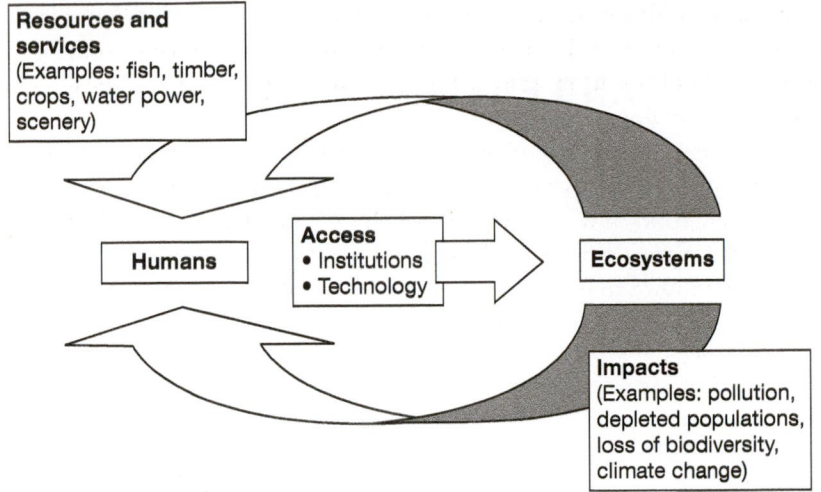

FIGURE 3.1
Human societies live among ecosystems. Three sets of connections shape humans to the natural world.

Once diagnosed, environmental problems challenge us to take responsibility, to *modify the way that humans interact with nature so that our behavior maintains the structure and functioning of ecosystems.* The emergence of a global economy with planetary impacts means that we have larger responsibilities than ever before. Responsibilities are not simply exercised by individuals, however. Instead, individuals' choices are shaped in **institutions**.

As we will see throughout this book, environmental problems often lead us to think in terms of institutional responses to solve them. The word "institution" has a special meaning to social scientists. Institutions are the rules, norms, and patterns of behavior that define how members of society interact with one another. The way that rules affect behavior is especially clear in games, which have many of their rules spelled out explicitly. Baseball, for example, is an institution that is defined by a set of rules detailing roles, expectations, and ways of keeping score. A batter will not swing at a ball thrown by the second baseman, for example, because doing so would not be in compliance with the rules of the game. Swinging at a ball thrown by the second baseman makes no sense. It isn't baseball.

Institutions, however, include many things more serious than games. Markets, governments, and communities are all grounded in important clusters of societal institutions. The concept of property, for example, is about more than just who owns individual pieces of land. Our notion of property also includes a culturally shaped set of rules and beliefs about what people can or should *do* with land, water, and other components of the natural world. Ideas about property affect the environmental choices people make. (See Box 3.1: Property and Environment, page 48.)

Taking responsibility for the environment means changing institutions. This is a huge challenge. But consider recycling. It was rare before 1970, yet it has

BOX 3.1

PROPERTY AND ENVIRONMENT

Property is an abstract idea, taken up only in boring law school courses—or is it? Consider this real estate transaction concluded in July 1636 in the lower Connecticut River valley. A Native American clan represented by Commucke and Matanchon—both sachems, or more colloquially, "chiefs"—sold a piece of land near their village at Agawam, near what is now Springfield, Massachusetts. The purchasers were a group headed by an English fur trader named William Pynchon. The land parcel is about 15 miles south of the oxbow in the Thomas Cole painting discussed in Chapter 2. Eighteen coats changed hands, with a like number of hatchets, hoes, and knives, according to the deed recording the transaction. The environmental historian William Cronon interpreted the misunderstanding that unfolded:

> The Indians conceived of this sale as applying only to very *specific uses of the land*. They gave up none of their most important hunting and gathering privileges, they retained right to the cornfields, and evidently intended to keep living on the land much as they had done before. The rights they gave Pynchon were apparently *to occupy the land jointly with them*, to establish a village like their own.... The Agawam villagers gave up none of their sovereignty over themselves, and relinquished few of their activities on the land. What they conferred on Pynchon was a right of ownership identical to their own: not to possess the land as a tradeable commodity, but to use it as an ecological cornucopia.[1]

The idea that different groups could share access to an ecosystem, an idea called a **usufruct right**—essentially a temporary use right—is hardly a strange idea. Fishing holes and golf courses are shared this way, as are the common areas in a condominium complex or planned residential community. Getting permission to use the family car on a Friday night is another example of a usufruct right; your parents do not turn over ownership of the car to you, any more than the Indians intended to turn over the land to Pynchon. Instead, you get the (temporary) right to use the car that night.

The Indian livelihood was a seasonally mobile one: raising crops in summer but moving to follow game, fish, and the ripening of wild fruits. This made sense in a landscape with large variations in temperature across the year, and it meant that the ability to use the products of land and water was what counted—a usufruct right. But a usufruct right is not what the English thought they were buying with their payment:

> What the Indians perceived as a political negotiation between two sovereign groups the English perceived as an economic transaction. . . .
>
> . . . As the English understood these transactions, what was sold was not a bundle of usufruct rights, applying to a range of different "territories," but *the land itself*, an abstract area whose bounds in theory remained fixed no matter what the use to which it was put. Once the land was bounded in this new way, a host of ecological changes followed almost inevitably.[2]

The English prevailed, building fences and excluding their Native American neighbors. The adoption of a private property scheme in which land was owned by individuals, in turn, led to a transformation of the landscape into permanent fields, roads, mills, and eventually to what we now call a developed economy.

The English had not arrived on the coast of Massachusetts, about 100 miles from Agawam, until 1620. How did their new ways come to be adopted? The reason was probably that the native society was too weakened by illness to protest effectively. When Europeans arrived in North America in the seventeenth century, they unintentionally brought diseases with them, germs against which New World peoples had no defenses. Epidemics of smallpox and other virulent infections swept through Native American populations, often killing entire villages and lowering populations drastically. As one sees in parts of the world devastated by AIDS today, removing a large fraction of the adult population can undermine vital institutions, including family, governments, and economic relationships. Cronon called the New England found by colonists a "widowed" landscape.[3] In these circumstances, the colonists' ideas of property and economic relations readily supplanted those of the native peoples.

"Property rights . . . shifted with ecological use," Cronon commented.[4] This is a two-way relationship: changes in property rights lead over time to changes in ecosystems. The very idea of "property" is a basic institution that shapes how humans treat the landscapes they inhabit. Because those landscapes are fundamental to economic relations, property rights often enter into the way humans interact with one another. This is why all law schools teach property law, and why students of environmental studies need to think about property as an idea.

---

1. William Cronon, *Changes in the Land: Indians, Colonists, and the Ecology of New England* (New York: Hill and Wang, 1983), 67 (emphasis added).

2. Ibid., 68 (emphasis added).

3. Ibid., 12.

4. Ibid., 63.

become a routine part of the way we treat the things we discard. Recycling rates remain much lower than they could be, but the idea that containers for recyclables should be located in places such as homes, airports, and residence halls, that people should separate recyclable items for the trash collector, and that people should follow rules about what is garbage and what is recyclable—all of this is now widely accepted. Institutions are hard to change, but some important institutional reforms have already occurred for environmental reasons.

In this chapter, we focus on the right arrow in Figure 3.1, investigating how humans cause environmental problems. The focus here is on the institutional arrangements that shape humans' access to ecosystems. As we will see, technological innovations and the economic changes enabled by new technologies can trigger environmental problems and institutional crises. They can also open pathways toward potential solutions.

Institutions shape access in two overlapping ways that are important to environmental studies. First, institutions tie together the activities and motivations of individuals. Consider littering. It is usually convenient simply to toss something no longer wanted on the ground, yet few of us do that. Why? And why don't we consider the release of carbon dioxide from our cars a form of littering? These questions suggest that our activities and motivations do not fit together as simply as we might assume, and they lead us to the problem of the commons.

Second, institutions limit the access of some people to ecosystems they might use. People must still live from ecosystems, however, and people without choices can be led to destroy the ecosystems on which they depend. In this way, the large and growing inequalities of the world's populations stand in the way of sustainable development. As the World Commission on Environment and Development concluded, "many problems of resource depletion and environmental stress arise from disparities in economic and political power."[1] Understanding these disparities is essential to solving many environmental problems.

We first take up the problem of the commons, and then examine the way that the control of access to shared ecosystems has played a central role in economic development, much of which is not currently sustainable.

A PARABLE OF THE COMMONS

In 1968 geneticist Garrett Hardin published an article called "The Tragedy of the Commons."[2] Today, almost two generations later, Hardin's essay remains one of the most frequently cited pieces of environmental writing because of its simple but powerful model for thinking about entire classes of environmental problems. Hardin's argument was not original, but he wrote clearly and forcefully at a pivotal historical moment, which influenced many people. Hardin also overlooked the role of inequality in creating environmental problems—a theme to which we will return below.

Hardin made his argument with a parable. He imagined a pasture that could provide benefits to a community of livestock herders. Every pasture's ability to nourish animals, however, is constrained by ecological factors. Herders raise livestock to earn profits by selling products such as wool, milk, and meat. But when too many animals graze on a given pasture, the quantity and quality of the grass that feeds the livestock deteriorates. In extreme conditions, overgrazing can permanently damage the ability of a pasture to support the animals. Generally, then, there is an **optimal** number of livestock that maximizes total benefits to the group. If too few animals are grazing, more could be fed on the pasture; if too many are grazing, the pasture will not feed any of them well.

Now suppose that no rules govern the use of the pasture by members of the community. Social scientists apply the term **open access** to this situation. Under conditions of open access, anyone may freely bring livestock to graze without social or legal constraints. This arrangement makes it apparently logical for individuals to increase their herd sizes whenever doing so would benefit them personally. Unfortunately, the drawback is that because the total benefits provided by livestock grazing are distributed among herders in accord with the size of each herder's flock, individual herders have incentives to bring new animals to the pasture as long as putting an additional animal on the pasture is profitable. But this means that herd sizes will increase beyond the level that is optimal for the community as a whole, perhaps even until the land becomes so degraded that the pasture is no longer able to support grazing. Unlimited use of the pasture is unsustainable in economic and ecological terms.

It can be very difficult for an individual to address this problem of the commons by acting alone. If Little Bo Peep decides to limit the size of her flock out of concern for preserving the land while other herders do not take the same precautions, Bo will be less well off than the other herders and the land will continue to suffer. So it does not make sense for her to deprive herself of some of her sheep. Conversely, if Ms. Peep decides not to limit the size of her flock while all other herders do limit their flocks out of concern for the land, then she will reap additional benefits—and the condition of the land will improve thanks to the efforts of others. So, in an open-access situation, no matter what the other herders do, Bo will always fare better by not limiting the size of her flock.

Hardin described this situation in the following terms. The rational herdsman, he argued,

concludes that the only sensible course for him to pursue is to add another animal to his herd. . . . But this is the conclusion reached by each and every rational herdsman sharing a commons. Therein is the tragedy. Each man is locked into a system that compels him to increase his herd without limit—in a world that is limited. Ruin is the destination toward which all men rush. . . . Freedom in a commons brings ruin to all.[3]

Hardin called this conflict between the common good and individual self-interest the **tragedy of the commons**. Hardin's analysis proposes a simple, powerful theory, one of a family of economic theories of rational choice. The tragedy is driven by the shepherds' being "rational" in a modern economist's sense of the term. All are trapped in a situation in which they feel compelled individually to do something that is unreasonable collectively. Hardin did *not* say that he observed the tragic scenario at specific places and times—his argument does not depend on observation. Instead, Hardin made predictions about social behavior using a simple, plausible model of human motivation. This is theorizing at work, with a simple model being asserted as an explanation of a wide range of phenomena. (See the "Two Approaches to Science" section in Chapter 2.)

REAL COMMONS, REAL TRAGEDIES

A good deal of human experience, nonetheless, seems to fit Hardin's scenario. The degradation of pasture lands under conditions of open access has occurred many times during human history. In the United States, for example, as railroads reached across the Great Plains after the Civil War, a ranching economy rapidly emerged. Large herds of cattle roamed the plains from Texas to the Dakotas, and each year, cattle drives organized by cow punchers led the cattle to railroad towns such as Dodge City, Kansas, eventually to be taken to stockyards and slaughterhouses in Chicago and elsewhere. This way of life looms large in Americans' views of their history, but in reality it lasted only one human generation. In part because the Great Plains were open access, with free grazing and transit routes available to all comers, the ranchers rapidly increased the size of their herds, which consumed the forage on the prairies far faster than the bison that had inhabited the plains for many centuries. (For a contemporary account, see Theodore Roosevelt's memoir, *Ranch Life and the Hunting-Trail*.) By the end of the nineteenth century, the cowboy and the open range had passed into history as the prairie was depleted and that business model of ranching failed. Cowboys lived on in the movies but not on the plains.

Similarly, when colonists arrived in the land they called New England in the seventeenth century, they initiated trade with the indigenous people in beaver pelts, which could be sold in Europe for high prices. Pelts had not been in such high demand before, and the Indians and English treated them as an open-access resource. The Pilgrims landed at Plymouth, Massachusetts, in 1620, and by 1700, beavers had been trapped to commercial extinction throughout New England.

(The animals were not biologically extinct, though, and their descendants eventually made a comeback in the late twentieth century.)

We can apply the tragedy of the commons to other environmental problems. We can think of pollution as being like grazing, a thought that may seem odd at first. When polluted water or industrial gases are dumped, they go into the water or air that belongs to everyone, like the open-access pasture. The polluter (who might be you, as you drive your old beater down a city street) gains the advantage of having a place to get rid of his or her waste, but everyone loses as the air or water is fouled.

Think about a crowded road at rush hour. You are an environmentalist and decide to ride the bus instead of driving, even though the bus stops a lot and doesn't come by very often. While riding in the bus, you notice that the traffic does not seem any lighter just because your car isn't out there. On a day when you'll be late for an exam if you wait for the bus, you decide to drive. Of course, when we all choose to drive, the result—in addition to dense traffic—is cumulative environmental degradation that can impair the health and well-being of all members of society. In many places, we would all be better off if we switched to public transportation, but making this change as isolated individuals is impractical and ineffective. So instead, we act in a manner that is individually rational and drive, even though we know it would be better if everyone would stop driving cars.

These examples of the tragedy of the commons can be seen as instances of a scenario: Technological change increases the ability of humans to exploit and intervene in once-natural ecosystems. When the situation is one of open access, that technological ability leads to overuse. The problem is that *no one who wants to utilize a given resource has incentives to curb his or her own use of that resource*, so that a growing ability to utilize the resource will be disastrous.

Railroads and the development of long-distance cattle drives made it possible to move large numbers of animals to market. Open access to the grasslands then triggered a tragedy of the commons. Similarly, transportation across the Atlantic made it profitable to trap beavers and send their skins to European furriers, and unrestrained trapping followed. Affordable automobiles and a pattern of urban development that made public transit appear to be more costly than cars driven by individuals stimulated traffic jams, smog, and sprawl. The stories are very different, but their structure has a shared thread.

COERCION

When beneficiaries of a resource are not or cannot be excluded, it becomes economically rational for each person to exploit the common resource (grass, beaver pelts, clean air) to obtain individual benefits, even when it becomes obvious that all the users together are overexploiting the commons and that greater shared benefits could be achieved through collective restraint. The response to the tragedy of the commons, Garrett Hardin argued, was "**mutual coercion mutually agreed to.**" But when an environmental problem erupts, the coercion imposed may not be mutual and it may not be agreed to. This can produce the dilemmas of inequality that we brought up in Chapter 1 in discussing sustainable development.

An important example comes from Great Britain, the land that gave us the word "commons" in the first place. Until the eighteenth century, grazing "in common" was both widespread and successful, not just in medieval England but also in most of Europe. In some cases, these grazing commons were maintained successfully for more than a thousand years, making them among the most sustainable livelihoods in the historical record. The herders in these pastures did not own the land where they fed their animals. Rather, as a matter of custom and culture, they were allowed to use the land for grazing, and collectively, they managed the land in a way that generally avoided the tragedies that Hardin saw as being almost inevitable (we will return to how they did this shortly). As a matter of law, however, the land continued to be owned by the nobility—the lords who legally held title to the land. These landlords permitted the herders and the farmers—the commoners who did the labor of the community—to cultivate small gardens and to graze animals on land that was not being cultivated or used for hunting.

In response to population growth and the economic changes that set the stage for the modern world, landlords faced a complicated combination of pressures and opportunities to reap economic gains by taking the common lands back for themselves and kicking out the herders. This process, called **enclosure**, unfolded over more than a century. Enclosing the commons meant building walls and fences around them and excluding the commoners who formerly used them. When the commoners protested and threatened rebellion, the landlords went to Parliament and had the enclosures written into law. The commoners were driven off, and many ended up in the cities, where they became the labor force of industrialization.

This was not an inevitable tragedy of the commons. Instead it was a tragedy of dispossession, the result of the exercise of power by the landed nobility. It was hardly what Hardin called "the remorseless working of things." Instead, the lords' appeal to Parliament is an example of what economists call "**rent-seeking**," a term that refers to the use of political influence by a group to secure economic benefits ("rents") by altering public policy. Colleges pressing for more advantageous student loans from the government are rent-seeking, and so are the owners of coal-fired power plants when they make contributions to politicians who vote against global warming legislation. Rent-seeking is inherent in self-government—in mutual coercion mutually agreed to—although some practices can be outlawed or limited. This is one of the dilemmas of democracy: those with more resources can usually tilt the rules in their favor, and voters face a constant struggle to restrain, or even to find out about, rent-seeking.

The present-day descendants of the English commoners have reason to think their fate may not be a bad one, however. They live in an advanced industrial society, wealthy beyond the imaginings of either the landlords or the commoners of two hundred years ago, with things like modern medicines, universal education, and environmental awareness. Elsewhere, however, the systematic exclusion of the powerless has not produced wealth over generations. Native Americans in North America, as well as indigenous people in countries burdened by colonialism in Latin America, Africa, and Asia, have much to resent. This is a problem of sustainable development, obviously: people who cannot meet the needs of the present are often too busy trying to survive to worry about assuring future generations the capacity to meet their needs. But there is a more immediate environmental problem as well.

If the humans affected by the decline of a natural resource are poor and powerless, they will be unable to stop the rich and powerful from taking or destroying resources that they value, or to sound an alarm that wealthier and more powerful people will heed. Exploitation of nature can proceed unrestrained because those who know the condition of ecosystems have no ability to bring their knowledge to bear. (See Box 3.2: Commons and Inequality in the Mississippi Delta, page 56.) This is a problem that can be seen from the Amazon to the Congo to Central Asia, where the overlords of the Soviet Union ruined lands in the name of economic progress for the people. We will return to this subject in Chapters 4 and 6.

UNCERTAINTY

The second problem with the important idea of mutual coercion mutually agreed to is determining how much restraint is necessary to keep a commons healthy. It is often unclear whether a given ecological resource is being overexploited. Sometimes, as in the case of the Channel Island fox, no one spots the problem until the damage or risks reach crisis levels. Sometimes the damage is visible initially only to scientists monitoring the population or resource. Fishermen and marine scientists have battled for years, for example, over whether particular fish species are being overfished. The fishers are naturalists: they have experience searching for fish, and they think, rightly, that they know a lot. The scientists take measurements and analyze the resulting data using complicated models. People reach different conclusions about whether to restrict fishing, and scientific uncertainty has often been a contributing factor in the failure to impose limits capable of maintaining fish populations.

Who's right? Wild populations fluctuate a lot from year to year, so short-term changes in the size or abundance of fish are hard to interpret. A decline in one season may be followed by profusion in the next. Measuring fish populations accurately is particularly difficult, given the location and behavior of most fish species. Yet over time, it has become clear that many high-value fish species have

BOX 3.2

COMMONS AND INEQUALITY IN THE MISSISSIPPI DELTA

How do commons, misunderstanding of science, and power intersect in our own society? Consider New Orleans. The damage inflicted by Hurricane Katrina in 2005 was worse by far because of the erosion of the once-extensive wetlands of southern Louisiana. Those wetlands would have caught the wall of water thrown up by the storm, moderating its fury as it blew into the city. The erosion occurred, over many decades, because there was money to be made (rents to be captured) by digging canals through the region's wetlands for navigation and for developing oil and gas resources. A small number of politically influential people made that money.

Moreover, for well more than a century, the construction of levees to protect towns and fields all along the Mississippi River from flooding drastically reduced the volume of sediments reaching the Mississippi delta (see the figure). These sediments had replenished the wetlands during past floods, but as the muddy

Sediment budget of the Mississippi River.

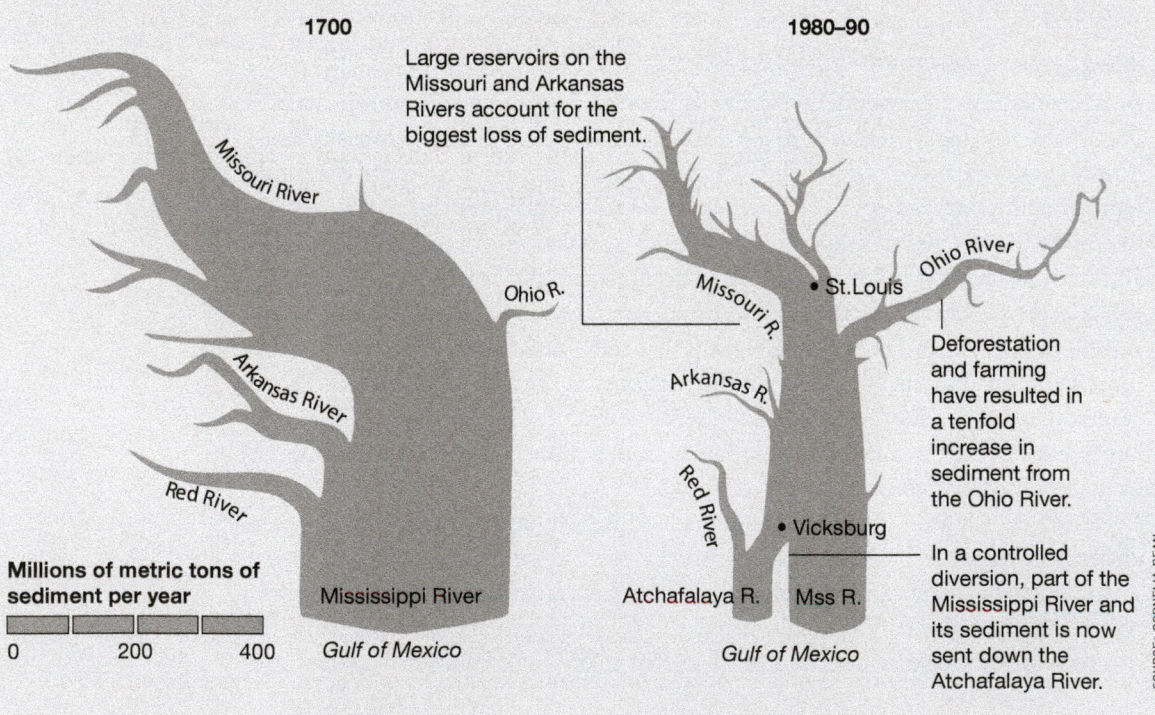

1700

Large reservoirs on the Missouri and Arkansas Rivers account for the biggest loss of sediment.

Missouri River

Ohio R.

Arkansas River

Red River

Millions of metric tons of sediment per year

0   200   400

Mississippi River

Gulf of Mexico

1980–90

Missouri R.

St. Louis

Ohio River

Arkansas R.

Red River

Vicksburg

Atchafalaya R.   Mss R.

Gulf of Mexico

Deforestation and farming have resulted in a tenfold increase in sediment from the Ohio River.

In a controlled diversion, part of the Mississippi River and its sediment is now sent down the Atchafalaya River.

SOURCE: CORNELIA DEAN.

Mississippi became less muddy, the wetlands at the mouth of the river dwindled. Even though scientists had warned for years of the consequences of weakening the wetlands, no one stood to gain from halting the erosion. A tragedy of the commons unfolded.

After the levees broke in New Orleans, refugees from the flooded lowlands included many of the city's poor, and they have waited years for assistance to rebuild. Many of those who profited from the economic development and ecological undermining of the Mississippi Delta have not offered to help. Instead, the assistance and encouragement provided by government at all levels induces people to rebuild where their flooded houses stood. That is, the new houses are exposed to the same risk of flooding when the next big storm hits. Tragedy is compounded by irony: neither the residents hoping to rebuild nor the governments that should be taking responsibility for humans in a perilous landscape seem to grasp the implications of environmental science. That science has been explained numerous times by experts, the press, and banks and insurance companies reluctant to put their money at risk. But these communities have long been home to their residents, and loyalties to place are reinforced by businesses and local governments that may see disaster relief from the federal government as a way to buffer short-sighted decisions.

been severely over-harvested, just like the beaver of New England more than three hundred years ago. Today, according to scientists, more than two-thirds of the world's commercially valuable ocean fish species are being overfished or are in danger of extinction.

This is a wider problem than fishing. Those with strong economic interests and political influence can derive benefits from arguing that the science is not yet conclusive. The reason is simple, though often overlooked. Scientific knowledge is by nature provisional. Science is an organized effort to get closer to verifiable explanations and predictions, a process of successive approximation in which errors are expected and debate continues even as understanding improves. Despite stereotypes about overeager bureaucrats, most government officials are reluctant to impose regulations on the basis of uncertain science. That reluctance is appropriate. The first scientists to see connections between certain widely used industrial chemicals and depletion of ozone in the upper atmosphere, for example, saw themselves as advancing a preliminary **hypothesis**—not a confident assertion, but a tentative explanation to be tested by further investigation. It took more than a decade before a seasonal hole in the ozone layer was detected by scientists. By then, a majority of scientists who knew of the original research had come to see the initial hypothesis as a well-established finding. Governments in Europe and the United States acted to ban some uses of those chemicals. The finding of an ozone hole above Antarctica in the late spring of 1985 then spurred a global agreement to phase out all uses of those chemicals.

The catch comes in deciding when a hypothesis has become a well-established finding. Those responsible for pollution or resource degradation are often able to delay or avoid taking responsibility if they can successfully argue that the science is "not yet proven." In many cases, moreover, an industry avoiding environmental responsibility is willing to spend far more money in contesting the science than innocent victims, or society at large, may be able to spend in demonstrating the reality of the harms. Even science can be drawn into rent-seeking. Industries responsible for large greenhouse gas emissions persisted in their claim that climate change was too unproven to justify action, long after a majority of the world's climate scientists had reached the conclusion that global warming was real and that human actions played a significant role in causing it. This verdict was acknowledged by the award of the 2007 Nobel Peace Prize to the Intergovernmental Panel on Climate Change, the international committee of scientists appointed by the United Nations to weigh the evidence.

The word "verdict" is an interesting one. To argue that the science is inconclusive can be a way of arguing that the polluting activity should be treated as innocent until proven otherwise. Because economic benefits can be maintained by doing so, many polluters clearly think of this as a sensible approach. So might the larger society. If you think about the similarity between littering and venting your car's exhaust to the atmosphere, you can see that all drivers might want to say that driving is innocent until proven guilty and that sentencing should be long deferred. Science can be conclusive among scientists, as is the case with global climate change. But shifting the presumption of innocence is not a scientific question. It is an institutional one. (See Box 3.2: Commons and Inequality in the Mississippi Delta, page 56, for other examples of institutional factors that can prevail despite scientific warnings.)

INSTITUTIONAL SOLUTIONS

Individuals cannot overcome the logic of an open-access commons by acting alone. What is needed, instead, is an **institutional solution**—a way to alter the expectations and rules that shape everyone's behavior. Only by adopting rules limiting open access can a small number of polluters or resource users be stopped from damaging the environment on which their fellow community members also depend, and only by shifting the expectations facing all the herders can the ruthless and self-defeating logic of the commons be changed.

For the past two generations, the institutional solution most persuasively advocated has been a "free market." The promise of the free market is that individuals competing with one another will find their behavior coordinated in ways that are mutually beneficial. The Scottish philosopher Adam Smith compared that coordination to the action of "an invisible hand." When a commons is poorly managed or development is uneven, however, this promise of mutually beneficial coordination is not being realized. Is it possible to *modify* the destructive rules of an open-access commons and convert it into a market that works to the benefit of its participants and the ecosystems they use? Or is it possible to *replace* the open-access commons with an institutional structure that protects the ecosystem while providing humans efficient access to its benefits?

Before we dig into these questions, it is important to understand that a "free" market is not free—it isn't an open-access commons. Instead, real-world markets are structured by rules, including property rights and contractual obligations. These rules deter theft, overuse, pollution, and other forms of behavior unacceptable to the community of users. These rules do not work all the time, and often the way they operate can be inequitable, as we have discussed above. But there *are* rules and they define markets and other forms of orderly access to the benefits of nature. From the perspective of the community, then, the point is not whether a market is free but rather whether social institutions enable their members to draw upon nature sustainably, efficiently, and equitably. In some important cases, as we will see below, it is possible to use a suitably designed market mechanism to foster all three of those ends.

In other instances, government regulations enable an economy to function while protecting a commons. Since 1970 the air quality in most American cities has improved dramatically, even as industrial activity and the use of automobiles has gone up. What happened? The story is complicated, but a key element is that both industrial processes and the vehicles we drive have been improved in response to legal standards adopted in the Clean Air Act. In states where air quality has been poor, these rules require that all vehicles must pass annual inspections, to make sure their air pollution equipment is installed and working properly. In addition, the rules limit pollutant emissions from factories and power plants. This is not a free-market solution, because inspections are not voluntary and automakers must install pollution-control equipment if their vehicles are to be sold in the United States. But it works reasonably well.

PRIVATIZATION

Air pollution regulation is an example of mutual coercion mutually agreed to—the imposition of binding rules restraining and shaping human inclinations. In his analysis of the tragedy of the commons, Hardin saw two alternative ways to implement mutual coercion. The first is usually called **privatization**, although this is a somewhat misleading term. Privatization is an action *by government* to divide commons among individual owners, with each asserting control over their own property. As with other forms of property, the rights of the owner are then backed up by the civil and criminal justice system of the government. Privatization is a form of enclosure, though it can be considerably more equitable than the Acts of Enclosure in Britain more than two centuries ago.

In the United States, the Homestead Act of 1862 privatized substantial areas of land west of the Mississippi. A homesteader could establish ownership to 160 acres of public land by growing crops for five years and building a dwelling on it. Only minimal fees were required.

In other circumstances, commons have been auctioned off. This has happened recently to portions of the electromagnetic spectrum used to carry cell-phone and other kinds of wireless communication, with many millions of dollars changing hands. Through these auctions, the federal government has sold licenses to specific telecommunications companies for the use of specific frequency ranges to broadcast signals through the air.

The owner of a privatized resource is no longer sharing that resource with others but has an economic asset. Ownership brings with it an economic incentive to take responsibility for assets. Most herders who own their own grazing land manage their land and flock to maximize the economic benefits derived from the land, because they bear the entire cost and gain the entire benefit of adding an animal. The land will only be overgrazed if the owner makes a mistake and adds too many animals. He or she has nothing to fear from the decisions of other herders, because they do not have access to the land. Often one sees neat front yards and dirty streets, especially in poor neighborhoods. The yards are privately owned, belonging to people who take pride in them, or at least may be ashamed to have an unkempt yard. The streets are public—a commons—with no one to care for them, particularly if the neighborhood has little influence over municipal street cleaners' schedules.

Privatization aims to solve the problem of the commons by eliminating it and replacing open access with institutions that protect and enforce private ownership. It is worth seeing how this fits with Hardin's formula. Mutual agreement can be reached in dividing up the commons, whether by willing purchase, as in the case of cell-phone spectrum, or through the sweat equity of the homesteader. There is mutual coercion in the protection of property rights. If Little Bo Peep loses her sheep, her neighbor may call the animal control officer to round them up before they eat up his grass. But a second, harsher element of mutual coercion is that careless stewards are allowed to impoverish themselves. People are forced to take responsibility and to suffer the consequences of not doing so, or may suffer bad luck with no help from the community. Finally, recall the Acts of Enclosure: privatization has sometimes been imposed through coercion that is neither mutual nor mutually agreed to.

Privatization is appealing in societies, such as in the United States, that are suspicious of public bureaucracy (but do not have similarly negative feelings about the private bureaucracies of corporations). Yet it is difficult or impossible to privatize some commons, such as the global atmosphere or flowing rivers. Bear in mind, too, that government is essential even in a privatized situation, because property rights must be defined and protected. Conservative critics of regulation agree that protection of property rights is an essential function of the state.

In addition, ecological economists and historians have criticized many specific instances of privatization. In the nineteenth century, for example, little land actually went to the homesteaders who are enshrined in American myth as pioneers of the West. The federal government first bought the land that now constitutes most of the West through the Louisiana Purchase of 1803. The nation then gave away the vast majority of that land, or sold it at rock-bottom prices, to large landowners, in an effort to stimulate economic development. In his classic analysis *The Development of American Agriculture*, historian Willard Cochrane concluded that out of the roughly 1 billion acres distributed, only about 15 percent, or 147 million acres, actually went to homesteaders.[4] Even though the United States has never had a landed gentry like the European nobility, Cochrane's analysis suggests that the new nation soon developed something similar, with 85 percent of public land going to railroads, timber companies, and other large landowners. These became the backbone of the uneven development celebrated in the painting *American Progress* that we saw in Chapter 2.

REGULATION AND PUBLIC TRUST

A second approach to mutual coercion is government **regulation**, or its cousin, **government ownership**. This approach is built around institutions that embody the following principle: if a commons belongs to the whole community, then the whole community, acting through its government, should take responsibility. A national park should have rangers to take care of it. The air over a city should be protected by rules that are obeyed by polluters and enforced by an environmental protection agency.

The idea of collective responsibility for a commons is known in legal terms as the "public trust doctrine." As the name indicates, the government is entrusted with the care of the commons, but the public owns it. This can mean that the government should be prevented from selling a public resource. In this way, the sale of one part of the Lake Michigan coastline in Chicago by the state of Illinois was canceled in 1892 by the U.S. Supreme Court, which relied on the concept of **public trust** and held that the coastline belonged to all and therefore could not be sold. In other cases, the government has successfully sold or given away resources, as when it auctions off electromagnetic spectrum, leases areas of the ocean floor to oil companies for offshore oil exploration, or permits people to dump their wastes into nearby rivers and streams. Of course, many public resources—land, water, air, and others—are managed by government agencies at various levels, from city playgrounds to the air over the United States.

Environmentalists tend to think in terms of regulation as a response to environmental problems, but it is sometimes possible to implement policies that look and feel like privatization. Before the 1990 Clean Air Act in the United States, power plants and many other large facilities had to comply with complicated, multivolume regulations to be able to emit sulfur dioxide, a potent form of air pollution (sulfur dioxide gas turns into sulfuric acid when it encounters water, as in a person's lungs). The regulations were aimed at assuring that the facilities had up-to-date technology, so the rules would often change. Under the law, which set goals based on protecting public health, the regulators could not take cost into account, and the polluting businesses complained continually of unaffordable rules that interfered with their ability to earn a fair return for their shareholders. The air over American cities had gotten cleaner, but the industries were dragging their feet, proposals to tighten environmental regulations faced continual opposition, and regulatory officials were buffeted by political conflict.

The 1990 amendments to the Clean Air Act rewrote the rules in a way that made it easier to reduce sulfur dioxide pollution in the nation as a whole, gave polluting businesses a great deal more flexibility, and lowered the burdens and conflict surrounding the remaining regulations. All of this was accomplished by creating a market-based form of regulation. The law assigned rights to emit sulfur dioxide to polluting industries—a new form of property. The government calculated how much sulfur dioxide emissions the atmosphere should absorb and then set that as an initial cap of more than 10 million tons per year.

This amount became a pool of total rights to be divided up among all the polluters. The individual polluters could decide whether to reduce emissions, which required investment, or buy rights to emit pollution on the open market, which also cost money. Either way, the right to emit sulfur dioxide become a cost for firms, whereas it previously had not been one. Firms whose emissions would exceed their cap could buy or trade for the additional rights to pollute from companies who had reduced their emissions and had extra rights left over.

Firms had already been subject to air pollution regulations, so emitting sulfur dioxide had already been costly in the sense that the company had to pay for pollution-control equipment. Under the cap-and-trade system, the rules were simplified. Instead of facing requirements to use specific kinds of pollution-control technology, the polluting facilities now only had to report their emissions and, if the emissions exceeded the cap, show that they had purchased the rights to emit those additional quantities of sulfur dioxide.

By limiting the total quantity of emission rights for the United States, the U.S. Environmental Protection Agency struck a kind of balance. This approach created a simplified regulatory regime that operates through a new form of private property backed by an enforcement program. The new system's rules intruded far less into the decisions made by businesses. This simplicity also benefited the public and the natural world. The total emissions cap provided a concrete definition of a public trust in the atmosphere. Some level of emissions is allowed, reducing air quality to a degree, but to a degree that Congress has deemed acceptable. Exceeding the total level impairs the public trust. The total cap has been substantially reduced since this policy was initially implemented in 1995, in effect transferring some rights to clean air back to the public. Passed with the support of President George H. W. Bush, the sulfur dioxide policy is an example of government regulation that works by using a market mechanism. It has enjoyed support among both businesses and environmental groups.

The cap-and-trade approach has been criticized, however. The criticism is important because this approach is now being considered for global adoption to regulate emission of the greenhouse gases responsible for global warming. The problem lies in the initial distribution of emission rights. This is the step, remember, in which government encloses the commons and transfers pieces of it as property to polluters in the form of the right to pollute. As with the enclosure of the English commons, one can ask who benefits from this transaction.

Business author Peter Barnes has argued that the air belongs to everyone, not just to polluters, and thus the sensible approach is not to *give* the rights to those who have been polluting in the past, but rather to hold an auction.[5] Once the polluters have purchased the emissions rights from citizens—represented by their governments—the polluters would then be allowed to sell them to each other. The proceeds of the auction, however, would not go to polluters, but to governments for investment in health care, parks, education, and other benefits to the larger society, to make up for the continuing losses from pollution. The earnings from the auction then forge a link between pollution-control policy and the public trust doctrine.

This argument sounds sensible, but it has an unsettling implication if applied to greenhouse gases on an international scale. The public trust argument implies that the right to pollute the atmosphere belongs to all the people of the world, rather than to those businesses that are now polluting the most. If those rights were divided in proportion to population, an auction of the right to emit greenhouse gases would be expected to transfer large sums from the developed economies and China, where greenhouse gas emissions are now highest, to the developing countries, whose large, poor populations have little access to energy from fossil fuels. Those transfers could dwarf all of the foreign investment and development assistance made up to the present.

The people in poor countries are in a position similar to that of the commoners in the time of the Acts of Enclosure. How much are the consumers and businesses of the rich world—the landlords of this story—willing to pay the poor and powerless, and under what conditions? Will development be unequal or sustainable? This is not an academic issue only, but one that lies on the agenda of the climate treaty negotiations already in progress. It may well delay global action on global warming, or even thwart it altogether. This is how important institutions for commons are.

COMMUNITIES

In addition to privatization and regulation, a third alternative can avoid the tragedy of the commons. This is to approach the commons through the **community governance** institutions that have grown up around them. "Commons" and "community" come from the same linguistic root, so there is a deep tie. In fact, community-based governance of commons has been the dominant form of natural resource management throughout human history. In that sense, Hardin's theorizing ignored a large body of experience that is useful in thinking about protecting the environment.

To begin, are there commons that do *not* produce tragic results? Students often share apartments or dorm suites, and so they come to share refrigerators, kitchens, and bathrooms. The maintenance of these parts of the habitat can be, shall we say, casual. But when parents arrive for a visit, the bathroom is often presentable and the overflowing trash receptacles have been emptied. How does this happen? It is impractical to divide up a shower, toilet, and bathroom floor, so privatization is not at work. There is no government regulation. Instead, our students report that something called "social pressure" is at work, leading them to pitch in and clean up. Hardin's analysis of individual rationality and collective welfare does not seem to have room for social pressure.

Beyond the shared bathroom, we see a growing number of American states and local governments taking action on climate change even though the federal government has rejected the Kyoto Protocol of 1997. The federal argument is that of the commons: if the United States incurs the costs of controlling greenhouse gases, then other nations will bypass us economically. Yet a coalition of states sued in 2005, in a case called *Massachusetts v. Environmental Protection Agency*, demanding regulation of greenhouse gases. Did these states not understand the inexorable logic of the tragedy of the commons?

Here is what the environmentalist and poet Gary Snyder said about Hardin's discussion:

> In the abstract the sharing of a natural area might be thought of as a matter of access to "common pool resources" with no limits or controls on individual exploitation. The fact is that such sharing developed over millennia and always within territorial and social contexts.... the commons is both

specific land and the traditional community institution that determines the carrying capacity and defines the rights and obligations of those who use it.[6]

Look carefully at the phrase "the commons is both specific land and [a] traditional community institution." Snyder is saying that something has been left out of Hardin's conception of the commons. To Hardin, the commons is a place where all herders are out for themselves. To Snyder—and to most of the people who have actually participated in commons grazing arrangements over the centuries—the commons is a place where rules, customs, and loyalties shape how people understand and pursue their self-interest. All of these arrangements make the commons something very different from the open-access setting for inevitable tragedy. As people make use of the natural world, they can affect ecosystems by altering habitats, over-harvesting, or pollution. But when people's actions are governed by social institutions that can recognize these problems and respond to them, the commons can be well managed. The sense of place discussed in Chapter 2 is a name for loyalties and sensibilities that can lead people to act to preserve a place. As we will discuss in Chapter 4, however, humans increasingly act on scales far larger than those that can be governed by place loyalties alone.

COMMUNITY-BASED GOVERNANCE

Social scientists have found many examples of commons that have been successfully managed for many human generations, including pastures in Switzerland that have been grazed for more than a thousand years, and irrigation systems that have been well maintained and used to share water for centuries. Political scientist Elinor Ostrom drew together hundreds of studies of well-managed commons and proposed eight "design principles." They are paraphrased in Table 3.1. We will call them **community-governance principles**.

1. *The boundaries and membership are clearly defined.* The first principle says that the well-governed commons belongs to a community that will claim it and defend it. Those who are entitled to use the commons are clearly specified, and everyone knows how they should use it. Along the coast of Maine, the lobstermen know who owns each lobster trap in a cove, and they know where the traps may be placed. Without formal government, community rule can be harsh. If you set a trap without the approval of the "harbor gangs," your float will be cut off and you will not be able to find your trap. Outsiders are thus kept out.

Private property is one way to deal with the question of boundaries and membership in a community. A private owner lives in a community where all agree to respect the property claims of others. But as the lobstermen illustrate, some

TABLE 3.1   COMMUNITY-GOVERNANCE PRINCIPLES FOR WELL-MANAGED
COMMONS

1. The boundaries and membership are clearly defined.
2. The rules fit local conditions.
3. The rules are open to modification by participants.
4. Monitoring is effective and affordable.
5. Enforcement relies on *graduated* sanctions.
6. Conflict resolution is fast and fair.
7. Control is local.
8. Local control is exercised within nested institutions.

Source: Adapted from Elinor Ostrom, Governing the Commons: The Evolution of Institutions for
Collective Action *(Cambridge, England: Cambridge University Press, 1990), chap. 3.*

examples of communal ownership do not entail public ownership or government control. The sense of place discussed in Chapter 2 is central to community-based governance.

2. *The rules fit local conditions.* Rules for using commons work only if they fit local conditions. In the Maine lobster fishery, the number of traps allowed in each area is controlled by a set of rules that are jointly determined by government regulators and local lobstermen, with careful attention to catch levels and the biological condition of the resource. They specify that the openings in the traps must be designed so that lobsters that are not ready for harvest can escape, to grow and be caught another day. In general, rules that work in one place may not be transferable to other places.

3. *The rules are open to modification by participants.* Commons that have survived over long periods generally have mechanisms by which rules can be modified by community members. Often, rules are suspended or modified when a crisis arises. Thus, a community that operates a shared irrigation system may decide to draw down the reservoir holding their shared water during a drought. This self-government may not conform to conventional notions of fairness, however. The journalist John Tierney put this more bluntly, writing of the Maine lobstermen, "The harbor gangs are built around the management principles of Tony Soprano."[7]

4 and 5. *Monitoring is effective and affordable, and enforcement relies on graduated sanctions.* The fourth and fifth community-governance principles are particularly important. A way must be devised to observe users' behavior and enforce the rules of use. When there are no consequences for abusing the commons, the logic of overexploiting a resource to obtain individual gain can be hard to stop. This is

why some states require that a sticker be affixed to your car's windshield when its pollution-control system has passed inspection. If your sticker has expired, you are likely to get a ticket. Other states require that the pollution-testing station send your test results to the agency that issues license plates, assuring that no licenses or renewals will be issued until the car meets pollution limits.

Both monitoring and enforcement can also be accomplished by nongovernmental means such as social pressure. This is how the bathroom in the student apartment gets cleaned up. This example also reminds us that monitoring and enforcement are often easier in small groups than in large ones. In small groups, the consequences of another person's misbehavior are direct, and it is usually apparent who violated the rules. The ability to catch violators red-handed, in turn, often makes it easier to enforce sanctions. If the members of the community know one another well, the prospect of getting caught can restrain misbehavior in different ways than is the case when enforcer and violator are strangers (see Box 3.3: Honor Code, page 68). More generally, developing low-cost, convenient means of monitoring environmental behavior, together with simple means of enforcing rules, is crucial. This is an element of environmental policy that gets little attention, however.

The history of well-managed commons also says that the sanctions or punishments for violating the rules of the community should be *graduated*. A first-time offender, or someone who breaks a rule to feed her needy family, should face a lighter penalty than a poacher who steals from the community repeatedly. This might seem to be a statement about fairness, but ideas about what is fair vary considerably across cultures. The fifth governance principle rests on a practical consideration. Commons are shared, by definition, and often they are used by everyone, as with the air we breathe. Rules that work under these conditions do not seek to drive out one-time or infrequent rule breakers, but rather to draw them back into the community and encourage them to follow the rules in the future. Making good behavior worthwhile is essential, and this is what graduated sanctions do.

6. *Conflict resolution is fast and fair.* Especially when monitoring and enforcement are done informally, the alleged rule breaker and the monitor who caught him or her can get into a dispute. Therefore, dispute resolution mechanisms are needed. When you get a traffic ticket, you have the right to contest the accusation in court. If Bo Peep's sheep are found in a neighbor's pasture, was it Bo Peep's fault or was the neighbor's fence broken? Some New England communities still have officials called "fence viewers," who have historically sorted out such disputes, although they get little work nowadays. More generally, ways of addressing conflict need to be quick and fair.

7 and 8. *Local control is exercised within nested institutions.* Finally, the rules developed locally by those who use the commons (principles 2 and 3) need to be respected by higher levels of authority. This principle was violated when the commons in Britain were enclosed, with the long-established traditions of shared grazing space

HONOR CODE

Commons—and the problems of commons—are found in circumstances that we think of as social and far from nature. The honor codes in place at many colleges and universities illustrate the importance of monitoring and evaluation in governing commons. If you think about it for a moment, you will see that cheating abuses a commons. Someone who gets away with a plagiarized paper or problem set gains a short-term benefit while distorting the grading curve, if there is one, or devaluing all grades even if there is no curve. An honor code relies on decentralized monitoring by fellow students, who are more likely to detect suspicious behavior than are faculty.

When the code works, everyone gains. For example, it is much easier to arrange for make-up examinations when people have confidence in an honor code, because exam questions do not need to be rewritten to prevent cheating. But monitoring needs to be backed up by enforcement, and many students feel hesitant about reporting violations when conflict may result. Turning to others for help in pursuing suspicions, such as residence hall advisers, deans, and faculty, is therefore important. By drawing on social resources beyond oneself, the commons can be safeguarded against the moral erosion associated with an excessive emphasis on individual rationality.

being overruled by Parliament. More broadly, higher levels of government have often ruined well-governed commons by disrupting local rules. The declaration of national parks in developing countries has sometimes been misused to remove people from lands that they have inhabited and used sustainably for long periods. Governments have sometimes done so in response to pressures from environmental organizations based in the United States, where national parks are assumed to exclude humans. In other cases, the central governments of developing nations have needed little encouragement to remove indigenous peoples so that others can profit from the land, the way the landlords in Britain did during the Enclosure Movement.

As this example shows, sometimes the overruling of local arrangements has been done in the name of a wider goal such as biodiversity conservation, rather than just to serve the powerful few. In other cases, actions can be taken in the name of conservation but with other goals lurking in the background. In still other cases, local conditions shaping access to the commons might exclude women or members of an ethnic minority, for example, and higher levels of government might intervene to provide more equitable access. In these instances, the paradox of uneven development again intrudes into the problem of the commons.

TRADITION AND POLICY

The principles of commons governance are not really about the *design* of policies, as Ostrom implied. Commons principles represent the accumulated experience of communities that have survived for many human generations. These communities appear to govern their use of commons with rules that fit these governance principles.

Community-governance principles generally cannot be "adopted." They must evolve. Remember Snyder's assertion that the rules for sharing the capacity of commons "developed over millennia and always within territorial and social contexts." The rules that emerge this way are **traditions**, regarded as legitimate because things have been done that way for a very long time. They have become institutions in the way we defined the term at the beginning of this chapter. Yet in contemporary society, we encounter much that is new, and more that is historically recent. The first mass-produced automobile, the Ford Model T, rolled out of the plant in 1908, only about a century ago, and much of the environmental and social change associated with cars—such as suburbs, freeways, and smog—have little historical precedent.

In addition, today's societies are highly mobile. The loyalties to place that reinforced and defined commons were weakened as people moved from agriculturally oriented communities to cities, and then from cities to suburbs, over the course of the twentieth century. Loyalties are undermined by rapid changes in the way people earn a living in places like the once-thriving manufacturing cities of the Northeast and Midwest. Such changes in the economy and technology create new commons, such as the very limited capacity of rivers to handle toxic waste. If these commons are open access, as they often are at the outset because the hazard is unrecognized, Hardin's tragedy can unfold.

Instead of tradition, contemporary societies have turned to adopting rules explicitly designed for governing, which are collectively referred to as **public policy**. That is, the community turns to government as the means of articulating and enforcing its will. The U.S. Constitution is an important example of this way of approaching the rules of a society. The Constitution created the government of the United States, empowering it to enact laws that eventually applied across a continent-sized territory while also reflecting the political will of a population dispersed across that wide land. This was an extraordinary political feat in the eighteenth century, before the invention of the telegraph or other means of communicating across such vast distances. Today, recognizing commons, the potential for tragedy, and ways of revising established institutions to cure shared problems can still be seen as important challenges for environmental governance and public policy.

The knowledge distilled in the community-governance principles summarizes a large body of social observations. It is an observation-based theory of commons, complementing Hardin's theoretical assertions derived from economic theory. Hardin argued that all commons will be mismanaged. The governance principles, on the other hand, show how some communities have in fact managed commons quite effectively. The governance principles can thus be seen as a richer description of what needs to be mutually agreed to if mutual coercion is to work.

RESPONSIBILITY, MISMATCH, AND ENVIRONMENTAL PROBLEMS

The ideas of commons and community provide a language that puts humans into the landscape so that we can describe how institutional arrangements and technology shape the access that people have to ecosystems. This is the access symbolized by the right arrow in Figure 3.1. In explaining access, we encountered the fact that access is unequal in all cases except for open access. Yet open access leads to the tragedy of the commons. As a result, the question of inequality arises in all well-managed commons. This may seem surprising, but social justice and sensible use of the resources of nature turn out to be deeply connected.

The language of human institutions in landscapes is complicated and subtle, and studying this book might be seen as an introduction to the grammar of that language. So far, you have learned some of the vocabulary of commons and community. Now we turn to a key statement about the way environmental problems are connected to human institutions: *when human responsibility does not match natural system boundaries or rhythms, environmental problems will result.*

This sentence asserts that environmental problems can be diagnosed by looking for mismatches between human and natural systems. This is a sweeping claim. In addition, *every* environmental problem involves a failure to foresee mismatches, a failure to take responsible actions once mismatches are identified, or both. Hence, solutions to environmental problems require eliminating or managing mismatches. Table 3.2 puts these generalizations into more concrete terms.

We have discussed spatial mismatches of responsibility above, including pollution and climate change. By putting dirty water into a stream, a polluter ignores or denies responsibility for the well-being of the stream's inhabitants and other

TABLE 3.2   MISMATCHES BETWEEN HUMAN USE AND NATURAL SYSTEMS

Mismatch	Examples of environmental problem	Possible solutions
Spatial (benefits are localized, costs are widespread)	Pollution; climate change	Regulation; technology; transparency
Temporal (benefits are immediate, costs are delayed)	Over-harvesting of fish or trees; buildings that can't withstand earthquakes	Binding agreements (insurance, liability laws) based on measurement and projection
Functional (benefits follow functional use, costs are defined as "external")	Water rights; sprawl	Enable larger range of stakeholders to participate; transparency; wider scope and accountability for planning

users. The logic of individual rationality can be tempting or overwhelming, and overcoming the problems of these commons requires changes in rules and expectations so that people will take responsibility more readily.

Sometimes transparency induces people to take responsibility. U.S. manufacturing firms are required to disclose the amount of pollutants they release into the environment. Even though the amounts released comply with environmental laws, the fact that the local community or the shareholders in the company know what is being dumped has proved to be an influential restraint in some cases, leading to voluntary reductions in the amounts being released.

Temporal mismatches occur when benefits can be taken in the short term but the costs of environmental degradation are delayed. A builder who sells a house may deny responsibility for the damage that comes with an earthquake or a flood a few years later. Fishermen sell their catch today, but they may not acknowledge the impact on fish populations in the future. Timber companies that log trees may not admit their role in the flooding that happens when rain rushes off denuded hillsides in the spring.

In many situations, problems can be foreseen. Where are earthquakes likely to occur, and how violent is the shaking of the ground likely to be? Are populations of young fish declining already? Are forest slopes vulnerable to rapid runoff and erosion once the trees have been stripped away? When environmental problems can be predicted, responsibility can be fixed—sometimes in a single actor or company, as in the case of a logging operation, and sometimes collectively. Laws that hold users of a commons liable for the damage that they cause, together with insurance to protect against known risks, are tools that can enable sound governance of commons across time. Neither insurance nor legal liability is a form of government regulation, and they can respond to the values of the community in a fashion that is different from public policy.

Recognizing that responsibility must be taken is not the same thing as taking it, however. As we discussed in connection with the cap-and-trade policy, above, the fact that the rich nations emit a disproportionate share of greenhouse gases suggests that those nations, including the United States, must now take a disproportionate share of the responsibility. A resident of the United States is responsible for annually emitting roughly 30 times as much carbon dioxide as a resident of Bangladesh, yet the nearly 150 million residents of low-lying Bangladesh are much more likely to lose their homes as global climate change raises sea levels and drives more frequent severe storms. Discussions are underway to turn the unevenly shared responsibility for climate change into actions that will reduce greenhouse gas emissions and pay for the costly adaptations needed by those most at risk. However, these are only discussions so far.

The third category of mismatch is functional. This happens when people simplify nature into a collection of resources. The water flowing in a river is economically valuable when used for irrigation or to generate hydropower, but the water also defines an ecosystem for aquatic life. If the natural world of the river is ignored, problems are likely to follow, affecting people such as those who depend on the river's fish, as well as the fish and other nonhuman elements of the ecosystem.

Is it likely that users will take such an oversimplified view of a river? Under some arrangements, including those established by the water laws of the western states, they are legally required to do so. The right to water is based on whether a landowner puts the water to what is officially defined as "beneficial use" by irrigating crops, providing water for grazing animals, or meeting urban and industrial demand. The value of keeping water in streams to support fish and other wildlife is legally excluded from the definition of "beneficial" in every western state but Washington. To the fish and other aquatic life, of course, the water is not just beneficial but vital. This is a striking example of a functional mismatch between human responsibility and the natural world.

Land use provides another important example. Suburban development works in a practical sense—new homes are built and sold, homeowners pay mortgages, and over time, nearly all homeowners take care of and improve their houses and property. Yet the development process pushes into the shadows many aspects of ecosystem change and community development. Lured by the pastoral ideal of the Hudson River School painters described in Chapter 2, home buyers are drawn to houses on their own plot of land. They look for good schools, an aspect of community quality, and they want their waste put into sewers.

The low density of most suburbs leads to an increasing dependence on automobiles in daily life, because mass transportation is unaffordable when a small number of passengers must be collected over a wide area. Often, the result is sprawl, a pattern of dispersed development with high traffic congestion and commercial development that requires lots of parking spaces and obtrusive signage to serve customers and attract people driving by. Sprawl ironically undermines the very attractions of pastoral, low-density land use that drew buyers to the suburbs to begin with, which may be one of the reasons why the difficult problems of sprawl have raised widespread environmental concerns, even in politically conservative communities. Sprawl is the result of a mismatch: partly as a result of successful rent-seeking by real estate developers over a period of many decades, the designs of new communities tend not to match the needs of its citizens after the community is built. Yet in many communities, planning is resisted because zoning and other limitations on land use reduce the profits that can be made by developers. Real estate developers and contractors have a strong interest in rent-seeking. When they succeed, the result is a short-term, narrowly self-interested functional understanding of the ecological and social changes implied by development. Development is not inherently bad, but when the institutions shaping development ignore the responsibilities that will be borne by communities in the future, the lasting environmental consequences are likely to be worrisome.

The idea that every environmental problem has within it a failure of foresight or responsible action is a hypothesis. It is a useful starting point for analysis, because it describes a wide range of environmental issues, providing a framework for inquiry and problem solving. It is important to remember that identifying problems and discussing responsibilities do not lead automatically to solutions. Powerful economic interests and deeply embedded legal traditions, such as the idea of private property rights, are involved, which make even the recognition of environmental problems difficult. It is often hard to design solutions and even harder to adopt and enforce them. Our point here is to show one way to begin solving the problems.

THE ROLE OF ETHICS

Toward the end of his article on the tragedy of the commons, Hardin made a striking statement about the relationship among ethics, humans, and landscapes (which he called "the system"): "The morality of an act depends upon the state of the system at the time it is performed."[8] The idea that what is right depends on the situation might seem unfamiliar to some readers. Some might wonder whether the Golden Rule or similar precepts from other religious traditions apply. (Our answer is yes, by the way. Deciding what it means to "do unto others as you would have others do unto you" is not automatic but depends on the specific stakes or consequences of a decision. As circumstances change, the interpretation and application of this generalized principle shifts, so that the concrete guidance of the Golden Rule does depend on the situation.)

Hardin applied his analysis of the commons to the rapid growth in human population, and he suggested that the most private of human choices—whether to have a child—was leading to shared tragedy. At the time Hardin was writing, moral and religious concerns about birth control and abortion were just as prevalent as they are today. Hardin was arguing that the ethics of human reproduction must change in response to a changing environmental situation. When human densities are low, in other words, the biological imperative to reproduce may be appropriate. But as human populations rise, he argued, social restraints become necessary and right, including coercive intervention in decisions by families to have children. China's one-child policy is an example of mutual coercion of this kind, although it was not mutually agreed to, particularly at the outset.

Do you agree that morality changes as the relationship between humans and landscapes changes? Serious students of moral philosophy have taken different sides in this debate. The fact that this debate continues illustrates both the fluidity of moral reasoning and the importance of deliberation in reaching moral judgments. At the same time, however, arguments over the need for and proper application of mutual coercion tend to echo a concept in toxicology that "the dose makes the poison." Pure water offers a classic example: If you get no water, you die. But even pure water can kill you if you drink too much in a short time. An amount of water that might keep you in good health for years if portioned out can be enough to kill you if you drink it all at once.

Ask yourself this question: What doses of restraint are needed to cure environmental problems and build a sustainable economy? Obviously, a lot of change is taking place in the world, something that is probably apparent in your own life. But when has the human situation changed enough that *new institutions* have been needed to govern human life? The Catholic Church has long maintained that the sanctity of human life means that contraception is a sin. Is that view wrong now that 7 billion people inhabit the planet? Are the signs of climate change so clear that we must no longer think of fossil fuel as only an economic commodity? These questions are examples of the ways in which environmental concerns encounter Hardin's idea that values need to change in response to changes in situation.

The need for moral deliberation crops up at important points in environmental studies precisely because humans' success in changing the natural world has altered our situations. In many cases, those situations press us to rethink the practical application of principles of right and wrong. Sometimes, the rethinking can go further, to reexamination of the principles themselves. Learning to recognize when such thorny matters arise is another important part of environmental studies.

As you can see in this chapter's discussion of community, government, and the economic transactions implied by herding animals on a common pasture, the subject of environmental studies leads directly to fundamental questions in the areas of biophysical science, social science, and even ethics and philosophy. In landscapes, we discover not only humans but human institutions.

In this chapter, we have argued that environmental problems are commons problems, and that they lead in many cases to the paradox of uneven development. They stem broadly from a mismatch between human knowledge and responsibility on one side, and the structure and function of the ecosystems that support human and nonhuman life on the other. When a mismatch arises, individuals often find they have incentives to act in ways that undermine or create disproportionate impacts on shared ecosystems. Overcoming such tragedies can necessitate changes in the rules and expectations governing the behavior of all the people and organizations sharing our common resources; these are institutional solutions.

Examples of such solutions can be found in the way societies have organized government, in economic systems, and in the traditions of communities that have lived in harmony with their landscapes for centuries, for the fact is that tragedies of the commons have been resolved by communities throughout human history. Hardin's vision of commons susceptible to tragedy is widespread in modern market economies but is strikingly rare among traditional cultures. Almost all of the cultures of the world that survived long enough to be studied by anthropologists and archaeologists were ones that worked out successful ways to allocate and share resources over periods stretching across many human generations. This fact suggests that there are practical as well as ethical reasons for us to work harder to govern commons effectively today.

Recognizing the mismatches generally requires the resources of the natural sciences (to describe the ecosystems involved) and the social sciences and humanities (to describe the history and institutions that shape human behavior). It also requires new ways of thinking that stretch across all of these traditional methods of examining and understanding the universe we inhabit.

In devising solutions for these commons problems, it is helpful to consult a set of principles that seem to characterize well-managed commons and the communities that own them. One of those principles is that problems and solutions vary from place to place. As a result, paying attention to local conditions and to the people who have a direct stake in ecosystems is essential. Translating the principles into practical public policies is hard work, demanding both an understanding of social and natural systems and the ability to bring about institutional changes through political leadership.

Solving individual environmental problems is crucial but not sufficient. This is the paradox of uneven development. The ways in which societies have combined technology, political power, and cultural and material incentives over the past three hundred years has now created a global economy that is changing the planet, often in irreversible and pervasive ways. This economy is only partly under the control of human institutions. Those of us living in the rich nations are the principal beneficiaries of uneven development. Still, like our fellow species, we humans remain vulnerable to its environmental and social disruptions. In the next chapter, we turn to a fuller picture of those global forces of uneven development, and then later in the book, we look at the challenge of searching for a path of sustainable development.

FURTHER READING

Cochrane, Willard W. *The Development of American Agriculture: A Historical Analysis.* Minneapolis: University of Minnesota Press, 1979.

Cronon, William. Chapter 6 in *Changes in the Land: Indians, Colonists, and the Ecology of New England.* New York: Hill and Wang, 1983.

Hardin, Garrett. "The Tragedy of the Commons." *Science* 162, no. 3859 (1968): 1243–48.

Ostrom, Elinor. Chapter 3 in *Governing the Commons: The Evolution of Institutions for Collective Action.* Cambridge, England: Cambridge University Press, 1990.

Roosevelt, Theodore. Chapter 1 in *Ranch Life and the Hunting-Trail.* 1888. Reprint, Lincoln: University of Nebraska Press, 1983.

Snyder, Gary. "The Place, the Region, and the Commons." In *The Practice of the Wild,* 25–47. San Francisco: North Point Press, 1990.

KEY TERMS

community governance	government ownership	mutual coercion mutually agreed to	regulation rent-seeking
community-governance principles	hypothesis institution	open access optimal	tradition tragedy of the commons
enclosure	institutional solution	privatization public policy public trust	usufruct right

Chapter Four

A WORLD WITHOUT EDGES

DISPROPORTIONALITY AND THE SHROUD OF ENVIRONMENTAL IGNORANCE

In this chapter we explore a question that is important but for which a good answer can be very difficult to find: What impacts do I have on the natural world? The point we want you to take from the pages ahead is why the answer to this apparently straightforward question, one that is of clear concern to environmentalists, is so elusive. Humans now live in a global economy of extraordinary complexity—one so complicated, in fact, that we routinely fail to understand simple things about the environmental profile of our economic activities. As a result, Americans and other people, especially those living in rich countries, cause substantial environmental damage that they don't see, as a by-product of their ordinary lives.

This ignorance has an important practical implication. Although we can modify our lifestyles to lower the indirect harm they cause, individual actions by themselves cannot overcome the institutional patterns woven into our habits. Using the concepts developed in Chapter 3, we will see that commons problems are deeply buried in everyday life. By gaining a better picture of the complex world we inhabit, we can begin to recognize the institutional challenges that arise within it. In this chapter we look at the environmental profile of the American economy, find some surprises, and gain historical perspective on how we migrated into what we will call a "world without edges." In that world, environmental impacts have become profoundly obscure because we no longer act or live in just a single place in the way most people did in the past.

THIRTY-FIVE POUNDS OF GUILT

Most readers of this book think of Americans as a people who consume a lot. Robert Ayres, a scholar of technological systems, has given us an idea of how much. In 1993, the U.S. economy required just under 21 billion metric tons of raw materials.[1] To get an idea of how much stuff this is, the twin towers of the World Trade Center destroyed on September 11, 2001, weighed roughly 1 million tons.[2] The national economy uses about 20,000 times that amount in a year, and this total does not count the water used to process these raw materials. As impressive as these figures may seem, they hide a surprise. Ayres found that a bit less than 3 billion tons was turned into durable goods, mainly in the form of new construction such as buildings and highways. And just over half a billion tons ended up as products sold to consumers. All the rest of the input materials became some kind of waste: sawdust, scrap metal, toxic wastes, or just garbage; materials dissolved in streams and rivers; or gases, such as carbon dioxide, emitted to the atmosphere.

In short, for each pound of stuff Americans bought, the production process created 35 pounds of waste that never got to consumers. This is an impressive number.

Learning Objectives
When you have finished studying this chapter, you should be able to

---

⬊ identify a disproportionality that you can see in your everyday life, and confirm that it is a disproportionality by doing a little web research;

⬊ see elements of your family's life that make them inhabitants of the "world without edges," and other elements that demonstrate their sense of place;

⬊ explain how something that adds to your individual feeling of independence is the result of an increase in your dependence on others;

⬊ articulate how a long-dead society such as that of the Maya of Central America or Gilbert White's eighteenth-century England can be seen as relevant to the globalized economy and culture of the present.

By reducing producers' emissions and wastes, it may be possible to make very large gains beyond those that come from focusing on consumers' wastes alone.

The 35 pounds of waste contain a deeper insight, one that requires a bit of explaining. Of the wastes generated by an industrial economy, perhaps the most worrying are the toxic pollutants emitted by manufacturers. The U.S. Environmental Protection Agency (EPA) requires that the emissions of all large plants be reported, and the data are available to the public in the EPA's **Toxics Release Inventory (TRI)**. (If you're curious, go to www.epa.gov/triexplorer/statefactsheet .htm.) The TRI gives us a way to see which industries are responsible for some of the most dangerous of the 35 pounds of wastes that go with each pound of stuff we buy. In the year 2000, the mining industry accounted for five-sixths, or 83 percent, of all on-site toxic releases.

That is, the various sectors of the economy are not equal in terms of their environmental harm. Often it is argued that pollution is a price that must be paid to provide jobs. But in 2000, mining accounted for only 0.00072 percent of U.S. jobs. Copper mining, to illustrate this disparity, accounted for 1/1,000 of the gross domestic product of the United States in 2000, but it was responsible for nearly 16 percent of all toxic releases by weight.

What's going on? *A large fraction of the environmental harm comes from a very small fraction of the economy.* We call this an environmental **disproportionality**. As we will see below, disproportionality is both widespread and important in thinking about how to manage environmental impacts. The idea of disproportionality—that some important characteristic is distributed very unevenly—is more common than we might think. Summer thunderstorms drop their water in small patches; one neighborhood may be drenched, while a few blocks away the lawns are only dampened by the passage of a light shower. There are many more marine species found in coral reefs than in other ocean ecosystems. And, as activists have been arguing for nearly a generation, trash collection facilities, power plants, and other undesirable urban land uses are found in higher concentrations where poor people live, a problem of **environmental injustice**, as the poor face disproportionate exposure to environmental harm.

Disproportionality in the *creation* of environmental harms, however, has been oddly overlooked. Instead, politicians and analysts tend to focus on average consumers and total human impacts. Politically conservative commentators tend to see the creation of environmental harms as an unfortunate but necessary side effect of prosperity; leftist commentators generally share that view, but see the creation of environmental damage as additional evidence of the drawbacks of capitalism. From these perspectives, we are all equally guilty of that 35 pounds of waste and environmental harm, and we can do little about it without harming the average person and the total economy. The latter is not true. To see why, we need to learn a bit of statistics.

OUTLIERS

In the sciences, it is often appropriate to assume that extreme cases are misleading because they are the exception to the norm and that they should be ignored. In these cases, scientists try to understand the whole and therefore focus on an entire collection of data. They look at measures of central tendencies such as averages and medians, and they examine the differences in central tendencies across groups. A small number of people may live to be older than 110, but the health conditions of a community are captured more effectively by average life span or infant mortality. Statistics textbooks warn that extreme results, or **outliers**, can distort the estimate of average values. As the textbooks add, however, simply being an outlier does not mean that a data point should be deleted from the analysis; instead, there sometimes may be good reason to make extreme values the *focus* of analysis. A health analyst might hope to learn something about longevity by studying the handful of people who live to 110, for instance, or may want to establish why the people in one community or region tend to die at younger ages than those in the rest of the country.

It turns out that disproportionalities abound in the way that humans interact with landscapes. In Box 4.1: The Gini Coefficient: Measuring Inequality (page 81), the standard method for analyzing inequality is explained; this method provides a way to quantify disproportionality.

IS (THAT MUCH) POLLUTION NECESSARY?

Information on the U.S. economy is collected using a Standard Industrial Classification (SIC) system, which is similar to the classification system used to identify books in a university library. These SIC codes not only label different kinds of businesses but also permit more and less detailed studies. A two-digit SIC code analysis groups together major industries, such as Chemicals and Allied Products (SIC 28), into a single line of data, and a four-digit report provides data on a single industry, such as Pesticides and Agricultural Chemicals (SIC 2879). Notice that the first two digits in this code are the same as the two-digit code for the broad industry grouping for chemicals.

At the two-digit level, the pollutant emissions in the EPA's TRI show a Gini coefficient of .755 when releases were analyzed as reported—by weight (Fig. 4.1). When releases are instead recalculated in terms of their estimated toxicities, the inequality rises to a Gini coefficient of .865. By both weight and harm, then, *a very large fraction of all toxic emissions come from a small number of industries.* This is perhaps not so startling. A coal-burning electric power plant will spew out more carbon

BOX 4.1

THE GINI COEFFICIENT
Measuring Inequality

Children in grade school are commonly taught to line up by size, with the tallest at one end and the shortest at the other. This helps in organizing class photos, for instance. One can do the same thing with other sorts of data. The Italian statistician Corrado Gini proposed a method that has become the standard way to measure inequality of all kinds. Let's see how to do this with income. First, estimate the total income of a community's households. Then, count the fraction of households whose incomes are below 10 percent of the total, below 20 percent, and so forth. Then plot the percentages on a graph like in the figure below.

If the 50 percent of the population with the lowest incomes have 30 percent of the total, the curve would look something like the dashed line. If the poorest 50 percent earned only 10 percent, the line would look more like the dotted/dashed line. In a community in which wealth is divided exactly equally—as it might be in a convent—one would find the perfectly straight black line.

The **Gini coefficient** is a numerical summary of a diagram like this one. The coefficient for the income distribution shown by the dashed line is the ratio of the shaded area to the area below the black line. You can see that if the distribution were perfectly equal (the black line), the shaded area would be zero, and the Gini coefficient would be as well. The more unequal the distribution—the more bowed-out the curve—the closer the coefficient is to 1 (or 100 percent shaded).

The distribution of incomes in all nations is unequal. In practice, coefficients range from around .2 for historically egalitarian countries such as Bulgaria or Hungary, to around .6 for nations where powerful elites dominate the economy.[1] The world's highest coefficient today is Sierra Leone's .62. Most present-day European countries and Japan have coefficients of .25 to .32. Most African and South American countries, and in recent years, the United States, have Gini coefficients in the range of .45 to .50. If we focus on wealth instead of income—that is, the accumulated size of bank accounts and other assets, rather than the incomes earned this

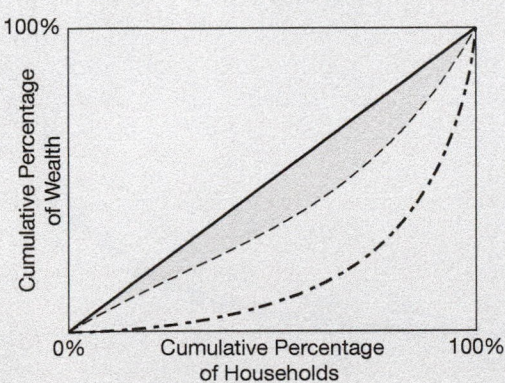

Measuring inequality with the Gini coefficient. A more bowed-out curve indicates greater inequality.

year—the Gini coefficient for the world as a whole is nearly .9. This is the coefficient we would find in a group of 10 people with a total wealth of $1,000 if one person had $991 and the other nine each had a dollar. Even so, more than 13 million people worldwide were millionaires in 2000.

The Gini coefficient can measure the distribution of any set of data. In the text, we use this approach to examine the distribution of emissions by industrial polluters. What we are asking you to look at, in that case, is the disproportionality of environmental impact across a set of polluting firms.

1. The wealth distribution figures in the discussion of the Gini coefficient are taken from James B. Davies, Susanna Sandstrom, Anthony Shorrocks, and Edward N. Wolff, "The World Distribution of Household Wealth," United Nations University World Institute for Development Economics Research, Helsinki, 2006, www.iariw.org/papers/2006/davies.pdf.

dioxide than a shopping center, no matter how one adjusts to make them comparable, such as by weight of wastes produced or the value of sales.

When one looks within a single industry, disproportionality seems more surprising. Why should one copper smelter be much dirtier than another? Yet consider the sixty-two plants that reported toxic emissions within the most toxic

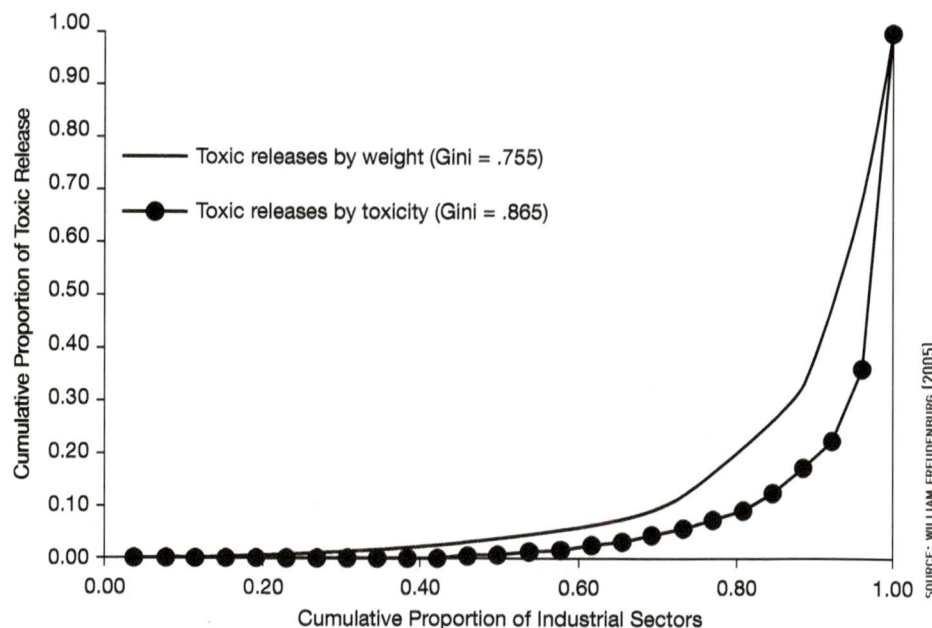

FIGURE 4.1
Toxic releases from major U.S. industrial sectors in 1993.

specific industry within the most toxic sector of the whole economy. This hazardous sector is SIC 333, Primary Nonferrous Metals, which produces basic metals that don't contain iron or steel. (In these statistics, "primary" metals are taken directly from nature, whereas secondary metals come from recycling.) In 1993, the distribution of toxic releases by those sixty-two facilities had a Gini coefficient of .975, an astonishingly high level to an experienced social analyst.[3] The disproportionality came, to a significant degree, from a single enterprise. The Magnesium Corporation of America plant in Rowley, Utah, accounted for more than 95 percent of the toxicity emitted from the entire 333 SIC code. Figure 4.2 shows the distribution with and without the Rowley plant. You can see how much more bowed-out the curve is when the plant is included. This is a graphic statement of how unequal the toxic emissions are within the group of primary nonferrous metal producers and how much of a difference a single plant can make in a population of sixty-two facilities.

In 2001, the EPA sued the Rowley plant's owner to force a cleanup of the facility, and the site was put on the national Superfund list, which identifies the most toxic spots in the country so that they can be remediated. The lawsuit was still pending as this book was being completed, so it is not clear when the Rowley plant (which is still operating) and its pollution will be cleaned up.

Release of toxic gases is a commons problem in the sense that we all breathe the air, but it is also a privatization problem in the sense that the polluting plants

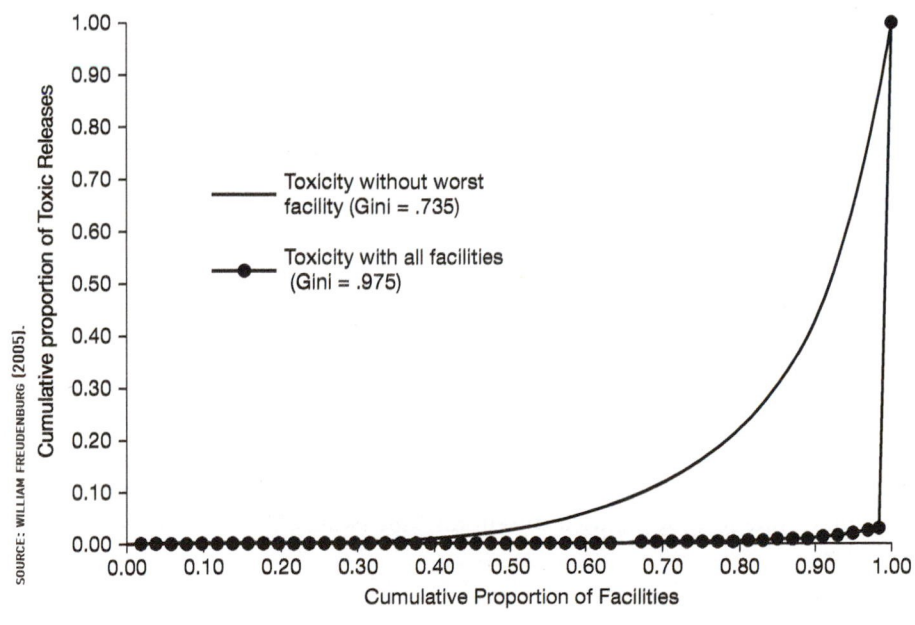

FIGURE 4.2
Toxic releases in the Primary Nonferrous Metals (SIC 333) sector of the U.S. economy, with and without the single worst facility, 1993.

are all privatizing a shared resource by using the air as a dump for their wastes. The Rowley facility accounts for a large percentage of the pollution emitted by its industrial sector. If the Rowley facility were to produce its products at a level of emissions that matched the industry's median, the total toxic burden from primary nonferrous metals would be reduced by more than 90 percent.

In many other environmentally sensitive parts of the economy we do not have data complete enough to compute a Gini coefficient. But strong evidence of disproportionality can be found nonetheless.

A key resource in the western United States, for example, is access to fresh water. In western states, the general pattern is that a single industry—agriculture—consumes roughly 80 percent of the water while generally contributing roughly 1 to 3 percent to the states' economies. Indeed, agriculture consumes about half of the water in the Valley of the Sun, where one finds Phoenix, a burgeoning metropolitan area in a desert setting where one doesn't think of crops so much as cactus.

As we saw in nonferrous metals, moreover, the disproportionality persists as the analysis gets more specific. Consider California. Does the water go into high-value specialty crops such as wine grapes or almonds? For the most part, the answer is no. Nearly half of all the water used in California's agriculture sector goes into growing grass and alfalfa—low-value crops that are used almost exclusively for hay. More than 60 percent of the agricultural water of the Golden State goes into four crops: grass, alfalfa, rice, and cotton. The largest percentage of these four crops are literally being grown in a desert—in the state's Central Valley, which receives only 5 inches of rain per year across its southern region. In addition, more than half of all water used in irrigation in the United States never makes it to the plants, being lost instead through leaks, evaporation, and runoff.

JOBS AND ECONOMY VERSUS ENVIRONMENT?

Deeply subsidized irrigation water is said to be vital for jobs in rural areas. In other words, wasteful use of water can help people. But studies in California by the Pacific Institute show that, over the course of a year, 1,000 acre-feet of water supports just twelve jobs in agriculture, or only one job if the water is used on some of California's high-volume, low-value crops that are the most water-intensive, such as rice. One acre-foot of water is a lot. It's about 300,000 gallons, or roughly the amount used by an American family over the course of a year in most of the country. A thousand acre-feet of water supports an average of a thousand jobs if used in schools, and twenty thousand jobs if used in industry.

In a 2001 study, the Natural Resources Defense Council, an environmental advocacy group, analyzed an alfalfa farm using 240 acre-feet of water in a given year (roughly 80 million gallons).[4] It generated $60,000 in sales and employed three workers. But a California semiconductor plant using the same amount of water generated 5,000 times that much in sales, $300 million, while providing jobs for up to 2,000 people.[5] Citizens might think of reasons why agricultural users should continue to use so much water—tradition, perhaps, or the argument that keeping land in agriculture is one way to keep it from being used instead for urban sprawl. But the argument that taking water away from these crops is bad for the economy—well, that argument simply doesn't hold water. The economy would probably be in much better shape if more of the water were to go to other uses. This might well include an option that is generally not defined as a "beneficial use" under western water law—namely, being returned to the streams, which would improve survival chances for fish and attract new economic activity in the form of tourism and recreation.

These examples strongly suggest that we should reexamine the conventional belief that environmental protection harms the economy and throws people out of work. The chemical and primary metals sectors of the economy, in combination, produced roughly 60 percent of all the toxic emissions in the nation in the TRI data summarized earlier, while producing 4.2 percent of the nation's economic output and providing just 1.4 percent of the jobs. Nor is it true that the most highly polluting facilities must be producing critical materials that simply cannot be produced any other way. Remember that one plant in Rowley, Utah, that accounted for more than 95 percent of the toxicity emitted for the entire industry. Recall that the plant was sued by the EPA over its handling of wastes, something that is hardly typical of the other 98 percent of the nonferrous metals industry.

In addition, little evidence supports the common assertion that stricter regulation of heavily polluting industries would drive those firms out of business. Analysts have found that high levels of pollution are found not among technologically advanced firms but among older ones that use outdated and generally inefficient methods. This makes sense, on reflection. In older industries such as primary metals, it can be difficult to raise capital to improve dirty plants, and easy just to stick with practices that have been in place for decades, keeping a clunker going until it sputters out, whatever the cost to workers or the environment. This hardly means, though, that such levels of pollution are economically necessary.

Consider a facility that emitted enough pollution to have a significant impact on the national total of hazardous emissions. In 1993, IMC-Agrico of St. James, Louisiana, was the country's leading emitter of toxics, by weight. This plant was producing fertilizer. Due in part to the publicity that accompanied its number one ranking in the TRI, the facility in St. James took basic but important steps to clean up its act. By 1998 the situation had improved, as the EPA found when it analyzed the plant's use of phosphoric acid in producing the fertilizer.

> Large quantities of gypsum are generated as by-products in the process. When rainwater comes in contact with the gypsum, stockpiled in uncovered outdoor stacks, it flushes out residual phosphoric acid.... Significant reductions in the amount of phosphoric acid reported to [the Toxics Release Inventory] have resulted from reducing the surface area of some stacks and from covering the stacks with grass-covered clay. Additionally, evaporation ponds built on top of inactive stacks were lined with a synthetic material, preventing some water from entering the stacks. A system was also implemented to collect water from within the stacks and recycle the phosphoric acid contained within.[6]

Planting grass on uncovered outdoor stacks of waste material could scarcely be considered economically ruinous. Nor was it technologically difficult. As a result of these straightforward measures, "IMC-Agrico company in St. James, Louisiana ... reported a 37.6 million-pound reduction in discharges to water"[7] by 1996, an amount that translates into nearly one-third of that facility's earlier total toxic release. Also, the firm's annual profits went up, not down.

In Chapter 3 we mentioned the sulfur dioxide emissions trading policy put in place by the Clean Air Act amendments of 1990. Before the policy was adopted, the utility industry had been arguing that cleaning up its emissions would be economically challenging—more than $1,500 per ton. But the actual cost soon proved to be much lower. In 1993 the federal government reported that auctions of emissions rights had brought prices of less than $450 per ton.[8] A citizens group called the Acid Rain Retirement Fund had raised donations sufficient to purchase the rights to more than 1,000 tons of sulfur dioxide, which they did not use, keeping the air cleaner.[9] Moreover, environmentally regulated sectors of the economy have done somewhat better against international competitors than have manufacturing firms in general, and the facilities producing higher levels of pollution have tended to earn slightly *lower* levels of profit than cleaner plants, even when looking within a single industry.

The story of IMC-Agrico is good news for Louisiana and the Gulf of Mexico, into which the polluted waters flow. The fact that cleaning up sulfur dioxide is much more affordable than expected is good news across the country. These stories, and the systematic analyses that support them, suggest that what we have in the American economy is a problem of "enclosure," or a tragedy of the *un*common. A few firms and activities create substantially more than their average share of pollution and other environmental problems; these harms are inflicted on nature and humans, but the profits go to a handful of firms. That pattern is reminiscent of the way the landlords of Britain enclosed the commons in order to retain the benefits from them for themselves, as we saw in Chapter 3. This is a pattern that can be seen whether we look at total environmental impacts, at environmental impacts per job, or pollution per dollar of output.

The deeper theme here has to do with the nature of effective environmental problem solving. First, disproportionality is present in many parts of the economy. Beware of jumping too quickly to averages and aggregate measures. Second, disproportionality matters. If there are only a few outliers, fixing the problem is likely to be much more feasible, and it is not likely to require shutting down an entire industry. Third, rather than being costly to a national or regional economy, *environmentally responsible business is good business*. Disproportionality has many syllables. It is a useful watchword.

Disproportionality in environmental performance did not just happen. It came about as an unplanned by-product of a global system of **economic specialization** governed by a specific set of institutional arrangements. The Rowley plant does not produce magnesium for local ranchers, but for distant markets. The large quantity of pollutants it releases has local impacts, but those impacts are driven by demand for products from far away. This global system of production is fairly recent, having arisen largely over the last two hundred years. So to see where disproportionality and the 35 pounds of guilt came from, we need to look at that history.

CROSSING SCALES

At the same time that there comes that increase in what the average man demands from the resources, he is apt to grow to lose the sense of his dependence upon nature. He lives in big cities. He deals in industries that do not bring him in close touch with nature. He does not realize the demand he is making upon nature.

—Theodore Roosevelt[10]

The linking of global to local has important and wide-ranging environmental effects. People in rich countries such as the United States now routinely influence environmental conditions far away through the choices we make about investment and consumption. Nearly always, however, we are unaware of the influences we exert. Think again about the 35 pounds of industrial wastes that are produced for every pound of products that we actually see. It is not so surprising that most of us think of ourselves as making local choices rather than contributing to distant changes. As we saw in Chapter 2, people's awareness of nature and environment is heavily affected by their sense of the places where they live, work, shop, and spend free time. Yet if people are having much more widespread impacts, then institutions that are grounded in place may no longer be adequate to guide the choices we make. Our choices now inescapably cross scales of space and time, and our everyday decisions, such as the food we buy or the way we heat our houses, have distant, sometimes delayed impacts.

To understand what has happened, we will adopt a term proposed by the scholar Don Gifford, the **world without edges**—the idea that the world we live in has become one in which the *boundaries* that once defined specific places *have dissolved* in important ways.[11] As with the idea of property, the world without edges is a set of institutional arrangements, which both reflects and drives ecological and social change.

Consider what happened to the economies of the Native Americans after colonists arrived in New England in the seventeenth century. Furs and timber became important to the Native Americans in ways that were new. They had used wood, of course, and their skills as hunters and trappers were far superior to those of the English. But the coming of the new settlers brought large changes, as the harvesting of animals and plants expanded dramatically. As we've mentioned before, beavers became commercially extinct by 1700, only eighty years after colonization began. By 1830, trees had been cleared to make way for farm fields and pastures all the way to the tops of hills, transforming the ecosystems that had been inhabited by Native Americans for ten thousand years. These changes were not caused only, or even primarily, by local activities. The furs and timber, like the tobacco and cotton grown farther south in colonial America, went into world trade. Trade is a set of institutions—the rules, markets, banks, and customs that shape the behavior of those who produce, transport, sell, and consume goods, and those who process the waste products. This behavior reshaped ecosystems, creating the specialized production systems, such as cotton plantations, in which we can recognize the disproportionality we see now in the industrial landscape.

Colonial landscapes became linked to European economies. Sales of beaver hats and the building of sailing ships that used American tree trunks in their masts and hulls affected how humans altered the environment of New England. European demand for tobacco and, later, cotton shaped how land was cultivated in the Mid-Atlantic and South. Slavery came to be established as an institution in the New World. In all these ways, the European settlers and their descendants created places rooted in ties to other places, far distant in comparison to the trade among neighboring communities that the Native Americans had known.

Further changes were to come. We will trace these changes by looking at the lives of three people who lived a century apart from one another. By examining their relationships to place, we will see that the world into which we were born is subtly but drastically different from the world in which the United States was founded in 1776. This change in culture and consciousness, in turn, has drastically altered human relationships to the environment. That is, the sense of place has a history, just as environmental awareness has a history, as we saw in Chapter 2.

GILBERT WHITE: A "STATIONARY MAN"

The first of these three people is the clergyman Gilbert White, who was born in the English village of Selborne in 1720 and died there in 1793. The church where White served as curate is still used for worship. It is located about 50 miles southwest of London—close enough that nowadays some residents of picturesque Selborne commute to work in the capital city, an arrangement unthinkable in White's time.

White was deeply attached to his home. His sense of himself was rooted in his hearth, his place. White was educated at Oxford, where he, like most college students of his time, prepared for the ministry. But he insisted on returning home to Selborne. When the Church of England assigned him elsewhere, White used his small salary to hire another man to serve in his place, so that he could stay at home and become the curate at the Selborne church. Here, he spent his entire life.

We remember White today for his writings as a natural historian, a way of talking about nature that he pioneered. He wrote about his ramblings and observations of the countryside in a book called *The Natural History of Selborne*, which first appeared near the end of his life. This book made White famous. The American scholar James Russell Lowell called it "the journal of Adam in Paradise."[12] Henry David Thoreau took this book to Walden Pond; Charles Darwin had it with him aboard the *Beagle*, on the voyage that laid the groundwork for the theory of evolution. Gilbert White was the first nature writer. His writing was a literary precursor of the Hudson River School painters and Transcendentalists a generation later. White marks the beginning of environmental awareness in its literary form.

Here is a passage describing White's studies of the echoes he found in the fields and woods around Selborne.

> A young gentleman, who had parted from his company in a summer evening walk, and was calling after them, stumbled upon a very curious one. . . . At first he was much surprised, and could not be persuaded but that he was mocked by some boy; but, repeating his trials in several languages, and finding his respondent to be a very adroit polyglot, he then discerned the deception.
>
> This echo in an evening, before rural noises cease, would repeat ten syllables most articulately and distinctly.[13]

Although he was well into middle age when he wrote this passage, White's enthusiasm for echoes and birds and the seasons of Selborne was as innocent and as full of curiosity as that of a child. With his Oxford education, White was also "an adroit polyglot,"[14] and he measured the duration of the echo by reciting Latin poetry. He repeatedly visited the spot where the echo could be heard, so as to see if the weather or time of day made a difference. White concluded that the evening air was "most elastic." He was an amateur scientist: someone who loved observing and thinking about what he saw around him. "Amateur" in this sense shares a root with "amorous."

In the preface to *The Natural History of Selborne*, Gilbert White called himself a **"stationary man"**—someone who did not stray far from his home in the country and liked it that way. Selborne was a village of fewer than four hundred people—about the population of a college dormitory today. White was of high status, well educated, and was paid an income for his ecclesiastical duties. It was not a large income but it gave him the leisure time to pursue his hobbies as a naturalist. He belonged to an elite in Selborne no more numerous than the resident advisers in an undergraduate dorm. He traveled occasionally to Oxford and London. Yet despite having some cosmopolitan ways, his life was led *in* Selborne to a much greater degree than is common today. Here is Don Gifford's account of White's household.

> [It] combined a high order of self-sufficiency with dependency on the outside services of the village butcher and miller, local farms for milk, butter, and cheese, wood- and peat-cutters, itinerant fishmongers, truffle hunters, and so forth. The village, in turn, was an almost closed agrarian economy....
>
> Throughout the year, White and...his parish were largely dependent on what was in season....White and his servants raised and stored and preserved a wide variety of fruits and berries and vegetables....White made his own malt and beer and some wine; he kept poultry, pigeons, bees, and at least three horses.[15]

White's was a world in which space and time were marked with less precision than we now take for granted. In one of the letters in the *Natural History*, White commented on the area covered by the village lands. Some, he wrote, "are of the opinion that the outline, in all its curves and indentings, does not comprise less than thirty miles."[16] The road signs of today, marked in tenths of a mile and sometimes mapped by global positioning devices, are far removed from the world of the stationary man. He didn't need to know exactly where he was in the modern sense of geometric coordinates laid out on a map, because he knew from direct experience, by walking and looking.

White was also bounded in time. As Gifford pointed out, the classical world of Greece and Rome, which White had studied at Oxford, was thought of as the golden age, a civilization that "established standards of absolute value against which the arts, letters, morality, and politics of the present should be measured and judged."[17] Today, techies wonder what the Next Big Thing will be, stock brokers are concerned with the ups and downs of tomorrow's market, and environmentalists try to predict the future of our planet's species and ecosystems. Our notions of

the significant achievements of humankind, such as athletic records or our grow-ing knowledge of tropical forests, tend to be much closer in time than the ancient Romans, and they point toward the future rather than the past.

Finally, White was bounded by the social order. He was a clergyman, a gentle-man who had enough income, time, and energy to make the writing of natural history into a hobby. He tested the echoes in Latin, a language few of his neighbors knew. The dream of equal opportunity for all, reflected in the multiculturalism that higher education has adopted in the United States and many other countries, was not part of Selborne, even though the American and French revolutions occurred while White kept his journals.

Being stationary did not mean that White was removed from the rest of the world completely. He had news of the French Revolution that began in 1789, and he noted in his journal a large volcanic eruption in Iceland in 1783. Each day, he drank tea from distant India. Yet to a degree far greater than in our own lives, White lived in a **self-sufficient** society. His food, his water, his work and religious activity, his social experiences, and most of all, his naturalist ramblings were all local. To him, being stationary was something to be treasured. His was a sense of place.

About two hundred miles away from Selborne lie the ruins of a cotton mill. A painting of the mill was made about a year after *The Natural History of Selborne* was published (Fig. 4.3). The Cromford mill was the first water-powered factory. It contained a water frame, a spinning machine that turned cotton fibers into thread. This was the birthplace of the Industrial Revolution. The builder of the mill, Richard Arkwright, died in 1792, a year before Gilbert White. These two men lived in the same society, in the same decades. White looked at the landscape and saw a nature that was, to him, mostly unchanged from Roman times; even the echoes spoke Latin to him. Arkwright looked at the landscape, but did not think in terms of the ancient world at all. Instead, he saw a source of energy and devised a way to harness it—an idea that soon changed the world.

White's lifelong passion to study and write about nature arose in a society just beginning to industrialize—just beginning to transform the human relationship with landscape. As we saw in Chapter 2, industrialization and the urbanization it brought in its wake gave rise to the romanticism of the Hudson River School painters in America. Perhaps surprisingly, the society that grew from Arkwright's industrialism was one ready to embrace White's celebration of nature.

HENRY DAVID THOREAU: CHOOSING TO BE STATIONARY IN AN INDUSTRIALIZING WORLD

Our second figure, Henry David Thoreau, was a nature writer, too. Born in 1817, Thoreau was just short of a century younger than Gilbert White. In the three generations that separated these two figures, writing about nature had become a pursuit whose literary value was indisputable. We remember Thoreau as a Transcendentalist, a friend of Ralph Waldo Emerson and other writers. Thoreau's writings continue to influence environmentalists today, and he is among the most widely read of the people in his literary circle. In his own time, Thoreau was also known as an antislavery activist, and his essay on political protest, *Civil Disobedience*, influenced many later leaders of political movements, including Martin Luther King Jr. and Mohandas K. Gandhi.

Thoreau published his masterpiece, *Walden; or, Life in the Woods,* in 1854. The book is an account of two years spent living on the shores of Walden Pond, a small lake in Concord, Massachusetts, about 15 miles northwest of Boston. Thoreau sought to live as much removed from civilization and society as he could achieve, building his own shelter and raising beans to feed himself.

Walden, like *The Natural History of Selborne*, is a landmark of nature writing. *Walden* is a meditation on solitude, on living as self-reliantly as one can. By the mid-nineteenth century, solitude had become so exceptional that Thoreau could make an explicit retreat from the spreading urbanization of Boston in a deliberate quest for an independent life. To become "stationary" had become a *literary* achievement, something done so as to discover experiences of a kind worth writ-

ing about. *Walden* was written in an apparently casual but carefully constructed style, as scholars have found by studying Thoreau's notes, drafts, and many revisions. The Hudson River School painters discussed in Chapter 2 combined scenic elements from different places in some of their canvases; they were Thoreau's contemporaries. Like those artists, Thoreau was making a scrupulously crafted statement. To his readers in the increasingly mobile societies of Europe and America, living in this deliberate fashion—in a bounded, stationary world—became a culturally significant idea to celebrate. As with *The Natural History of Selborne, Walden* did not become famous until its author had died in 1862, when Thoreau was only 44.

Thoreau wrote this about his isolation: "I have my horizon bounded by woods all to myself; a distant view of the railroad where it touches the pond on the one hand, and of the fence which skirts the woodland road on the other. . . . I have, as it were, my own sun and moon and stars, and a little world all to myself."[18] It is surprising to think that a nineteenth-century landscape with a train in it could seem isolated and pastoral. Thoreau's train is like the one in Jasper Cropsey's 1865 painting *Starrucca Viaduct*—a plume of smoke in the distance, a benign machine in a beautiful natural landscape that overshadows the small human in the foreground (Fig. 4.4).

Unlike Gilbert White, Thoreau could become a stationary man only temporarily. While he developed his literary and political career, Thoreau made a living manufacturing pencils, which were destined to be sent far away. Nature, too, had

FIGURE 4.4
Jasper Francis Cropsey, *Starrucca Viaduct* (1865).

changed. In the New England woods where Thoreau sauntered, all the large animals had disappeared, even the deer that have become common enough again to be a nuisance in suburban Concord today. Along with the economy and nature, culture also was changing. The classical allusions and Latin recitations of Gilbert White were mostly gone by Thoreau's time, replaced by Transcendentalist celebration of nature. The rain, Thoreau wrote, "being good for the grass, is good for me."[19] This is a sentiment also found in the poetry of Emily Dickinson, another prominent Transcendentalist.

Today, seeking solitude is something we do deliberately, like Thoreau, rather than as a matter of course, as Gilbert White did. To "get away from it all" is to retreat behind boundaries one has chosen. It is a temporary retreat for nearly all of us. What one sees in the two generations that separate *The Natural History of Selborne* from *Walden* is a world in which changes were coming with ever greater speed, an accelerating locomotive that carries us still.

DON GIFFORD: AN URBAN NOMAD

The third person to consider is Don Gifford, who taught English, American studies, and environmental studies at Williams College in Massachusetts. Gifford was born in 1919, two centuries after Gilbert White and a hundred years after Henry David Thoreau. A student of literature and history, Gifford argued in an essay published in 1990 that the "stationary" life has lost its meaning. Our reality, he wrote, has become a "world without edges."

> White could confidently locate the center of his world and trace the circumferences of its edges. Thoreau had some trouble with his centers—was it Concord or Walden Pond or the Celestial Kingdom?—and the edges were for him less securely determined, thanks to the swirl and fluidity being created by railroads, telegraphs, steamships, and all their hustling progeny.
>
> If I look up from my worktable and try to define the edges and circumferences of my worlds, the walls of this study dissolve. . . .we are urban nomads.[20]

White lived in a world that he literally paced off, walking through fields and gardens that nourished him, in forests whose wood kept him warm. Thoreau created a life in the woods, for a time, where he could insist upon self-sufficiency. But for a person living in the late twentieth century, that was no longer possible. Gifford was, like all of us living in America, tied to an urban, industrial, global economy.

This has a profound implication for our perception of place, an implication signaled in the word "nomad." Instead of the particular glories of Selborne or Prospect Park (discussed in Chapter 2), Gifford saw a world in which all airports look alike and all motel rooms are interchangeable. He suggested that our notions of place have become uncertain, in the sense that we do not know the relationships that tie us to the rest of the world, even if we know our current geographical coordinates.

THE ILLUSION OF INDEPENDENCE

One sign of the dissolving edges and uncertain boundaries in the places we inhabit is the shift in the knowledge we think is valuable. Identifying birds and trees was something White or Thoreau did habitually. In our society those skills are much less common. Our students are often surprised when we ask them to notice what kinds of clouds are in the sky. This, too, is something few of us do.

Does this mean our students are lazy and unintelligent? Not at all. They, and we, live in a society in which we learn what is significant about the sky by watching a computer model called a weather forecast, which is superimposed on a satellite radar image. This isn't an inferior world, necessarily. Certainly, we obtain more detailed and more reliable estimates of the weather this way than Gilbert White could routinely obtain. The death toll from Hurricane Katrina would have been significantly higher without advance warnings from satellites. But the weather forecast comes from a world that we cannot even see from our homes or work places. The weather report uses satellite images collected more than 100 miles overhead, computers running in distant places, and a commercial television network or the Internet to deliver it. It comes from the world without edges.

How is our world different from Gilbert White's? Markets and technology are the obvious answers. These are names for a large collection of nonlocal institutional arrangements, including the Internet, which operate at a scale far beyond the experience of any single person, even though they are all human creations operated by people. The world without edges creates a stage for new human experiences, including the news from the Persian Gulf or Paris, fresh fruit in February, and environmentalists who care about and provide support for landscapes they have never seen.

Gifford offered a thought-provoking comment: "We enjoy a radical independence that Gilbert White and his contemporaries never could have imagined. But . . . we find ourselves in the midst of independence that is, in fact, the fruit of our dependence."[21] The weather forecast gives us independence by making it easier to plan what clothes to wear. Our automobiles give us independence by enabling us to travel distances every day that Gilbert White would have found unimaginable. The melons that one can buy in a Minnesota supermarket in midwinter give us the independence to eat what we like no matter the season.

But each of these is actually the product of our *dependence on technologies and organizations that extend far beyond our experience*, and of which we are largely ignorant. Rising gasoline prices and war in the Middle East in the first years of the twenty-first century reminded drivers that their cars are fueled by sources that Americans cannot easily control. Few supermarket managers know where the food they sell comes from, so their customers don't know either. When tainted spinach makes people sick, it can take weeks to unravel the supply networks and pinpoint the problem. So pervasive is the world without edges that we aren't even aware we live in it, until we think about the worlds of White and Thoreau and realize that those worlds subtly and profoundly different from our own. Instead of independence, we live within a web of dependences, in a state of **interdependence**.

The world without edges is also a stage for massive *misunderstanding and ignorance*. The world without edges is a place where we inevitably act on distant landscapes in the routines of daily life. This means that we who live in developed economies cannot know what it is we do to the environment, at least not in the sense that White knew how his clothing was made or his food was grown. This profound ignorance can be mitigated by strategies such as buying sustainably harvested plywood or shade-grown coffee. But, as a practical matter, it cannot be avoided. It is a direct consequence of our wealth and comfort.

DISCONNECTION AND COLLAPSE

The world without edges is more than a barrier to environmentally responsible consumption, however. It is a social order in which disproportionality can rise and fall out of sight of citizens or their community's leaders, a world in which choices can be made for nations or the world even if decision makers do not grasp the dimensions of their choices. Surely those in charge know better! But history and archaeology suggest otherwise.

The **Maya** are a people who lived—and still live—in southern Mexico, Guatemala, and Belize. For nearly a thousand years—more than 4 times longer than the United States has been in existence—a Mayan civilization prospered in these rain forest landscapes. Agriculture flourished, supporting a society that built cities with populations as large as fifty thousand people, grand temple structures up to eighteen stories in height, and a culture with an intricate religious, artistic, and political life (Fig. 4.5).

Then, beginning around AD 850, in a period of not much more than a century, Mayan civilization collapsed and its grand cities were abandoned, leaving behind ruins decorated with elaborate carvings that are still being discovered and translated. Populations declined drastically, although they did not die out. (When the Spaniards arrived five centuries later, they found people speaking Mayan languages, which survive today in indigenous communities.)

Archaeologists are still learning about Mayan society. It appears to have been organized around religious and political institutions. The priest-kings of the Maya conducted elaborate ceremonies of self-sacrifice, and they collected taxes from farmers who had learned how to grow corn, beans, and other crops in the rich, humid landscape. Those taxes supported an elite class of warriors, priests, and traders. Oddly, no depictions have been found of royalty performing economic functions; attention was focused instead on religious ritual and warfare. The murals and inscriptions do show a society that changed, gradually at first, with a larger and larger fraction of the population entering the elite class. These were people who did not contribute to the agricultural economy, although they lived on the surpluses it generated.

FIGURE 4.5
Mayan temple at Tikal, in Guatemala.

With the growing burden of supporting the elites, economic conditions tightened and wars became more frequent and intense. Larger and more complex temples were constructed, as if the rulers turned to more elaborate appeals to the gods as conflict and competition increased. The investment of an increasing share of the society's resources in assets and activities that did not increase agricultural yields seems to have triggered a downward spiral of rapid decline.

Why did the rulers ignore the erosion of the economy? An answer comes in a broad theoretical proposition advanced by the anthropologist G. L. Cowgill: "We can never simply assume that stress or the threat of stress will automatically or even typically generate social or cultural development. . . . We always have to ask, *who* is experiencing the stress, *who* is in a position to do something about it, and *why* might they see it to be in their interest to do what they do?"[22] The elites in the Mayan cities could continue their way of life as the farming communities beyond the city walls declined, according to the archaeological clues left in the artifacts. Insulated from the material stress of the society as a whole, and living in a cultural setting that emphasized religious ceremonies and symbolism as a response to material troubles, the Mayan elite apparently did not see what was coming until it was too late.

Of course, our own society pays a great deal of attention to its material underpinnings. And yet, think of the world without edges in which we invest in a wide range of activities of doubtful environmental sustainability—from using irrigation water to grow hay, to overharvesting fish and timber, to continuing to develop patterns of urbanization that rely heavily on automobiles, even though fossil fuel supplies are dwindling fast. In Chapter 3, we called the disconnections fostered by the world without edges *functional* commons problems. Perhaps we should ask who is experiencing stress and who is in a position to do something about it—and whether warnings of collapse can be found in the answers to these questions.

The idea of disproportionality and the puzzling world without edges take environmental concerns beyond the sense of place discussed in Chapter 2. Environmental problems include local ones such as littering or polluted water. But other problems cross scales of space and time. Some of these are global, such as climate change. Many more arise from the web of interdependence that links all of us: consumption, investment, and participation in mass and worldwide information exchange.

The entanglements of the world without edges draw us into commons problems far from our experience. The generation of 35 pounds of waste for each pound of stuff we buy raises the question of whether the wastes are handled properly, and whether it is rational for so much waste to be produced in the first place. How would we know? If there is disproportionate hazard incurred by some things that we buy, can we avoid them? If one company produces 20 times as much toxicity per pound as other companies in the same industry, how can we tell if the products we buy contain material from that company? How often do we even know whether a product contains a particular material, from whatever source? More generally, how might people build institutions, along the lines suggested in Chapter 3, to govern the complexity of the world without edges?

As we said at the beginning of this book, no one knows the full answer to this last question. Still, solutions have been found for many commons problems. Implementing them in the world without edges remains difficult, and successful governance is the exception rather than the rule. But in many instances one can see how progress might be made through approaches such as identifying disproportionately polluting facilities and closing or cleaning them up.

FURTHER READING

Gifford, Don. "How We Are Housed." Chap. 6 in *The Farther Shore: A Natural History of Perception, 1798–1984*. New York: Vintage, 1991.

KEY TERMS

disproportionality	Gini coefficient	outliers	Toxics Release
economic	interdependence	self-sufficient	Inventory (TRI)
specialization	Maya	stationary man	world without
environmental			edges
injustice			

Chapter Five

THE ARCHITECTURE OF THE PLANET

CLIMATE, LIFE, AND THE PROVINCES OF NATURE

As William Shakespeare wrote, all the world's a stage. In this chapter, we describe the large-scale structure of Earth, the stage we share with other forms of life. Humans are modifying the architecture of the planet, with consequences that affect all living things, including ourselves.

In Chapter 2, we discussed the idea of place. Places are like rooms on the stage set of the planet. All places are distinct, though doors connect the places to one another. In the natural world, living things sometimes migrate or spread from one place to another, but a place is usually associated with the life-forms found there. For example, the grasslands of the Great Plains are home to bison and wild-flowers not found in the Appalachian Mountains of the eastern United States. Places have edges.

In Chapter 4, we suggested that humans live increasingly in a world without edges, in which trade, travel, and information tie distant places together. These human activities challenge the edges that define places, and they alter the natural world. This chapter sketches the world we are altering.

In some ways, Earth is not a single place, but rather a collection of different kinds of places, including ones beyond our direct experience. The planet as a whole, for instance, has been seen only by astronauts and is sensed almost entirely via scientific instruments. Other places are not remote, exactly, but our perceptions of them are formed by mass media rather than direct experience. The developing world is such a place for most Americans, as are deserts, permafrost regions near the poles, or rain forests. What we are trying to grasp in this chapter is a world of distinct places, all of them being changed by human institutions operating in a world without edges. It is on that stage where the grand challenges of sustainability are being played out.

The world's stage is set mainly by climate, the long-term patterns of the weather. Climate is the product of two factors: the long-term patterns in the circulation of the atmosphere and oceans, and **topography**, the very slowly changing shape of the land surface. Weather is driven by the Sun's energy and the planet's rotation. The way weather patterns are draped over the surface of the globe, in turn, defines a global climate, the architecture within which life unfolds.

Before the Industrial Revolution and the emergence of the world without edges, each human society lived almost entirely within a single province of the natural world. The agrarian society of Gilbert White or Confucius was populated by stationary people. Those societies changed relatively slowly, on a time scale of many human generations. The invisible present was known, in the sense that traditions and legends contained guidance and warnings about drought and other stresses. Now, in the dynamic global economy of the world without edges, human actions short-circuit the time scales and spatial isolation of traditional societies. Our species now acts in ways that threaten the stability and structure of the global architecture. We need to grasp the invisible present and the world without edges, because in their obscure reaches, we find both immense environmental challenges, such as changing climate, and the potential to meet and surmount these challenges.

⤷ disconnect in Consumer culture ⟶ Production

Learning Objectives
When you have finished studying this chapter, you should be able to

---

↘ look at a regional weather map on the web, or a television weather report, and recognize the rotation of the planet by the motion of storms;

↘ observe and appreciate how the vegetation changes as you travel from place to place;

↘ make an intelligent guess at the climate to be found in Chile or South Africa or Australia, and to relate those climates

to what you would see in different parts of the United States;

↘ explain the risks to biological diversity of the destruction of rain forest or coral reefs;

↘ suggest how bacteria, which reproduce over a time scale of minutes, might come to be resistant to antibiotics used to cure disease.

LIFE, AND AN APPLE

The natural world that humans inhabit and depend on is populated by living things. Scientists call this the **biosphere**. Here is how ecologist Edward O. Wilson described it.

> The most wonderful mystery of life may well be the means by which it created so much diversity from so little physical matter. The biosphere . . . makes up only about one part in ten billion in the earth's mass. . . . If the world were the size of an ordinary desktop globe and its surface were viewed edgewise an arm's length away, no trace of the biosphere could be seen with the naked eye. Yet life has divided into millions of species . . . each playing a unique role in relation to the whole. [1]

FIGURE 5.1
The skin of an apple has about the same relative thickness as Earth's biosphere.

The biosphere is draped over the planet. How thick *is* the biosphere? Living things can be found at high altitudes and down in the depths of the ocean. In the Himalayas, birds have been sighted at an altitude of more than 4 miles above sea level.[2] The deepest part of the ocean, in the Marianas Trench east of the Philippines, is about 5.5 miles below sea level. Adding these two together, the zone of life is

no more than 9.5 miles thick. Earth's radius is 4,000 miles, so the biosphere is 9.5/4,000, or about 0.24 percent.

How thick is the skin of an apple? The apple skin in Figure 5.1 is about 0.2 millimeters thick, which is 0.25 percent of the apple's 50-millimeter radius. In relative terms, then, the shell of the biosphere is about as thin as the skin of an apple. Within the thin shell of the biosphere, we find weather and all living things, the large-scale structures of the planetary ecosystem.

HEAT + ROTATION = WEATHER

Weather and life are dynamic processes, and both are powered by solar energy. The atmospheric changes we call weather occur mostly below 30,000 feet—within the zone where life is possible without technological support. If you fly on a commercial airliner at an elevation of 35,000 feet or more, you are above the biosphere, flying through bitterly cold, thin air. Few of us pause to reflect on the harshness of the environment beyond the plastic windows of an airplane.

Both atmosphere and ocean are fluids, able to flow internally in response to forces—if one pours water into a vase or pumps air into a balloon, the water or air flows in smoothly and takes the shape of the container. In this respect, fluids are unlike solids, which expand or contract but whose internal organization does not change, just as a skillet left on the stove retains its shape when heated and when it cools. Fluids in motion are very complex, and in a basic sense their physical behavior is not completely understood. This is an area in which field science is still important, in part because the mathematical theories that come from experimental science remain weak.

Field science is made up of many, many observations. How does one make sense of them? Models are often useful, even when they are very crude, in helping to organize observations. Bear in mind, though, that organizing observations is not the same thing as making predictions. Predictions can be made when observations can be put into mathematical terms. This is how space scientists calculate the orbits of satellites or how weather forecasters estimate the movement of storms. Many models useful for organizing observations are not easily restated in mathematical terms, however, including the one we are about to examine. A simple, obviously unrealistic model of Earth's atmosphere provides a useful way to think about the centuries of observations of the world's wind patterns.

Consider what happens when you put a pan of water on the stove. As the water heats, it begins to move in response to the addition of energy. One might think of this as parallel to what happens in Earth's atmosphere and ocean. The Sun is like the

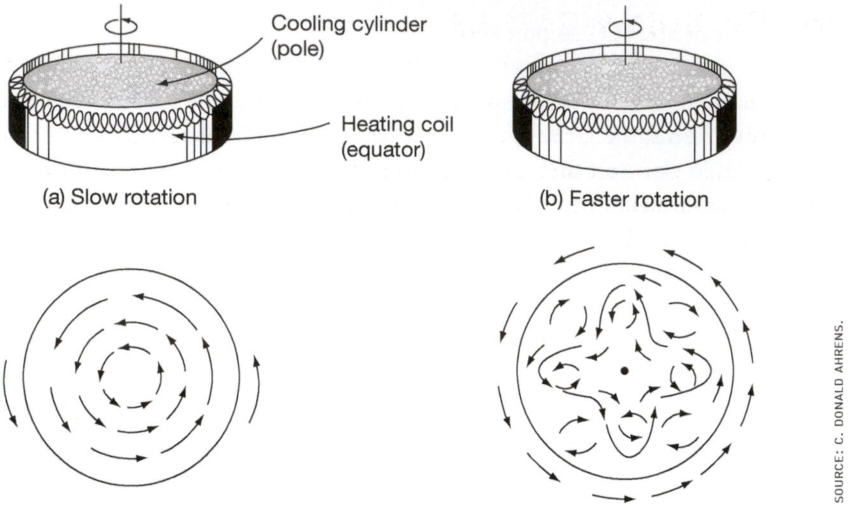

FIGURE 5.2
Two-dimensional model of Earth's atmosphere.

Cooling cylinder (pole)

Heating coil (equator)

(a) Slow rotation

(b) Faster rotation

SOURCE: C. DONALD AHRENS.

heat under the pan, and the atmosphere and ocean are like the water in the pan. In this scenario, weather is the "simmering" of the atmosphere by the "flames" of the Sun. To see how this idea plays out, we want to think about water, solar energy, and the planet's rotation.

Start with a pan of water. But instead of heating it from the bottom, put the heat along the outer rim and cool the center. This is a two-dimensional model of Earth, looking down from above one of the poles (Fig. 5.2). As it does in this model, the real planet absorbs more heat from the Sun along the equator than at the poles.

Next, rotate the pan. When the rotation is slow, the water flows in a circle. As the rotation speed increases, the fluid begins to circulate in eddies and loops. Depending on the speed of the rotation and how much heat is flowing from the rim to the center, one may see several cells (closed loops) forming. These flows can be stable, though the structures may rotate around the turning axis. A similar effect takes place on our three-dimensional planet, as we'll see shortly. In the real world, too, these cells persist, producing steady patterns of weather. Because Earth spins on an axis that is not at right angles with its orbit around the Sun, the angle at which the Sun's rays strike the surface of the planet changes over the course of the year. This variation is what gives rise to the seasons. So the steady pattern on Earth is not constant weather but an annual cycle. This repetitive pattern of weather that lasts over decades and centuries is what we call climate. It becomes part of the invisible present. Like the autumn leaves in North Adams, Massachusetts, that we mentioned in Chapter 2, the life-forms we observe reflect the environmental conditions experienced by living things.

WATER, SUN, WEATHER

Nearly all the water on Earth is salty ocean water. The oceans constitute 97 percent of the water on the planet. The other 3 percent is in freshwater, about two-thirds of which is in polar ice and glaciers. Most of the remainder is underground. Only 0.3 percent is in lakes and rivers—a drop in the global bucket.[3] But the freshwater in lakes and rivers is more than 10 times the quantity in the atmosphere. Water in the atmosphere constitutes only 10 parts per million of the water on the planet. One perspective on the role of water in the atmosphere is that a water molecule stays airborne, on average, for about eight days, whereas water in the oceans stays there, on average, for about twenty-five hundred years, a ratio of 1 to 100,000.[4] That raindrop on your face is made up of molecules that went into the air about a week ago, but the ocean spray on your face at the shore contains molecules that flowed into the sea before the first emperor took the throne in China, and centuries before Jesus preached by the Dead Sea. The fraction of the world's water moving through the atmosphere at any given time is tiny.

Yet that tiny amount of water is doing something crucial in shaping the environment. It is transporting heat from the Sun and, on the way, producing weather. The Sun's rays evaporate water into the atmosphere, and the water vapor carries that heat until it cools, transferring the energy to other molecules in the air, or until the water turns into rain or snow. This small fraction of the world's water that is in the atmosphere—10 parts per million—shapes the natural world in a way that seems disproportionate to the large reservoir of water from which it is drawn. But as we see in many environmental situations, small concentrations can have large effects. We will see this in global warming, because the change in greenhouse gas concentrations that is now having an impact on climate is only in the tens of parts per million. We see this in the concerns about toxic wastes, which can pose hazards to human health at levels of a few parts per million or less. And we also see this in the struggle against terrorism, because roughly 10 parts per million of the global Muslim population have been identified as members of al-Qaeda. As we have seen in the frustrations of this struggle, dealing with very low concentrations can be a tough challenge.

Weather is unstable on a scale of hours or days, and measures such as annual **precipitation** can vary by a factor of 10 from one year to the next. But weather over decades and longer periods does have stable spatial patterns.

Climate patterns depend in part on geography, as can be seen in Figure 5.3. Parts of the western United States, including Nevada and Arizona, are in a rain shadow: the moisture-laden storms from the Pacific Ocean drop nearly all their water to the west of the Sierra Nevada, and the rainfall is low farther east. The interior Amazon basin of South America has very high precipitation, as moisture from the Atlantic Ocean is carried eastward, falls as rain, and then evaporates back

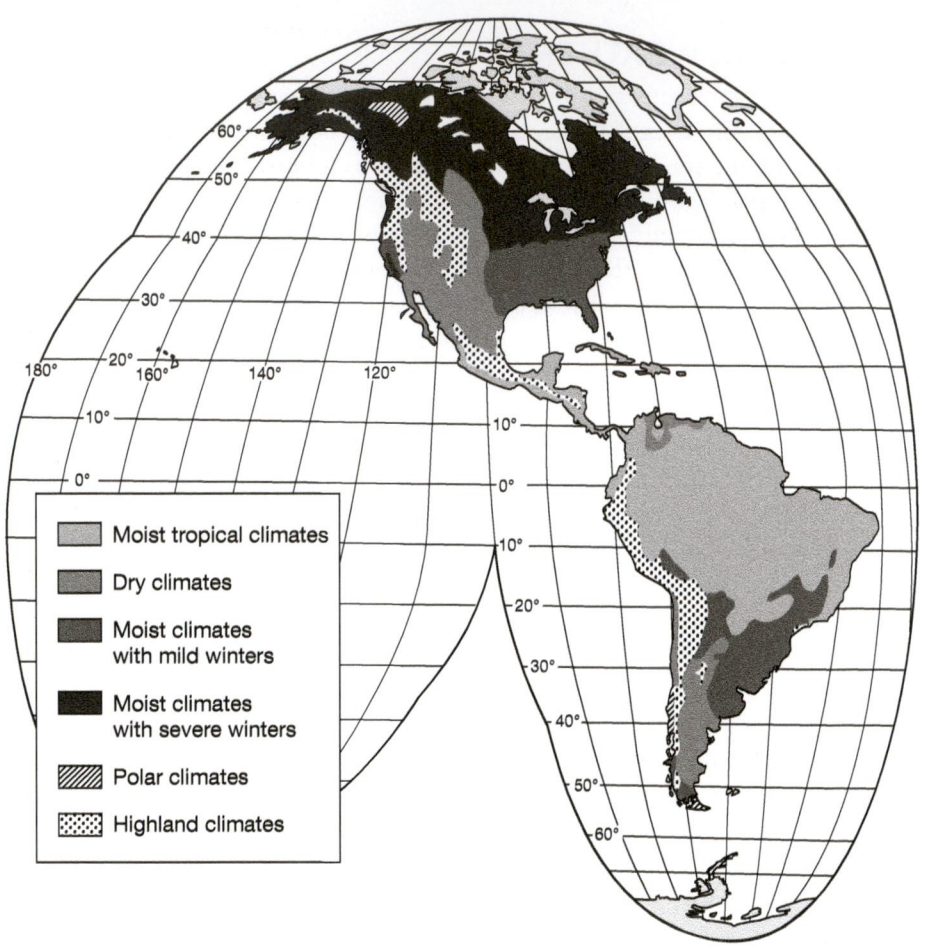

SOURCE: C. DONALD AHRENS.

FIGURE 5.3
Climates of
the Western
Hemisphere.

up into the atmosphere to fall as rain again, a cycle that continues until the moisture reaches the high wall of the Andes Mountains. The rain forests of the Amazon
are fed by this climate. The varied climates of the Western Hemisphere display a
complexity we often find in field science.

Yet simple patterns are hiding within that complexity. We can see them by
looking north and south, following the water as we move away from the equator
toward the poles. Figure 5.4 adds together precipitation from east to west, so that
all the observations from a given latitude line are summed up—all the rainfall
in a year along the equator added up, all the rainfall at 45° north (the latitude of
Portland, Oregon), and so on. This process of summing up a world's worth of data
leaves only the north–south pattern. Actually, what is added up isn't just rainfall
but net precipitation. Water evaporates from every part of Earth's surface, but in

some places more evaporates than falls, and in other places the reverse is true. This is the basic pattern represented in Figure 5.4, which shows the net inflow of water at the surface (precipitation minus evaporation) decreasing as the curve goes toward the right. Thus, the equator (latitude 0°) is wet, whereas the zone at roughly 20° north or south is so dry that more water evaporates (from the oceans, mainly) than falls as rain.

This figure brings out a simple fact that is not so easily visible in the mottled climate map. The north–south distribution of rainfall shows large-scale structure, roughly as large as oceans or continents.

Now let's look for that pattern on the planet. Figure 5.5 is a satellite image of an autumn day in the Western Hemisphere. One can see storms just north of the equator, and clear skies over the deserts of the southwestern United States. This was a normal day in the tropics, with a large hurricane in the temperate zone (Hurricane Rita in this instance). If one looks at Earth from space day after day over the year, one can see a persistent necklace of storms around the equator. These storms provide the rain that feeds the rain forest. This band of storms moves northward in the summer and southward in winter, but stays within the zone of the tropics. In fact, high rainfall defines the tropics. Clear skies are seen at about 23° north and south, and changing weather formations are found in temperate zones, where rain falls, but not as frequently as in the tropics.

The north–south distribution of storms, in short, corresponds to the net precipitation curve. This is what one would expect, because that curve comes from measurements made beneath the cloud patterns one sees from an orbiting satellite. Sometimes, tropical storms "boil" off the tropical storm belt. When the surface water in the ocean is warm—that is, in late summer—these escaped storms can get stronger and become hurricanes. Hurricanes threaten the eastern third of the United States in the summer and early autumn, from June through October.

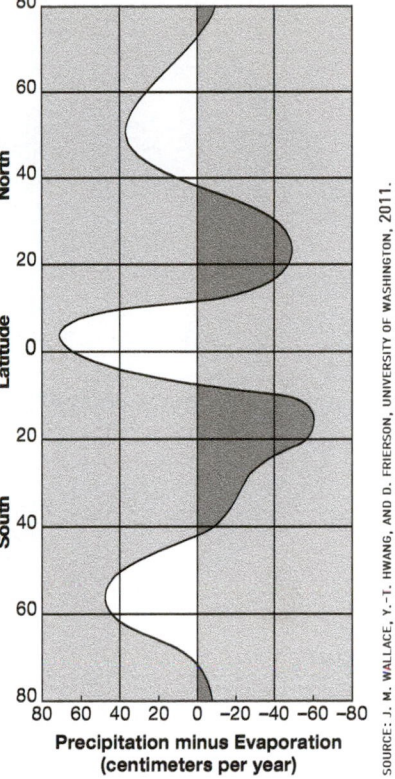

SOURCE: J. M. WALLACE, Y.-T. HWANG, AND D. FRIERSON, UNIVERSITY OF WASHINGTON, 2011.

FIGURE 5.4
Net precipitation (precipitation minus evaporation) by latitude.

CLIMATES AND GLOBAL CIRCULATION

Over time, weather becomes climate. If we look at where the skies are clear on most days, we find deserts. One can pick out the pattern on a world climate map (Fig. 5.6 on page 109). A global band of deserts is located at about 23° north and

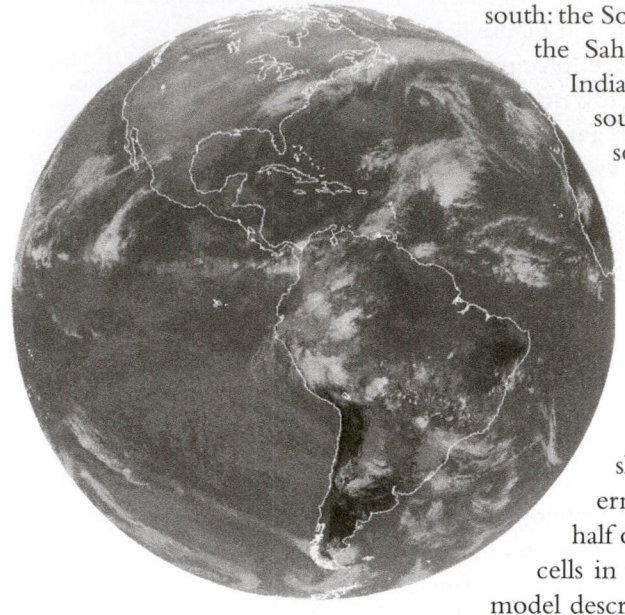

south: the Sonoran Desert of Mexico and the United States, the Sahara of northern Africa, and the Rajastan of India in the north; the Nazca of Peru, the Namib of southwestern Africa, and much of Australia in the southern hemisphere. These deserts lie within a narrow band of latitudes both north and south of the equator. To either side of these dry lands are regions of high moisture—a band of rain forests at the equator, and well-watered temperate zones toward the poles.

Now, let's take another look at what happens in the atmosphere. Figure 5.7 (on page 110) shows a three-cell model of the atmosphere. It shows three stable circulatory cells in the Northern Hemisphere, and three more in the southern half of the globe. These are similar to the circulatory cells in the rotating pan in Figure 5.2. The three-cell model describes the basic patterns of **prevailing winds**—the long-term average of the wind directions over time.

FIGURE 5.5
Satellite image of the Western Hemisphere, September 23, 2005.

The sun shines most intensely at the equator, evaporating water and raising it up as water vapor into the atmosphere. As the moisture rises, it cools and condenses into clouds. With further cooling as the air rises, droplets form, and they may be heavy enough to fall as rain. Some of the water vapor stays aloft and travels away from the equator, in a loop as wide, north to south, as the continental United States.

The winds in the tropics and subtropics—from Mexico or Florida down through the Caribbean in the Western Hemisphere—tend to blow out of the northeast. These winds carried along the sailing ships that linked Europe to its New World colonies in the Atlantic basin. Those winds came to be known as trade winds for that reason. You can see that the trade winds on both sides of the equator carry moist air from the Atlantic westward into the Amazon, providing the water that returns to the sea through the Amazon River basin.

The direction of flow of the trade winds is a result of the rotation of the planet, similar to the eddies in the heated and turning pan of water in Figure 5.2. The air in the trade winds is dry, because the moisture has been condensed as the tropical air rises into the cold upper atmosphere. The lands under this dry air tend—no surprise—to be deserts.

A second cell is located over the temperate zones; this is the weather we see in the continental United States, Chile, or South Africa. In this cell, the winds come out of the west, so storms come in mostly from the west in the lands of

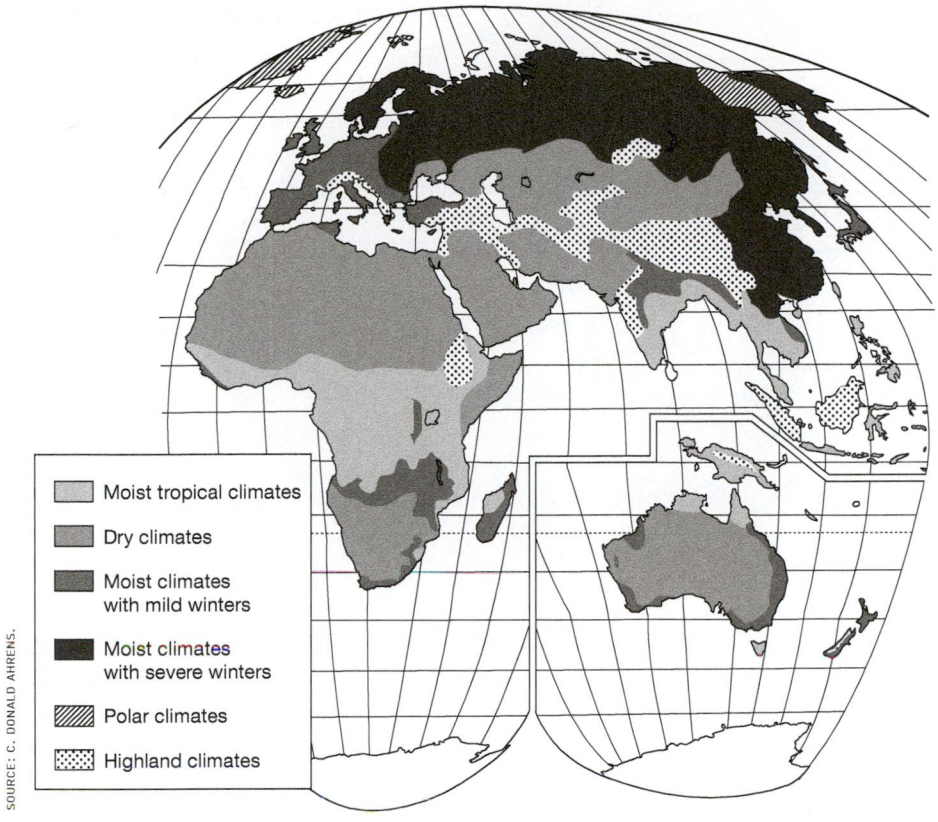

SOURCE: C. DONALD AHRENS.

Moist tropical climates

Dry climates

Moist climates
with mild winters

Moist climates
with severe winters

Polar climates

Highland climates

FIGURE 5.6
Climates of the
land areas on
Earth.

the temperate zones. As Figure 5.8 (on page 110) indicates, this fact was already known in the nineteenth century. This westerly tendency of the winds is also a result of Earth's rotation.

In the seams between the cells are **jet streams**—high-altitude rivers of air that steer storms and speed airliners along when they can hitch a ride. The jet streams flow at high velocity, often more than 100 miles per hour, so a plane flying inside one can travel over the ground a lot faster than it would in still air, or a lot slower if it is going against the jet stream. The jet streams form at the boundary between cells, a bit like a rip current in the ocean water as it runs along the shore. The cells themselves, and their boundary jet streams, move northward in summer and south in winter. This is why northern portions of the United States get blasts of "arctic air"—air coming, literally, from the north polar cell—in the wintertime. The jet streams that mark the boundaries of the cells meander back and forth, sometimes looping far south or north like a jump rope. If one looks

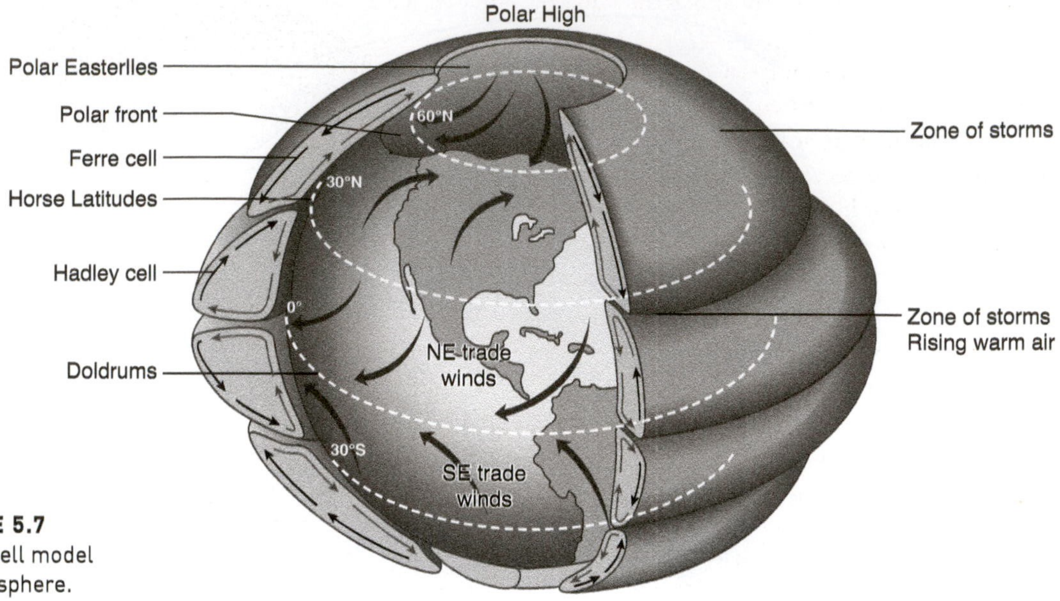

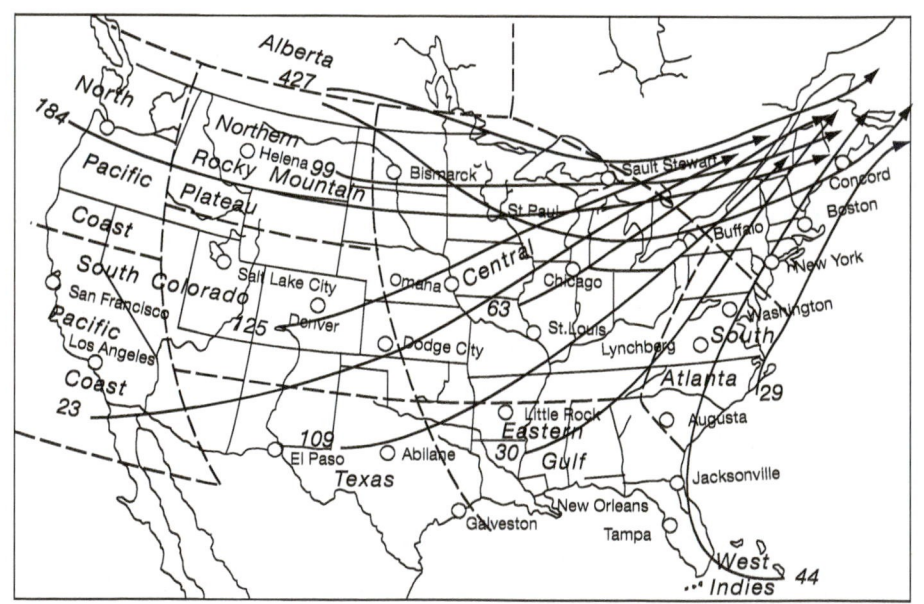

FIGURE 5.7
Three-cell model
of atmosphere.

Polar High

Polar Easterlies

Polar front

Ferre cell

Horse Latitudes

Hadley cell

Doldrums

Zone of storms

Zone of storms
Rising warm air

60°N

30°N

0°

30°S

NE trade
winds

SE trade
winds

SOURCE: C. DONALD AHRENS.

FIGURE 5.8
The number of
storm tracks
observed between
1884 and 1893.

SOURCE: MARK MONMONIER.

at this pattern from directly above the equator, one can see a banded pattern (Fig. 5.7).

This banding is similar to what we see on Jupiter (Fig. 5.9). On that planet, too, the combination of rotation and uneven heating produces a wind pattern that can be seen as bands. Jupiter does not have landmasses like Earth does. Indeed, Jupiter does not have a surface in the same sense at all, because the planet that we see is composed almost entirely of cold hydrogen and helium. With its featureless topography, Jupiter's atmospheric circulation has a clearer structure than Earth's.

The three-cell model of Earth's atmosphere is a product of natural history and field science. It was put together from centuries of observation, mainly by sailors, with additional work by scientists over the past two hundred years. We think the three cells result from the joint effects of rotation and heating. Yet, because fluids are very complex, we still have no theory grounded in the physics of the molecules in the atmosphere to tell us why, in principle, there are three and only three cells, or why each hemisphere has two jet streams. (We *have* learned that the bands on Jupiter and Saturn are the result of atmospheric cells like those on Earth.) Nonetheless, by combining field science observations and the mathematical theories we do have, it has been possible to assemble quite detailed models of Earth's atmosphere. Projections of global warming are made using these models.

FIGURE 5.9
The visible banding of Jupiter's atmosphere.

DIVERSE LIFE

The map of climates in Figure 5.6 shows the natural world divided into distinct provinces. With that map, it is possible to see why life should be highly diverse. Most living things spend their lives in a single place—a spatial region that is small compared to these climate patches. Plants and nonmigratory animals normally live within one climate zone. Migratory animals move seasonally, taking advantage of opportunities in more than one climate zone, but they rely on the regularity of the seasons in each one. Over many generations, life-forms that take advantage of the particular features of a climate are more successful than those that do not. As climates have changed in the past, some life-forms, such as the dinosaurs, have gone extinct. As with Gilbert White in Selborne, the natural history of Earth is mostly the story of "stationary" animals and plants defined by their environments.

This intricate pattern of living things is what Edward Wilson called "the wonderful mystery of life" in the passage quoted at the beginning of this chapter. How does this wonderful mystery come to be? Scientists have worked for about 150 years to elaborate on the theory of evolution advanced by Charles Darwin in 1859. (See Box 5.1: Natural Selection: Joining Natural History to a Simple, Powerful Theory, page 113.) Darwin's ideas came as a shock in Victorian England, because evolution proposed an explanation for the extraordinary complexity and grandeur of the biological world that did not require divine intervention or, indeed, any kind of intelligent design. (See Box 5.2: Darwin and Natural Theology, page 115.) Although the controversies have not died out, biological science built on evolution has become the authoritative scientific account of life. This understanding has produced, among other things, advances in medicine that have been hailed as miracles in their own right. In addition, evolutionary biology provides persuasive explanations of such facts as the chemistry of all living things appearing to be nearly identical in its fundamentals.

The idea of evolution by natural selection (see Box 5.1, page 113) provides a way to understand why maple trees throughout the northern temperate zone drop their leaves in the autumn. The winters are harsh in temperate climates, with fewer hours of daylight and temperatures falling below freezing. Trees that are deciduous (drop their leaves each autumn) tend to survive the seasonal fluctuation in their environment. They do not choose to be deciduous; they do not learn to drop their leaves. It is the trees that are deciduous (for whatever reason) that survive more often in these climates, and thus are able to produce new generations of trees. Over time, the population of trees that survives in the temperate zone has developed into many deciduous species, such as the maple.

Now, apply this kind of thinking to life in general, and one can see that different plants, animals, and microbes will, over many generations, tend to occupy different climate zones. Where do the individual differences that are chosen by natural selection originate? Charles Darwin did not know, but he could readily see variations within populations and that some of these variations seemed to be inherited, such as the shape of birds' beaks or the colors of flowers. We now know that many individual differences are programmed in the **genes** of living things, and that random changes, or **mutations**, in the reproductive genes can produce individual differences that can be inherited. Nearly all mutations are harmful or confer no benefit, so the evolutionary process is normally slow, operating on a time scale of many generations.

BOX 5.1

NATURAL SELECTION
Joining Natural History to a Simple, Powerful Theory

The scientific explanation for why different species are found in different habitats was first articulated by English biologist Charles Darwin in *The Origin of Species* (1859). This book is a great landmark of the human intellect, and the focus of a great and continuing controversy. With a simple but deep analysis, Darwin created a powerful theory to explain the underpinnings of natural history. Behind the enormous complexity of what we see in the natural world, Darwin discerned a remarkably simple order. The name he gave this order was **evolution**.

The concept is not hard to grasp. Within a population of living things, some reproduce more than others of the same kind or species. The reasons for this combine circumstances and individual differences. A given plant may happen to be eaten by a grazing animal, and that plant may not reproduce. Tasty plants may be eaten more often than prickly or sour ones. An individual's characteristics are partly **heritable traits**—that is, they can be inherited by that individual's descendants. Over the generations, traits that favor reproductive success will, by definition, survive better. That is, the characteristics that survive are carried by a larger number of descendants. This is the process of **natural selection**: a population is selected by natural forces such as the ability to elude predators, survive drought, or reproduce.

Those traits that improve the survival of individuals become more common in the population over time. Species that are ill equipped by their heritage to survive dwindle in number and eventually go extinct. It has been said, jokingly, that extinction is nature's way of doing a cost-benefit analysis. Species that were once well adapted to their situations may find their environments have changed, or that the ways they fed and reproduced no longer work as their population size grows. The phrase "survival of the fittest" was coined to describe this situation, but it says less than it seems: those species that survive turn out, by definition, to have been "fit." Predicting fitness in advance is often hard to do.

Natural selection is a process by which different species can emerge. For example, imagine a species of grass that grows at the base of a mountain. Each year, the individual plants bear seeds, but the seeds aren't the same. A few might grow better in wet soil and others might grow well in cooler weather. The grass seeds are blown by the wind and eventually settle on a patch of ground. Most do not land in conditions where they can grow, but some do, and these germinate and grow into new plants. Over time, the original population spreads, extending its range. But the survival of the seeds is not uniform. The few seeds that can grow well in wet conditions survive better when they happen to land in a swampy area, and others die without producing seeds of their own. The seeds that can flourish in cool weather might do better at higher altitude. Of the seeds that happen to be blown up the mountainside, these survive and bear seeds more readily than seeds that do not have the trait of growing well in cool weather. Over many generations, the grass spreads, but the populations that spread are no longer identical to the original one. What was one species has become two or more: one species colonizes a swampy area, another grows well up on the slopes, and the original species continues to occupy the base of the mountain.

For this explanation to work, there must be some source of variation within natural populations, and there must be enough time—enough generations—for the selection process to produce different species. Look in a crowded classroom and you'll see a lot of visible variation in our own species. Darwin did not know where the variation comes from. The idea of a genetic code that directs the growth and development of individuals and varies, very slightly, from individual to individual was not developed until the end of the nineteenth century. But Darwin, a talented natural historian, could see, as we can, that the variation exists.

Darwin lived and worked during a time when geologists became more convinced of the great age of most of the rocks on Earth's surface. The realization that the natural world was hundreds of millions of years old meant that there had been much more time for life to **adapt** to the varied conditions of the globe than had been thought. Darwin grasped the significance of these two observations and proposed that there would be both time enough and sufficient variation in populations for the slow process of natural selection to explain the origin of species.

BIOGEOGRAPHY AND DIVERSITY

As natural historians and field scientists have studied ecological settings, they have discovered an important pattern. Table 5.1 shows bird species as an indicator of biological diversity, or biodiversity, which can be defined as the number of species in a relatively small geographic area. What this table indicates is a pattern found for other groupings of living things. *As one moves toward the equator, levels of diversity rise.* This pattern is a basic finding of **biogeography**, the study of the distribution of life-forms across space and through geological time. (As you might expect, at a

BOX 5.2   DARWIN AND NATURAL THEOLOGY

Darwin's theory of evolution was controversial because it provided an explanation of the complexity and diversity of nature without needing a supernatural designer. Of course, the idea that nature is sacred is one we encountered in the Hudson River School paintings. Those canvases were being produced during the years that Darwin was writing his masterpiece, *The Origin of Species* (1859). The Hudson River School painters depicted human figures as tiny in relation to the landscape as a way of expressing the majesty of divine creation, an idea seen in Frederick Church's *Heart of the Andes* (in this painting, the human figures are standing around a wooden cross in a clearing to the left of the waterfall). At roughly the same time, Emily Dickinson had written, "Nature is Heaven."[1] Darwin replied, in essence, that nature is chance variation and patience—although it is not lacking in wonder and surprise to the human mind (see Box 5.3: Genetics and the Invisible Present, page 117). Darwin's ideas were heretical as well as scientifically revolutionary in their import. In fact, Frederick Church suddenly stopped making paintings like

Frederic Church, *Heart of the Andes* (1859). To the left of the waterfall in the foreground, there is a clearing with a cross standing in it, and a person next to it, virtually invisible in this image.

this one, and the paleontologist Stephen Jay Gould has speculated that Darwin led Church to lose his faith in natural theology.[2]

Even today, a large majority of Americans tell pollsters that they do not believe in evolution, and vigorous attempts have been made to rewrite textbooks to treat evolution as "only a theory." Such suspicion of uncomfortable scientific ideas is neither rare nor unimportant, even in a society that celebrates its open-mindedness in many things. The fact remains that Darwinian evolution has been reaffirmed by scientific studies for more than a century, and the modern economy relies on evolution's application to biology for developments ranging from new medicines to ways to protect crops from pests. Even those who say they do not believe in evolution enjoy the fruits of Darwin's ideas when they eat corn or take an antibiotic medication.

1. Emily Dickinson, *Complete Poems of Emily Dickinson,* ed. Thomas H. Johnson (Boston: Little, Brown, 1960), poem 668.

2. Stephen Jay Gould, "Church, Humboldt, and Darwin: The Tension and Harmony of Art and Science," in *Frederic Edwin Church* (exhibition catalogue), ed. Franklin Kelly, with Stephen Jay Gould, James Anthony Ryan, and Debora Rindge (Washington, DC: Smithsonian Institution Press, 1989), 94–107.

* As you move closer to the equator, species diversity increases

particular latitude, larger areas contain more species. Table 5.1 on page 118 compares areas of roughly equal size.)

In New England, one would likely encounter only a few dozen tree species in a patch of forest 1 mile in diameter. A forest in Panama or New Guinea may have several hundred tree species in a patch of land that size. What is true of trees is even more so for small life-forms. In a tropical climate, thousands of species of insects can be found in an area the size of a college campus, whereas we would find perhaps a few dozen species on a U.S. campus (not counting the specimens in the natural history museum and laboratories, of course). Overall, ecologists have estimated that *10 to 100 million species* of life-forms are large enough to see. Of these, fewer than 2 million have been identified. This is a great frontier of field science to be explored. If you're adventurous, you can find and name species. One alumna of Williams College in Massachusetts named an Australian leaf-eating beetle after the school's founder, as a gift for the college's 200th anniversary.

Why are there more species as one moves toward the equator? Scientists aren't quite sure. It is a fact of field science for which there is, as yet, no satisfying account in a mathematical theory. For example, we don't have a formula to translate latitude into an estimate of diversity of species. What we do know is that John James Audubon could attempt to paint every species of North American bird, and in the

BOX 5.3

GENETICS AND THE INVISIBLE PRESENT

Each living thing contains an "invisible present" of its lineage. Its genes incorporate the editing of natural selection over hundreds of millions of years. Each of us contains, genetically, an invisible present of our parents and ancestors. Although cultural legacies are not carried by genes, it is possible to think about how you also carry the invisible presence of some of your teachers, coaches, friends, and enemies, not in a genetic sense but in the formative memories, habits, and fears that contribute to your personality and identity.

All living things reflect this process. Some species became separate populations rather recently; for example, dogs and wolves are descended from common ancestors that lived during the last Ice Age. Other species diverged long ago; the common ancestor of bacteria and elephants is hundreds of millions of years in the past. Yet the genetic and biochemical processes in all living things and across species are very similar. For example, the mitochondria that process the energy used in living cells work the same way in bananas as in the campesinos who cultivate them. The similarities are so uniform and so widespread that it is overwhelmingly likely that all life has a common origin. The tree of life diagram (see figure) shows how different groups of species are related to one another, measuring the distance down the branches by their genetic relatedness.

We are literally related to all living things. As the cell biologist Ursula Goodenough pointed out, "the flow of genes from the common ancestor has been a constant flow, diverted into countless culverts but flowing steadily, from the beginning to the present. We are all, we creatures who are alive today,

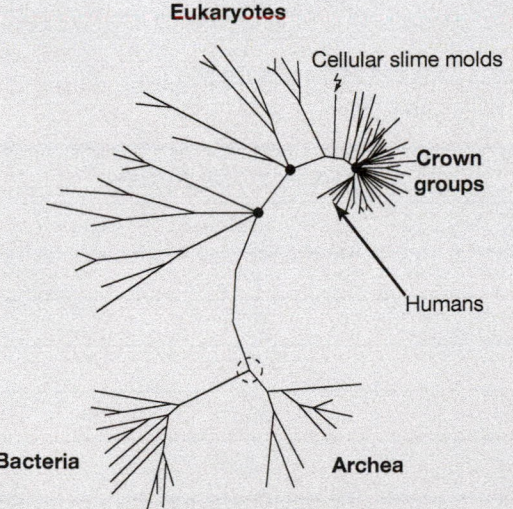

Eukaryotes

Cellular slime molds

Crown groups

Humans

Bacteria

Archea

SOURCE: URSULA GOODENOUGH (1998).

The tree of life. This diagram shows the degree to which different groups of living things are related to one another. The clusters of bacteria and Archea (a microscopic form of life) are so far distant from the others that they form distinct kingdoms. Plants and animals are in the "crown groups," of the Eukaryotes, which are the forms of life made up of multiple cells organized in complex structures such as skin, bones, or eyes.

equally old."[1] She argued that we should take that relatedness seriously, particularly when human actions endanger an entire species. For once a species goes extinct, it cannot be brought back. That the family of living things has become so diverse, Goodenough suggested, is reason for awe: "The outpouring of biological diversity calls us to marvel at its fecundity. It also calls us to stand before its presence with deep, abiding humility."[2] As this reverent language implies, Goodenough suggested that contemplating biodiversity is like entering a cathedral of life, and the tree of evolution is its rose window. She wanted us to see in evolutionary biology the grandeur of nature that the Hudson River School artists put in their canvases. However, now the brushstrokes are in the style and language of our time—the language of science.

1. Ursula Goodenough, *The Sacred Depths of Nature* (Oxford, UK: Oxford University Press, 1998), 85.
2. Ibid., 86.

1840s he published a set of 435 plates titled *Birds of North America* (Fig. 5.10). Such a task would have been infeasible in a tropical setting rather than the temperate zone in which Audubon worked.

Scientists do think the great complexity and intense competition in tropical forests and coastal ecosystems plays a role in species proliferation. Consider Figure 5.11, which shows tree species distribution in the tropics as a function of rainfall and height.

TABLE 5.1   NUMBER OF BIRD SPECIES AT DIFFERENT LATITUDES, IN PLACES OF ROUGHLY SIMILAR AREA

Location	Number of bird species
Greenland	56
Labrador	81
Newfoundland	118
New York State	195
Guatemala	469
Colombia	1,525

From Wilson (1993).

In the tropics, even when there is year-round warmth, trees tend to drop their leaves in dry seasons. As one compares places with different averages of yearly rainfall, trees that continue to photosynthesize do better. With more rain, trees can grow taller, and the ones that compete more effectively get a larger share of sunlight. Think of this diagram as a field science summary, like the net precipitation diagram in Figure 5.4. Again, there is a simple pattern hidden in the intricate texture of natural landscapes.

In tropical regions, soil is generally poor because it is very old and has not been stirred up by glaciers during recent ice ages, as the soil in many temperate zones has been. Tropical soils have also been washed by rains for millions of years, and the water-soluble nutrients have thus largely been leached away, except in areas close to volcanic activity. But sunlight is intense. Accordingly, the basic resources available to trees are sunlight and water. Trees compete with one another by growing taller and spreading their leaves to form a canopy that intercepts the sunlight. Under the top layer, other plants grow where they can catch the light leaking through the canopy. As you can see, where the rainfall is high, the trees that do well are the taller ones, though their height is limited by the water available from rainfall. This is the result of natural selection. In competing for

FIGURE 5.10
John James Audubon, *Snowy Egret (Leucophoyx thula)*, 1840.

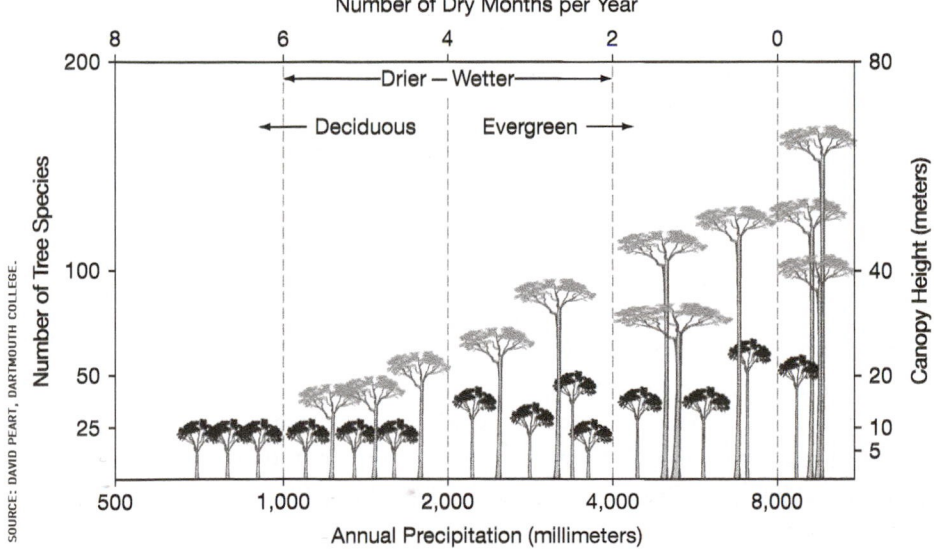

SOURCE: DAVID PEART, DARTMOUTH COLLEGE.

FIGURE 5.11
Diversity and tree height in rain forests.

FIGURE 5.12
A stand of old-growth rain forest in Sumba, Indonesia.

sunlight, the tall plants do better than the shorter ones, and the tall plant species come to dominate the forest (Fig. 5.12).

The woody structures formed by the plant life, in turn, create an intricate geometry that forms the environments of other living things. Trees live a long time, and they are complicated objects. A forest is, from an ecological perspective, a different kind of place than bare ground or a grassland or a lake.

Forests contain a wealth of **microhabitats**—small spaces in which the conditions of life are different from those around them. The crook of a tree may catch rainwater and form a miniature pond, an environment different from the bark nearby or the soil at the base of the tree. The understory plants that grow beneath the canopy make the spatial environment more complicated than a field of grasses. Vines and hanging plants such as orchids create more nooks and crannies within which animals can hide and hunt in different ways. When a tall tree falls in the rain forest, it creates a clearing where sunlight can reach the forest floor. This environmental change, in turn, means that different life-forms can survive there until tall trees replace the fallen tree and close up the clearing again.

Differences in habitat can arise in other ways, too—through place-to-place variations in elevation or mineral distribution in soils, or because living things themselves change the physical and chemical conditions of micro-habitats. Sometimes an insect infestation kills only some species of trees in a forest, creating a pattern of clearings different from the effect of a violent storm or of human loggers. In these varying settings, different forms of life can find ways to make a living and form distinct communities. Over many generations, those communities can nurture distinct species.

In this way, forests foster diversity. A similar kind of richness of microhabitats can be found in coral reefs, another ecosystem type in which living things create structures that persist over long periods, providing stable spatial structures of high variety (Fig. 5.13). In short, life-forms interact with climate to become an evolutionary force in their own right.

Wet tropical environments, notably rain forests and tropical coral reef systems with their many and varied microhabitats, contain tremendous diversity, as we have mentioned. Tropical rain forests cover about 6 percent of Earth's land area, but they contain more than 50 percent of species. Their profusion of varied microhabitats has produced many **endemic species**—life-forms that only live within a limited range, such as a dragonfly found only at a single lake or a flowering plant that lives only on the ridges of three neighboring hills. Endemic species are vulnerable when land is cleared by logging or for farming, which has been happening a lot in recent decades as the human economy has grown and the once-distant pull of the world without edges reaches all but the most isolated communities.

FIGURE 5.13
Corals on sea floor.

THE WORLD IS ECOSYSTEMS

"The world is places," Gary Snyder wrote. We can now read his statement in a different way than we did in Chapter 2, where we emphasized the way that human experience and traditions have been rooted in places that are small by the standards of modern travel, trade, and communications. We can see now that the planet being altered by humans is a collection of ecosystems of various spatial dimensions. (See Box 5.4: Net Primary Productivity, page 122.) The natural world is places, each of them shaped by climate and the distinctive ecological communities in them. Many of these ecosystems have now been claimed and modified by humans.

Table 5.2 on page 123 sets out the principal places—in this broader sense of ecosystems modified by humans—that we will explore in the rest of this book, together with the kinds of commons dilemmas that humans face in modifying them for our own benefit.

BOX 5.4

NET PRIMARY PRODUCTIVITY

All life on Earth relies, directly or indirectly, on energy from the Sun and the ability of plants to **photosynthesize,** the process that transforms the energy in sunlight into the chemicals and energy needed by all living things. The distribution of photosynthesis taking place on Earth is one way we can see the planet's architecture. Biologists measure photosynthesis by **net primary productivity (NPP)**, which is the amount of solar energy captured in this lowest (primary) level of the food chain.

NPP can be estimated from data collected by satellite instruments. The figure estimates NPP based on data collected between 1982 and 1998, which were then used to estimate the steady-state NPP of the planet. The individual cells or pixels in this image are a quarter of a degree on a side. Note that the NPP is only shown on land, even though the world's oceans contain large populations of **phytoplankton**, microscopic organisms that contain chlorophyll and carry out photosynthesis.

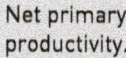

Net primary productivity.

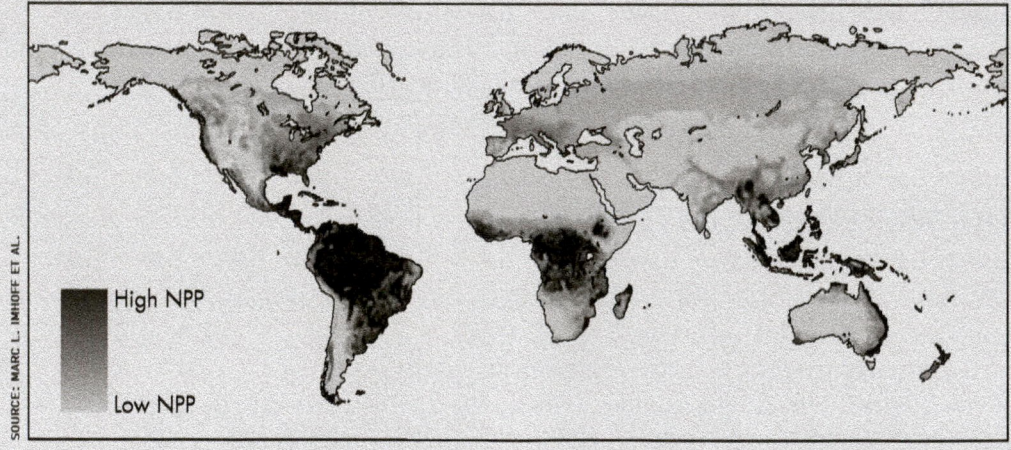

SOURCE: MARC L. IMHOFF ET AL.

High NPP

Low NPP

TABLE 5.2   ECOSYSTEMS BEING MODIFIED BY HUMANS IN WAYS THAT RAISE COMMONS DILEMMAS.

Ecosystem type	Commons dilemmas and grand challenges	Questions about responses being attempted
Planet as a whole	Climate change Loss of biodiversity Global-scale economy, driven by demand from rich nations	Is global-scale governance feasible? Can development be made sustainable?
Critically damaged ecosystems	Severely polluted landscapes Seascapes and landscapes destroyed by overharvesting of fish and trees	How can outside assistance be provided? Can we sense limits before destruction becomes irreversible?
Protected areas (parks, coral reefs, biological preserves)	Ecosystems are being protected from human pressures, often by removing traditional human inhabitants	How do we insulate protected areas from human-caused changes? What should be the role of humans in these ecosystems?
Human settlements, including industrial landscapes	Ecosystems transformed by human activities Rapid population growth in cities in poor countries Changes in consumption of energy and materials	How do we recognize, enable, and manage rapid changes? How can social capability be built for communities being torn apart and reassembled by these changes?

We view these ecosystems from a distinctly human perspective. They range in scale from a patch of prairie the size of a city block being restored between intersecting train tracks in Chicago all the way up to the planetary atmosphere. From the perspective of human history, these ecosystems have been modified relatively recently. Even the long-settled cities have grown dramatically within the last 150 years and now face problems such as toxic wastes on a far larger scale than they did before the Industrial Revolution. Judged against the time scales of evolution, these environmental challenges are so new that there has been little time for adaptation. For that reason, the speed of these changes is reason for concern. The ways of the natural world cannot react fast enough to avoid major damage, so we must rely on human institutions.

The scales and tempos of human life do not match those of the natural ecosystems we are claiming and transforming. Climates do change, but the ice ages that punctuate the cycles of geological time come and go over tens of thousands of years. Life-forms do change, but natural selection takes many generations to do its work. The limitations of human perception that we have identified with the names "invisible present" and "world without edges" show that greed and laziness are not needed to explain a lot of environmental degradation. More often, disproportionality is a sign of a common resource, such as a flowing river, that has been taken over by a polluter, and a society that acquiesces in the polluter's seizure of something that belongs to all. More generally, the human institutions we have today are marked by frailty, shortsightedness, and tendencies toward self-serving error. So when we say we must rely on human institutions to steer our way toward a sustainable future, we also imply that grand challenges lie ahead. These are the subject of the next part of this book.

FURTHER READING

Goodenough, Ursula. *The Sacred Depths of Nature.* Oxford, UK: Oxford University Press, 1998.

United Nations World Water Assessment Programme. *Water in a Changing World.* Third United Nations World Water Development Report, 2009. Available at www.unesco.org/water/wwap/wwdr/wwdr3/tableofcontents.shtml (accessed 5/09).

Wilson, Edward O. *The Diversity of Life.* Cambridge, MA: Harvard University Press, 1993.

KEY TERMS

adaptation	gene	natural selection	phytoplankton
biogeography	heritable trait	net primary	precipitation
biosphere	jet stream	productivity	prevailing winds
endemic species	microhabitat	(NPP)	topography
evolution	mutation	photosynthesis	

PART II

Grand Challenges

Chapter Six

THE MOST SUCCESSFUL SPECIES?

HUMAN DOMINATION OF ECOSYSTEMS

Perhaps the most striking disproportionality in environmental studies is the role of humans on Planet Earth. Standing on a national park trail, contemplating the fury of the sea, or looking down from an airplane at the open reaches of the Great Plains, the mind is drawn strongly toward the vision of the Hudson River School artists, whom we met in Chapter 2. Such encounters prompt us to see humans as insignificant figures set against the awesome grandeur of the natural world. Cities are smashed by an earthquake or the storm surge of a hurricane; droughts drive famines. But as we will see in this chapter, contrasted against the other species of the biosphere, and set within the scale of the planet's architecture, humans have become extraordinarily influential. Environmentalists tend to see that influence as a hazardous thing, but the power of humans is simultaneously the wellspring of responsibility, if we can learn to govern better the reach of our species.

CARNIVORES ARE (USUALLY) RARE

Life, like weather, is powered by solar energy (see the "Net Primary Productivity" box in Chapter 5). The basis of life is photosynthesis, the transformation of sunlight into stored chemical energy by plants. Animals and virtually all other living things, in turn, are nourished by plants or plant-eating animals. Light energy from the sun is absorbed by chlorophyll molecules in plants and transformed into chemical energy. That energy is used to convert water, carbon dioxide, and nutrients drawn from soil and air into compounds usable by living things. Peach pits, seaweed, and the energy that a Venus flytrap uses to engulf its prey are all products of photosynthesis. The fact that plants collect a lot of sunlight is reflected in the green landscapes we see in much of the world. Photosynthesis also generates oxygen, so the air we breathe continues to sustain life.

More than three-quarters of the light energy absorbed by plants is used for **metabolism**, the collection of processes that maintain life. The energy that grasps the fly in the Venus flytrap and the energy expended to dissolve the victim are examples of metabolism at work. The remaining solar energy produces the tissues of the living plant, including roots, flowers, and leaves. (A large fraction of the world's photosynthesis is done in single-celled organisms living in the oceans and freshwater. These microscopic phytoplankton contain chlorophyll, but they are too small to have leaves or other differentiated parts.)

Plants are consumed by animals, in turn, until at the top of a **food web**, or linked network of prey and predators, we find **carnivores**—the animals that eat animals,

Learning Objectives
When you have finished studying this chapter, you should be able to

↘ distinguish between the environmental impact of a vegetarian and that of a meat eater;

↘ point out differences between a lightly managed ecosystem, such as a forest or wetland, and a tightly controlled one, such as a farm or home aquarium;

↘ explain the significance of agriculture, which occupies less than 2 percent of the U.S. labor force today but plays a key role in the economy, social institutions, and in human history;

↘ reflect upon the "macroparasitic" character of many governments, and the promise of democracy to act in a manner that is not parasitic;

↘ identify some of the ecosystem services that you rely on in daily life and inquire into how the continuing supply of those services can be assured;

↘ explain why an increase in the price of gasoline may or may not reflect exhaustion of Earth's supply of oil;

↘ describe the concept of the "ecological footprint" and explain the limitations of the measurements we now have of human impacts on the natural world.

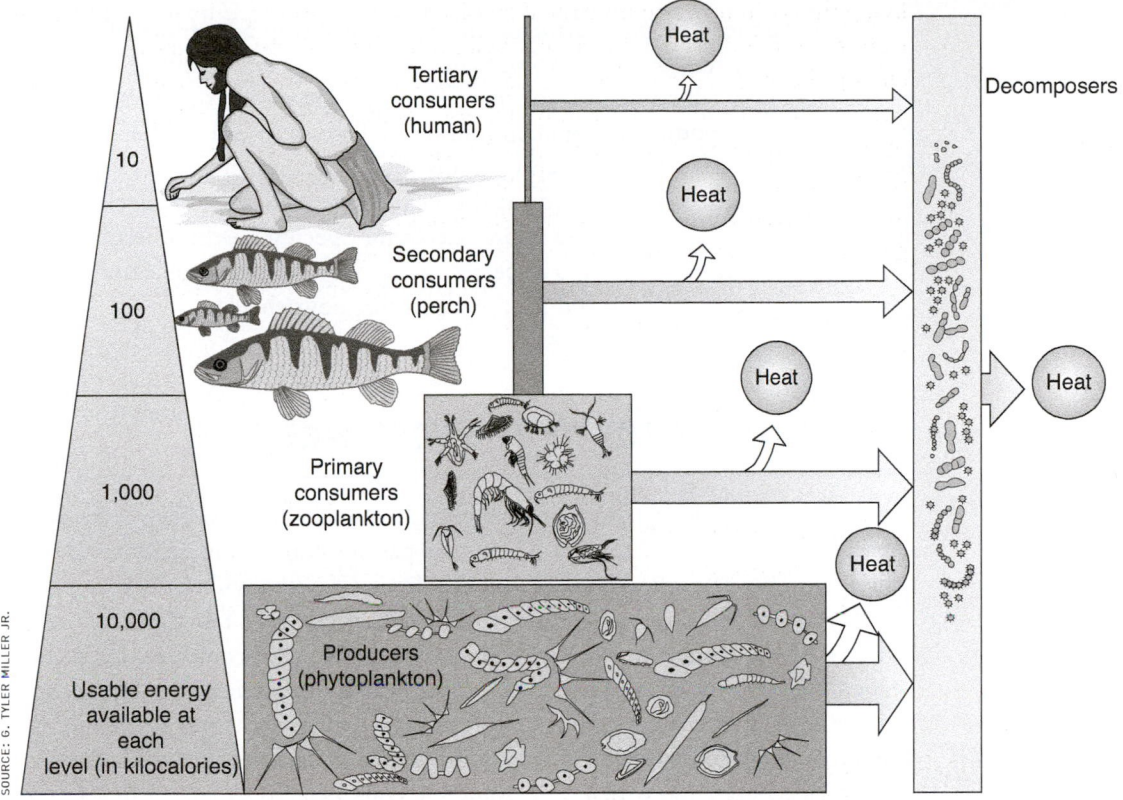

SOURCE: G. TYLER MILLER JR.

FIGURE 6.1 Generalized pyramid of energy flow in the biosphere.

and thus eat plants twice or more times removed. Along the way, a large fraction of the energy in the food is burned up in the metabolism of the consumers. For example, mammals, including humans, burn a lot of calories to maintain a steady internal temperature. This energy is not available to grow more muscle or hair.

The net result is that it takes more and more sunlight to feed the biochemical machinery of the animals higher up the food web, shown in Figure 6.1 as a pyramid. Thus, it is easier to feed a population of vegetarians than one of meat eaters—one needs less sunlight to provide the food supply of organisms that eat only plants. Turn this principle around, and one can see why adding meat to the diet is a status symbol for poor people: only the well-to-do can afford the sunlight (as well as the land, water, and animals) to raise meat to eat. It is one measure of how rich a society the United States has become that many people now choose not to eat meat. Eating meat has lost its significance as a status symbol in the developed societies of the world, and a turn toward vegetarianism has become widespread within the past generation. As this example shows, humans are versatile: we can eat both plants and animals—we are **omnivores**.

Following the same logic, plant-eating animals make up a lot more of the biomass in an ecosystem than carnivorous animals, because each carnivore eats many **herbivores** over a lifetime. It is for this reason that nut-eating squirrels are commonly seen, but spotting a mountain lion or a killer whale can be the event of a lifetime. Yet there is one meat-eating animal that is not uncommon. Do you see one now? Look in the mirror (or at the next nonvegetarian you meet). This is why this chapter is called "The Most Successful Species?"—the question mark is a reminder that our success as a species carries momentous implications for all animal species, including us, and for the plants that nourish all.

RISING TO THE TOP OF THE FOOD WEB

How did our species get to be so successful? Remember the idea of carrying capacity discussed in Chapter 3. There, the pasture could only accommodate a limited number of grazing animals. It is that limitation, you will recall, that shapes the tragedy of the commons. Over the course of the past ten thousand years, humans have managed to evade that limitation as far as our own species is concerned. We seem to have *enlarged the carrying capacity* of environments to accommodate more humans. This came about through two major waves of change: the invention of **agriculture** several thousand years ago; and the Industrial Revolution, which began more than two hundred years ago. Some elements of these revolutionary transformations are likely to be unsustainable, as we see now in the challenge of global warming, driven by the burning of fossil fuels for industry and by the release of greenhouse gases from agriculture.

Although humans are thought to have emerged into our current genetic lineage several million years ago in eastern Africa, recorded human history extends back only about eight thousand years. For more than 99 percent of our time on Earth, humans have been a hunting and gathering species, similar in some ways to the troops of chimpanzees that still inhabit our common African homeland. Like several other mammal species, humans seem to have hunted effectively in groups, so that large or swift prey could be brought down by teams of people. Humans may have been more effective than other hunting species because we developed language and were able to coordinate our actions better and learn faster. It is likely, in short, that we became the most successful species long before history was invented.

From Africa, humans spread into other parts of the world, reaching Australia by about forty thousand years ago. Our ability to cross water in boats, to make clothing and build shelter so that we could survive cold and harsh environmental conditions, and to control fire—these and many other adaptations were important to the diffusion of humanity across such a wide geographical range. In this way, too, we were a species capable of a rare kind of success, able to expand carrying capacity by living in habitats where we required technology such as heated shelters and warm clothing to survive. We were able to modify landscapes and to heat shelters through the controlled use of fire. In some places, such as Australia and parts of North America, hunting and gathering persisted until European colonization. But large farming economies existed in what would become the midwestern United States, as well as in the Andes and in Mexico, long before Europeans arrived.

CONTROLLING THE FOOD SUPPLY, CHANGING THE LAND

Agriculture brought the greatest proportional expansion of humans' carrying capacity. One anthropologist has called agriculture the "greatest technical achievement in the human record."[1] Carrying capacity is a function of technology, specifically food-production technology. This means that, as humans shifted to different food-production methods, they also changed the size of the human population that can be fed, so long as environmental conditions enabled food production to continue at a high level.

Enlarging the human carrying capacity has meant replacing ecosystems that were managed lightly or not at all with agricultural ecosystems controlled by people (see Box 6.1: Agricultural Ecosystems, page 132). How have people reshaped the carrying capacities of the landscapes they inhabit? The story began long before there was writing, so there are uncertainties and vigorous debates among archaeologists and anthropologists. One version is based on the work of noted scholar William McNeill, one of the first contemporary historians to develop ecological ideas in the portrayal of human history, and geographer Jared Diamond, particularly in his wide-ranging book, *Guns, Germs, and Steel*.[2]

The domestication of grains, beans, and root crops such as yams, together with a suite of tamed animals that included chickens and pigs, seems to have occurred within a fairly brief time—but in far distant places (Table 6.1 on page 133). The differences in the groups of species domesticated by humans, together with their common nutritional capacity to feed a population far larger than could be supported by hunting and gathering, seems to suggest the independent invention of agriculture in different parts of the world. The most recent Ice Age ended not long before the first signs of cultivated crops in the archaeological record, so the climate changes that accompanied the retreat of the ice may be related to the invention of agriculture.

BOX 6.1

AGRICULTURAL ECOSYSTEMS

The practice of agriculture includes dramatically different human relationships with the land, as we can see in our own country. America was a developing country in the colonial period. The American Indians were already growing crops in coastal and southern New England when the English arrived in 1620. Indeed, the Indians were able to help feed the starving Puritans in that first winter. The Indians practiced a shifting agriculture, farming land for several years and then moving on as the nutrients in the soil were depleted. This style of agriculture was practiced together with hunting and gathering. In the tropics today, this method of making a living is sometimes called "slash and burn," a derogatory term that might lead one to overlook the fact that this has been, over wide areas and long periods, a sustainable means of gaining a livelihood.

Agricultural practices replace the natural ecosystem with one of human design. This usually means that the number of species of plants and animals in a cultivated landscape is far smaller than were found on that same land before humans took it over. The species that live on the farm are chosen for their value to people, rather than because they are able to survive against the pressures of natural selection. Chickens that can be carried off easily by hawks, lettuce that is relished by deer, fields of grain that cannot exist without protection by scarecrows—all of these are

Gulf of Mexico hypoxic zone, July 21–27, 2008.

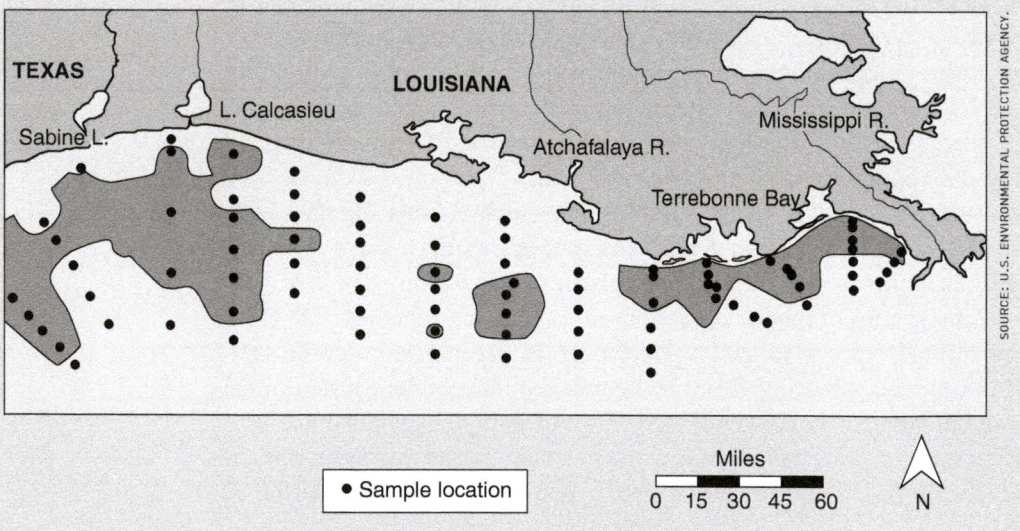

species that are abundant because people nurture and protect them. The plants and animals of a farm are generally removed when they are young (think of lamb or spinach). Young organisms mostly grow rapidly, and replacing them with seedlings or newborns resets the biological clock each year, enabling people to gain more from their cultivated ecosystems.

The ecosystem of the farm is often precarious. Soil that is turned over each spring is carried off more easily by rain or wind than soil anchored by the roots of trees and brush. Erosion can be accelerated by poor practices such as plowing in patterns that do not slow the runoff of rainwater. In contemporary farming in many nations, fertilizer is added to the soil, and pesticides are used to kill off insects and other species that compete with people for farm produce. When fertilizers and pesticides are in turn carried off by rain into waterways, they pollute water and foster the growth of organisms that choke streams or lakes. The Mississippi River's waters carry nutrients and pesticides into a large zone in the Gulf of Mexico where the nutrients suffocate the marine life by lowering the oxygen level of the water. The resultant "dead zone" (the shaded area in the figure) is a desert—it is "hypoxic," as scientists say, meaning without life-sustaining oxygen. You can see by the state boundaries of Louisiana and Texas that the hypoxic zone in 2008 was quite large. In some years, the area of the dead zone is as large as New Jersey.

TABLE 6.1   DOMESTICATION OF PLANTS AND ANIMALS BY HUMANS

Area	Plants	Animals	Estimated date of domestication (years before present)
Southwest Asia (Near East)	Wheat, pea, olive	Sheep, goat	10,500
China	Rice, millet	Pig, silkworm	9,500
Mesoamerica	Corn, beans, squash	Turkey	5,500
Andes and Amazonia	Potato, manioc	Llama, guinea pig	5,500
Eastern North America	Sunflower, goosefoot	None	4,500

Source: Based on Jared Diamond, Guns, Germs, and Steel *(New York: Norton, 1997), 100.*

Agriculture brought about a very large expansion in the number of people who could be supported by a given land area. Most of the material in most plants, such as leaves, stems, and wood, cannot be digested by people. In a small number of plant species, however, we can consume significant parts—including seeds such as the grains on an ear of corn, flowers such as broccoli or artichoke, roots such as potatoes and the manioc that feeds many in Africa today—and thus these are the plants we have promoted and protected through agriculture. Humans have also managed to breed a small number of animal species successfully under domestic conditions, including turkeys, cattle, and guinea pigs. Domesticated plants and animals, cultivated intensively over a relatively small land area, enabled humans to obtain many times more food per acre than was possible through hunting and gathering. As shown in Table 6.1, the main food sources of today were being produced regularly long ago. As Jared Diamond has pointed out, "By Roman times, almost all of today's leading crops were being cultivated somewhere in the world."[3]

As Europeans colonized the New World from the sixteenth century onward, they often occupied lands where the native peoples had been living partly or wholly by hunting and gathering. In this way of life, fire is often employed as a management tool: undergrowth is burned so as to promote the sprouting of green shoots, which attract game animals so they can be hunted.

When the English arrived in New England, they learned to grow corn and vegetables from the Indians who lived along those shores. The English added grazing animals and replaced hunting and gathering with a sedentary way of life based on private ownership of land, as discussed in Chapter 3. This unleashed a set of forces that reshaped the landscape (Table 6.2). People did different things on the land: they plowed and grew crops over a wider area; they grew animals and used their manure to fertilize the land; they killed wolves that threatened their domestic animals; they divided the land into farms owned by individuals, and they built towns where there was little farming in densely settled areas, but crops could be brought to marketplaces along roads. All this replaced the forests of New England with the pastoral landscape portrayed by Thomas Cole in *The Oxbow* (see Chapter 2).

As the historian William Cronon documented in *Changes in the Land*, the new agricultural practices changed the landscape—but not sustainably.[4] The ideas and institutions brought by the English led to much more intensive use of the land. New methods of cultivation and domestication of animals produced more abundant food supplies, although the higher density of animals and crop plants also favored the spread of disease and pests. There were other, less visible costs as well. Because the English concept of property meant permanent cultivation of a given plot of land, soil was not allowed to recuperate and farming did not last as the sole support of its human population. Now, almost four centuries later, New England has returned to a lightly used, forested landscape. There are many more

TABLE 6.2   FORCES CHANGING THE LANDSCAPE OF NEW ENGLAND, 1620–1830, AS A HUNTING AND GATHERING ECONOMY WAS REPLACED BY INTENSIVE AGRICULTURE

Changes

- Domestication of grazing animals
- Plowing of cleared land
- Ownership of domesticated animals (instead of hunting wild animals)
- Killing of wolves (carnivore competitors)
- Fencing of land
- Separation of different land uses, especially pasture from nonpasture
- Building of roads to link farms to markets

Consequences

- Much higher levels of food and other materials used by humans (benefit)
- Higher densities of domesticated animals (benefit)
- Habitats favoring weeds, pests, and diseases, mostly introduced species (cost)
- Overgrazed lands (cost)
- Depleted soils (cost)

Source: William Cronon, Changes in the Land *(New York: Hill and Wang, 1983), chap. 7.*

people living there now, and they mostly import their food—as one can see by the brightly colored fruits and vegetables available in food markets there in late winter, when local fields are covered in snow.

SETTLEMENT, SOCIETY, COMPLEXITY

Agriculture brought more than larger food supplies. It stimulated a qualitative change in human society: **permanent settlements** and the rise of **specialized roles** in society. Consider the rain forests and coral reefs discussed in Chapter 5. Permanent, complex structures seem to foster high levels of biological diversity. Something similar appears in cultural development. Hunting and gathering is a nomadic way of life, as people follow seasonal abundances of plants ripening in different microhabitats and exploit the life cycles of migrating fish or birds or game. With cultivated plants and animals, however, year-round settlement becomes imperative in order to tend the crops as they grow, to store the surplus after the harvest, and to guard crop lands and storage facilities that are fixed. The demographer Joel Cohen commented, "The plow tied the farmer to a place."[5]

Permanent settlements (Fig. 6.2), in turn, created the possibility of social differentiation, as humans could take on different, complementary roles in societies fixed in place. In a nomadic way of life, work was shared within a family or clan grouping as the family moved with the seasons. Because wild edible plants or animals were widely scattered, the groups that could work together were small ones, as the thinly spread resources could only sustain a small number of humans in a given area. Populations fed by farmers, by contrast, could grow to much higher densities because of the much larger food supply that could be produced in a given area of land.

In societies that were no longer nomadic (called sedentary societies), people could harvest enough food to see them through the succeeding year. When agriculture became more efficient, enough food was produced that some families could stop farming, if they could get the rest of the community to feed them. This made room for a **division of labor**, including the development of governments. Some scholars have pointed to irrigation as a key innovation, which may have fostered the invention of government and legal institutions.[6] An irrigation system is a commons, often too large for an individual or clan to manage, but requiring

FIGURE 6.2
Site of Uruk in Iraq. Uruk was one of the first cities with a population of more than 10,000 living in an area of several hundred acres—about the size of a college campus, but more densely settled.

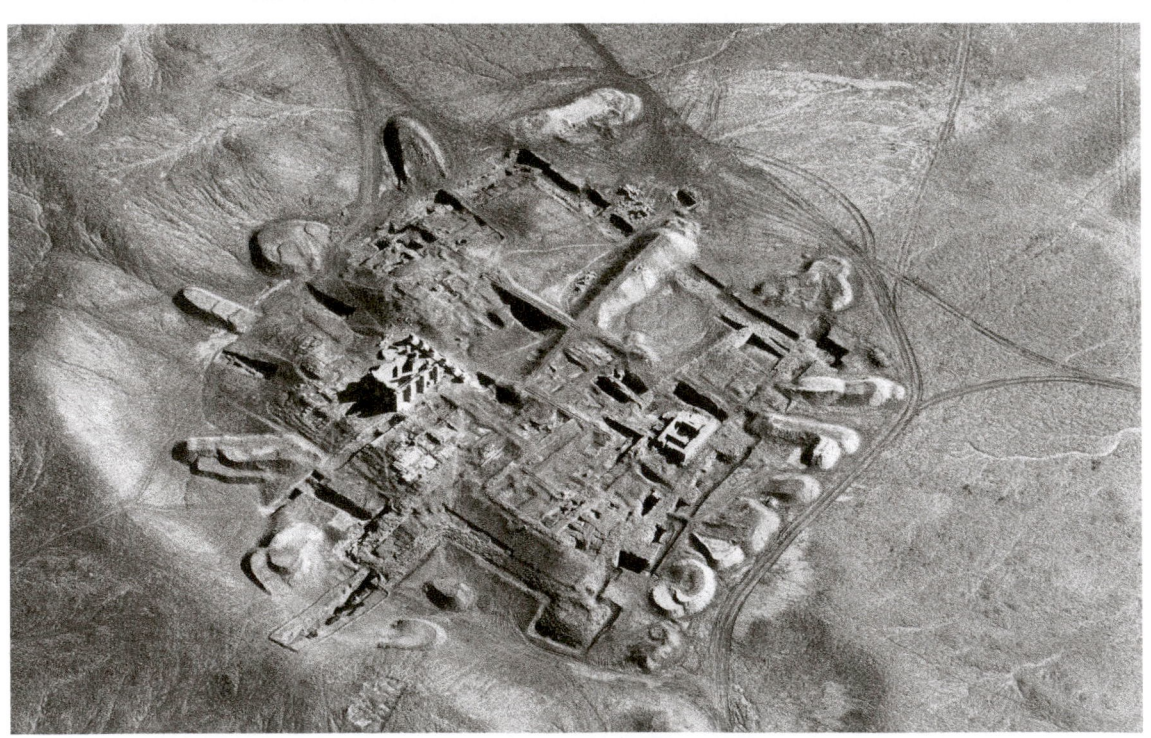

maintenance over long periods if the gains of reliable water are to be realized. The taxes and labor needed to create an irrigation system thus seem to be social requirements, which were supplied through government. The rules for contributing labor and for using the shared resources of the irrigation system, in turn, might have served as the prototype of law.

Priests, merchants, warriors, and scholars appeared, all supported by the farmers whose agricultural output was taxed, usually within the context of religious observances. With these more complicated interactions, writing and counting became functionally necessary. History, religion, music, dance, art—in a word, culture—could be recorded and passed from one generation to the next by means other than rote memorization. Agriculture, forming the material base of what we recognize as civilization, may be the most far-reaching collection of technological capabilities of our species.

From an ecological perspective, these social specializations served to make their communities stronger and more resilient in the face of war, drought, or other stresses. Warriors with specialized weapons and tactics could defend and conquer more effectively. Administrative capabilities enabled the redirection of resources to handle routine needs such as religious ceremonies or education, or for reacting to unexpected events such as damaging storms or epidemic illnesses. Perhaps most important, the much higher productivity of farming gave agricultural societies the numbers and military capability to displace hunter-gatherer societies, so that farmers could consolidate their control of fertile areas, such as river valleys.

Societies with these capacities continued to suffer sickness, pestilence, and famine, but they survived more often. Their institutional arrangements have been enshrined and explained in a dazzling variety of texts and legal traditions, in literatures and liturgies in every tongue and music. In the Old Testament, for example, Joseph won his freedom by interpreting the Egyptian pharaoh's dream as a revelation to store grain after abundant harvests, so that a society may survive the lean years likely to follow. As we saw in the governance principles for managing commons in Chapter 3, the variety of rules, customs, and institutions documented by social scientists is a coat of many colors, but in its various forms it serves needs that arise in all human communities and the commons they must share among their members.

GOVERNMENTS, PARASITIC AND RESPONSIVE

Civilization, with its characteristic social differentiations, highlights the central roles of inequality and power in human history. Humans, in dominating nature, also came to dominate other humans. The relationship between these forms of domination was captured in a provocative ecological interpretation of the human past put forth by the historian William McNeill. He argued that history could be understood as the interactions among parasites. Agriculture attracts pests and predators. The high concentration of food in a crop is a biological opportunity, not only for the cultivators but for other species and kinds of people. Rats and roaches have been companions of humans for a long time. Some nomadic peoples, including the Mongol hordes of central Asia and the Viking raiders of Scandinavia, survived for many generations by raiding settled societies, building a livelihood on militarized predation.

Disease organisms, too, do well in dense human settlements. Not only do humans and their domesticated animals provide a large habitat for the pathogens, but high densities foster transmission of microbes from one person or animal to another. This means that pathogens can survive even if they kill their hosts. Microbes reproduce rapidly, often on a scale of hours, whereas the host populations of animals or people reproduce far more slowly. The host population adapts slowly in comparison to the generational time scale of the pathogen. As with the long-lived trees of the rain forest, the characteristics of the hosts become significant in an evolutionary way to the microbes.

The result is a trajectory of adaptation that has been seen repeatedly. Disease-causing organisms that can propagate without killing their victims, or by killing only those that can be relatively easily replaced, such as young children, tend to survive in larger numbers. Those strains of bacteria and viruses that irritate but are not fatal—such as the rhinoviruses that cause the common cold—become the microbial analogue of rats, entrenched and hard for people to avoid. Until the development of medicines based on a fundamental understanding of the biology of pathogens, beginning in the nineteenth century, endemic diseases that fit this profile were part of the human condition.

McNeill called disease-causing organisms and pests microparasites: species, almost all smaller than humans, that erupt into human history as plague or pestilence, or form an irritating backdrop to everyday life, and are more debilitating than computer viruses or acne. The microparasites, McNeill suggested, have human and social analogues, which he called **macroparasites**: criminals, marauders, unjust rulers, and dysfunctional institutions. "When one man or group of men seize goods or compel services from other human beings, they . . . may . . . be called macroparasites," McNeill wrote.[7] All these, he pointed out, can be seen in history, and they exact a price on those who cannot resist—including farmers, who cultivate the crops on which all depend.

Governments, in particular, may be seen as macroparasitic in much of human history. Whether particular rulers were enlightened or not, government implemented a rough bargain. Rulers provided defense against enemies, settlement of conflicts so that economic activities could be carried out, and control of criminals. In exchange, they levied taxes, sometimes in the form of labor rather than money or crops, and they conscripted young men to serve in military forces. As with diseases, a form of macroparasitism evolved that was resilient and durable. Governments ruled, by the threat of force, populations that were productive enough in their agriculture and trade that they could provide both food and weapons to a military ruling class. These societies could expand by conquest. Those that learned to rule large territories effectively could persist for centuries. In short, in these empires, the total carrying capacity was increased, despite taxes and conscription that reduced the economic returns of farming. These societies worked, in a material sense, with governments serving a symbiotic role and not just a parasitic one.

A key social invention of such durable empires was **bureaucratic organization**: networks of officials who could implement the orders of rulers over large territories before the advent of rapid communication technology. Bureaucracy, in turn, made possible taxes instead of plunder—a form of parasitism that did not kill the victims but burdened them. Such forms of military and bureaucratic rule arose in China, the Roman Empire, and the New World more than a thousand years ago, before these regions were in regular contact—a pattern that suggested to McNeill that the logic of this form of government had more to it than the ambitions of kings, but was grounded as well on a macroparasitic ecology of power.

In few cases were those who paid the taxes or provided the foot solders consulted by their rulers. It was not until the eighteenth century that democracy on a mass scale took root. Unlike the earlier Greek and Roman models of democracy, in which only a small fraction of the population could actually vote or influence decisions, mass democracy has evolved into a system of government in which nearly all adults residing in a nation-state are eligible to participate in making binding choices. Voting, constitutional government, political parties, and the institutions of self-rule were significant cultural mutations in governments, arising just as the Industrial Revolution brought another major increase in the human carrying capacity. We will return to this part of the story in the last part of this book, as we discuss the responses to the grand environmental challenges that are our focus now. You might pause to notice, however, that both classical environmentalism and community-based governance operate through self-government—that is, through *collective* action that is not parasitic but can serve the interests of the group.

In comparison to the hunter-gatherer society built around clan groups bound by family ties, the empire based on agriculture was a highly differentiated world, where a person's place was defined by power relationships cemented into institutional forms, such as aristocracy and serfdom. The status hierarchies were typically rigid and characterized by exercises of authority that today seem unbearably arbitrary. This was a world with edges that were social as well as geographical. With the emergence of a world without edges, built on an industrialized, globalizing economy, we have seen the erosion of place in all the traditional senses: spatial, social, and functional. The rise of Tiger Woods, a gifted athlete born to a Thai woman and an African-American man and who became an icon of corporate advertising, is one example of the remarkable cultural changes still under way.

The search for sustainable development is one element of an attempt to create a community without edges—a set of loyalties and rules and practices by which we can guide our actions for the benefit of the community as well as ourselves. Today the presumption is that a community must be democratic to be functional—that those who are guided by the community's norms must have a stake in them, so that they will mostly guide themselves, instead of looking over their shoulders in fear of detection and enforcement. Humans have not lived as members of communities larger than China's nearly 1.5 billion, however, and the world population is now more than 4 times that large. One of the grand challenges, then, is the creation of effective governing institutions for a sustainable planet, institutions that will need to be both global in reach and as local as restraints on littering if they are to work. The ability of McDonald's or the Catholic Church to do something along these lines suggests that we need not be daunted.

BETTING ON LIMITS

The enormous success humans have had in expanding Earth's carrying capacity, particularly as our numbers and our material consumption have increased over the past 350 years, raises a question: Can we keep going, or are we going to encounter limits to Earth's carrying capacity, as all other forms of life do? This turns out to be a complicated question, and one that is of fundamental importance in classical environmentalism. If scientists have unambiguous evidence that humans are exceeding the capacity of nature to support our species, then an alarm must be sounded and heeded.

Since the beginning of contemporary environmentalism in the 1960s, natural scientists have sounded repeated warnings. Many of them have been ignored. An interesting wager, made in 1980 between ecologist Paul Ehrlich and an outspoken economist named Julian Simon, illuminates the puzzles that humans face when we wonder if we can continue to expand the capabilities of nature.

Ehrlich, an entomologist and environmental activist, gained wide fame in 1968 when he published *The Population Bomb*, a book that warned of impending famines as the populations in countries such as India were projected to exceed food supplies.[8] Ehrlich was warning of the approaching limits in carrying capacity, doing so at the same moment that Garrett Hardin's analysis of the tragedy of the commons was gaining wide notice.

Simon, an iconoclastic economist until his death in 1998, published an article in 1980 that was skeptical of environmental claims. The article summarized his argument, which appeared the following year in a book entitled *The Ultimate Resource*.[9] This argument states that humans and human institutions have been smart enough to anticipate environmental limitations and to implement innovations that have staved off disaster—and no sound reason has been put forward to make us think things have changed. In particular, Simon asserted that markets were efficient ways of sensing and responding to impending limits, so long as government stayed out of the way. This opposed the recommendations for strong intervention that biologists including Ehrlich and Hardin made.

Julian Simon's original article was subtitled "An Oversupply of False Bad News," in which he specifically criticized Paul Ehrlich as a wrongheaded doomsayer.[10] This attracted Ehrlich's attention. He, along with two other scientists, decided to take up a bet that Simon had offered. As resources like gold or platinum get scarce, Simon noted, they become more valuable, prompting a search for cheaper substitutes. If natural resources were indeed becoming scarcer, their prices should rise; but if human innovation found substitutes quickly, the prices of natural resources might fall instead. Simon declared that he was willing to bet that prices of a specific set of resources would fall rather than rise.

Ehrlich put down $1,000. In 1980 that sum would have been enough to buy small quantities of five industrially important metals: chromium, copper, nickel, tin, and tungsten. If by 1990 the five metals were worth more than in 1980, after adjusting for inflation, Simon would pay Ehrlich the difference. And if the five metals cost less than $1,000, Ehrlich would pay the skeptical economist.

"It is such an obvious proposition in a finite world: things run out," the journalist John Tierney wrote in a story about the wager.[11] Yet it was Ehrlich who wrote the check to Simon, and it was for a cool $576. The prices of all five metals had declined over the ten-year span of the wager, losing half their value. Simon's wager was not a bad return on a thousand dollars. What had happened, Tierney wrote, was technological innovation. Chrome-plated fenders on cars were replaced by cheaper molded plastic; fiber-optic cables started to replace copper wires for carrying the signals of the infant Internet; tungsten cutting tools met competition from less expensive ceramics; and an international tin cartel collapsed when so much tin was mined that it proved impossible for the sellers to control prices.

In a word, the market seemed to work just as Simon argued it would. The high price of metals such as tungsten can offer an economic opportunity. A cheaper substitute or a more effective one at the same price will earn its inventor and producer money, sometimes a lot of money. These opportunities are being explored now in many places throughout the world without edges. Human ingenuity may not be the ultimate resource, as Simon argued, but it is a large one.

Care is needed, however, in interpreting the results of this wager. In the late 1990s and early 2000s, the prices of most metals and other natural resources rose sharply as rapid economic growth in China and India stimulated increased demand.

Although technological change serves to reduce some prices (the *technology effect*), demand growth led by higher production and consumption pulls prices higher (the *scarcity effect*). Whether prices rise or fall over time depends on the net result of these opposing effects, among others. Many economists believe that resource prices are likely to rise in the future as available resource supplies become scarcer relative to growing demand.

Interestingly, a recent analysis by environmental economists at the College of the Holy Cross points out that Ehrlich would have won the resource price wager if the bet had come due in 2007 rather than 1990.[12] Resource prices are highly volatile, with random fluctuations around a long-term trend. Luckily for Simon, 1990 brought a short-term lull in metal prices, which reversed after he had collected from Ehrlich. The study found that prices for the five metals in the bet rose between 1900 and 2007 with random peaks and valleys. The market, it would appear, is not the smoothly operating mechanism that Simon envisioned.

Keep in mind that scarce metals of the kind Simon and Ehrlich bet on are non-renewable resources, so the idea of carrying capacity doesn't really apply. Because only a finite amount of tungsten or platinum is contained in the earth, its price has more to do with economic demand and the ease or difficulty of finding the ores and producing the metal than it does with whether we are running through all the tungsten in the world. (Where might we get tungsten when the ores have run out? In garbage dumps, where the concentrations of some rare metals is already comparable to the levels in commercial ores.)

Something similar seems to be happening with oil prices, which increased more than tenfold between the late 1990s and 2008, only to fall as the worldwide recession of 2008–10 took hold (Fig. 6.3). Are we running out of oil? Maybe, or maybe not. On the one hand, conventional oil resources are sufficient to meet current demand for at least several decades. On the other hand, oil demand has grown rapidly in recent years to fuel economic growth in places such as China and India. Only a finite quantity of oil exists, and there is no doubt that we are burning oil far faster than it is being created by natural processes. The question, then, is when and at what price substitute technologies (e.g., synthetic oil from coal, electricity from sunlight, palm oil from plantations installed in cleared rain forests) will become available in the future.

Optimists envision a market response that will give rise to a smooth transition with relatively low energy prices. Pessimists worry that the financial and environmental costs of the transition may be far higher than the optimists hope. The trouble is, no one has a crystal ball to predict the future, despite the best efforts of futurists and economic modelers. Anyone who could predict the future could make a great deal of money by buying and selling contracts for future delivery of petroleum at stipulated prices. Such futures contracts are traded in commodities

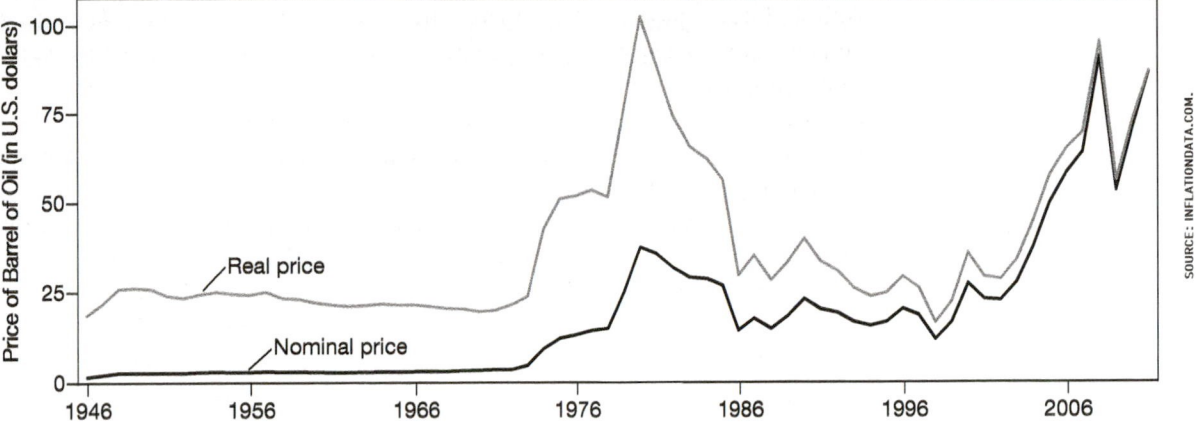

SOURCE: INFLATIONDATA.COM.

markets that resemble stock markets. Futures contracts are a commercial form of the Simon–Ehrlich bet. Bettors are found on both sides of oil futures, however, indicating that no one is sure what will happen.

Environmental limitations come in various forms. Garrett Hardin's story about the pasture has to do with exceeding the ability of nature to produce things that people need, such as grass in the pasture for their animals. But humans today also are depleting nonrenewable resources such as natural gas or platinum, which are finite and for which there is no long-term carrying capacity. Although resource prices might rise in the future if demand outstrips available supplies, we have seen how markets and new technologies can sometimes gracefully and flexibly resolve concerns about resource scarcity. When prices go up, businesses and consumers find substitutes, such as burning wood or paying closer attention to caulking and insulation when heating oil becomes more costly, or installing solar panels. It is too simple, then, to assert that the physical scarcity of natural resources automatically translates into higher prices and lower living standards. Even with given resource limits, a sustainable future can be achieved through the interplay of markets and well-designed public policies.

FIGURE 6.3
World price of a barrel of oil. The nominal price is the price paid at the time; the real price is adjusted for inflation (in 2012 dollars).

ECOSYSTEM SERVICES

The bet between Simon and Ehrlich focused on resources traded in markets. But this experiment missed a crucial aspect of human needs and the natural world. To understand what is omitted, we need to take up the idea of **ecosystem services**.

We're familiar with the fact that we obtain many goods from nature, including crops, seafood, timber, and medicines. Today, most of these can and do come from distant locales, from the world without edges. But as Stanford ecologist Gretchen Daily observed,

> Natural ecosystems also perform fundamental life-support services without which human civilizations would cease to thrive. These include the purification of air and water, detoxification and decomposition of wastes, regulation of climate, regeneration of soil fertility, and production and maintenance of biodiversity.[13]

Thus, we not only need nature's goods, such as food, we also need nature's *services*, such as the slow release of water from mountain snowfields and glaciers.

Some of these services, such as the pollination of food crops by honeybees, can be privatized. The owners of almond trees, for example, rent beehives for several weeks each year to make sure their trees are properly fertilized so that they will produce nuts in large quantities. In 2007, entomologist May Berenbaum estimated that pollination services by honeybees accounted for most of the bees' contribution of $14 billion per year to the American economy.[14]

Most ecosystem services are less easily explained, and the idea is a bit abstract (for a related, but oversimplified portrayal, see Box 6.2: Ecological Footprints, page 145). Think about brushing your teeth this morning. You used water from the tap. Where did it go? Not far away, because water is too heavy to transport very far before being released to a river or the sea. Now, think about the water you will use to brush your teeth tonight. Where will that water come from? For most of us, our water supply is also located not far away, although Southern California and some major cities such as New York collect and transport freshwater from distances of more than 100 miles. You wouldn't want to use the same water that you spat out this morning. As you read this, do you know where your morning toothpaste is? The provision of clean water is an example of an ecosystem service that is both vital and so much taken for granted that it is invisible. Tap water is also nearly free. Few Americans are aware of paying bills for drinking water; its cost is less than 1 percent of the cost of bottled water that may not be as carefully tested. More important, when water is drawn into a city's mains, what residents are paying for is the cost of the piping and the treatment plant, rather than the water itself.

Alarm bells should sound. When things that people value are taken from a shared natural world without charge—that is, from a commons—abuse is likely, a basic dynamic we have already seen. And when users cannot perceive the pressures they put on ecosystems, then the abuse is silent even to the abuser. Few of us have any way of knowing whether we are polluting a stream when we flush a toilet.

BOX 6.2

ECOLOGICAL FOOTPRINTS

Among all the living things that depend on consuming the products of photosynthesizing plants (the class of organisms that biologists call heterotrophs), humans constitute about 0.5 percent of the biomass, or about 1 part in 200. Yet humans **appropriate** at least one-third of all the net primary productivity on land. This disproportionality gives this chapter its title.

Trying to estimate how much of the planet's capacity to support life is under human control is the objective of several overlapping lines of research. None has come to be the accepted standard in the same way the gross domestic product—for all of its shortcomings—has become a standard measure of economic output.

The one with the most intuitively appealing name was developed in the early 1990s by regional planners William Rees and Mathis Wackernagel. They devised the ecological footprint, an estimate of "how much land and water area a human population requires to produce the resources it consumes and to absorb its carbon dioxide emissions, using prevailing technology."[1] The appealing notion of a footprint evokes the picture of a preindustrial city drawing its sustenance from the

Size of countries has been stretched or shrunk to indicate their effective consumption levels in 2005. This provides a measure of the countries' ecological footprint.

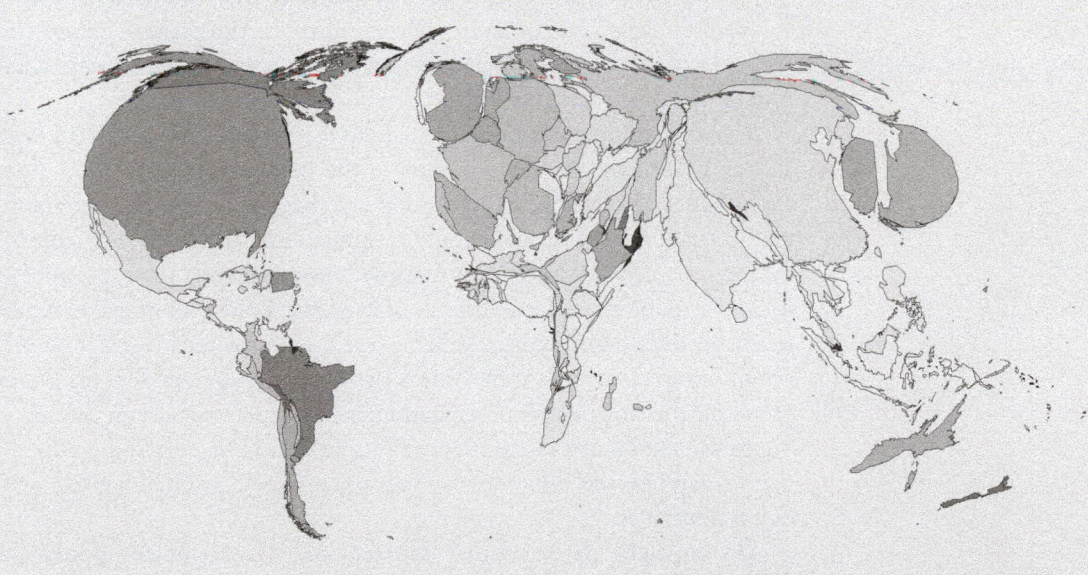

farmlands around it. The ecological footprint attempts to adapt this picture to cities and nations deeply enmeshed in a global economy. The footprints of a number of countries and cities has been analyzed. Using this approach, the average footprint of a person living in a high-income country such as Australia is 8 times as large as that of someone living in a low-income country such as Ethiopia. The map shows this disparity visually by distorting the area of each of the world's countries in proportion to its consumption levels; such maps are called **cartograms**.

The total footprint of humans on Earth, using these methods of analysis, grew to exceed Earth's "biocapacity" in the late 1980s. It is far from clear whether biocapacity is the correct measure, however. The portion of net primary productivity (see the "Net Primary Productivity" box in Chapter 5) being appropriated by humans is well less than 100 percent. Nonetheless, as you'll see later in this chapter, measures of the impact of humans in the natural world are impressively large.

1. Global Footprint Network, Footprint Basics—Overview, 2012, www.footprintnetwork.org/en/index.php/GFN/page/footprint_basics_overview/

Because ecosystem services are all but invisible to people who live in the world without edges, scientists have devised various ways to explain them. Figure 6.4 was used by an international group of 1,600 scientists participating in the Millennium Ecosystem Assessment, which was completed in 2005. The assessment was the first major step of an international process sponsored by the United Nations. Its aim was to draw government and public attention to the planet's ecosystems—and to the services they provide to people—in a way similar to the international scientific process that has brought sustained attention to climate change.

The first thing to notice in this flow diagram is that it is both wide-ranging and complicated. The various parts of the diagram provide a functional map of the world without edges. On the left are the landscapes: five different groups of ecosystem services that contribute to human life. Consider how this might work on the scale of a community. A forested hillside above a town produces wood and foliage through the photosynthesis of its plants, holds soil in place, and contributes to the production of oxygen. When a tree is cut down, its wood joins the stream of provisioning services as it become a chair, a box, paper, or fuel for a fireplace. The plants on the slope catch precipitation and slow its flow down the hill, contributing to the regulatory function by adding to flood control. And children playing in the woods, like the hunters who shoot game in it, are making use of the forest's cultural services.

On the right are boxes that represent the five major components of life and society that would be desired by a very large fraction of the human race. The flood

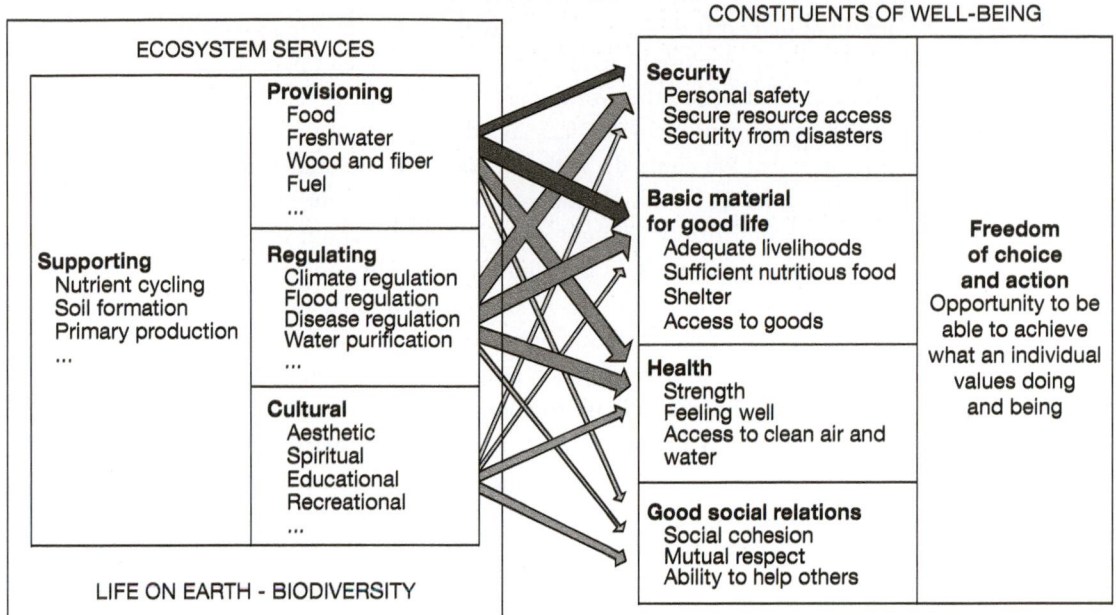

CONSTITUENTS OF WELL-BEING

ECOSYSTEM SERVICES

Supporting
Nutrient cycling
Soil formation
Primary production
...

Provisioning
Food
Freshwater
Wood and fiber
Fuel
...

Regulating
Climate regulation
Flood regulation
Disease regulation
Water purification
...

Cultural
Aesthetic
Spiritual
Educational
Recreational
...

LIFE ON EARTH - BIODIVERSITY

Security
Personal safety
Secure resource access
Security from disasters

**Basic material
for good life**
Adequate livelihoods
Sufficient nutritious food
Shelter
Access to goods

Health
Strength
Feeling well
Access to clean air and
water

Good social relations
Social cohesion
Mutual respect
Ability to help others

**Freedom
of choice
and action**
Opportunity to be
able to achieve
what an individual
values doing
and being

**Arrow's color
Potential for mediation
by socioeconomic
factors**

▢ Low
▢ Medium
▢ High

**Arrow's width
Intensity of linkages between
ecosystem services and
human well-being**

═══ Weak
═══ Medium
▢ Strong

SOURCE: MILLENNIUM ECOSYSTEM ASSESSMENT.

FIGURE 6.4
Ecosystem
services and their
relationship to
human well-being.

control service provided by the forest is a component of the physical security of the town. In most years in the hills of the western United States, wildfires burn large areas of land. People who live below burned-over hills know to fear mudslides and floods if there are heavy rains or a rapid melting of snow when the soil is still bare. If a lumber mill in the town uses logs from the hillside, the lumber produced contributes to the material needs of those who buy it, while the wages earned by the mill workers can underpin the social cohesion of the community. The forest's wild birds may bring beauty, song, and other benefits that are enjoyed by many humans, although they may also be infected with the West Nile or avian flu virus, so doctors in the town might be on the lookout for unusual illnesses among the town's residents.

These little stories illustrate two things. The first is the way that ecosystem services are woven into the fabric of daily life. All the things we do—from eating to exercising, working at a lathe, or taking a walk with an old friend—can bring us into contact with the natural world. In most of these interactions, though, we are unaware that it is nature we deal with. The water seems to come out of the tap, not from a river close by. The job comes from a company, not from a productive landscape that makes the work economically profitable. In each of these ways, the availability of ecosystem services and the quality of the environment are not just amenities that we could do without. Nature is essential, yet for the most part, it is something we do not think about as we go about our lives.

The second reality brought out by thinking about ecosystem services is how many of them are not sold through markets. The municipal public works department may tend the walkways in the wooded park above town, but it does not send residents a bill for flood control or water purification. Most of what can be sold is in the form of goods, such as logs from the forest, whereas services such as recreational value are difficult to charge for.

Why does this matter? It matters because of the insights that arise from the outcome of the Simon–Ehrlich wager. When efficient markets are operating, buyers and sellers recognize the approach of scarcity through prices, and they then respond, sometimes by innovating in a way that actually drives down the price of a natural resource, at least in the short run. But goods and services that are not bought or sold in the market have no established price, and thus one of the important signaling mechanisms of scarcity or decline in quality is absent. Unfortunately, the incentive to innovate and to find alternatives may not be there either. As we noted before, the absence of a market suggests an ecosystem service supplied through a commons that may not be well governed.

A third point is not immediately apparent: ecosystem services often come from ecosystems we do not see or do not pay attention to. The state of those ecosystems may be changing slowly, in the invisible present, and we usually learn of their condition only when something has gone wrong, as when the West Nile virus appears in a wild bird. Even as hunter-gatherers, humans relied on many of the same ecosystem services that we still rely on today. Clean air and water, disposal of our wastes, and maintenance of biodiversity remain crucial to human success in any landscape where humans are present, and many of these services must be supplied from nearby. The interruption or loss of those services drastically reduces the carrying capacity of a particular place. This has been true for all living species for as long as there has been life. What is new is the way humans in rich societies do not need—or feel they do not need—to know this truth. The freedom of choice and action that fosters this ignorance is the independence that Don Gifford discerned in the world without edges—an apparent independence that grows out of interdependence.

If ecosystem services can be abused when they are provided from commons, such as rivers, we need to think about institutional ways of dealing with commons through privatization, government regulation and ownership, or community governance. We will come back to these ideas again. The point now is to see the importance of ecosystem services: some of them are necessary to our well-being, and some are more precarious than we realize.

In summary, ecosystem services are

↘ necessary to life and well-being.

↘ usually local.

↘ often found in commons.

↘ hidden in the invisible present.

ARE THERE PLANETARY-SCALE LIMITS?

That there may be local problems with ecosystem services is not really a surprise in the sense that, in some cases, people have degraded environments so much that the local economy and habitability have been seriously impaired. The decline of the Mayan civilization in Central America was described in Chapter 3, and the more recent example of the disappearance of the Aral Sea in central Asia is sketched in Box 6.3: A Warning? The Aral Sea (page 150). The destruction of that sea, a lake nearly as large as Lake Superior, has tragic consequences for the communities that relied on its ecosystems. But does this suggest that humans are exceeding the capability of the planet to support our ways of life?

That question was surveyed in 1997 by a group of American scientists led by ecologist Peter Vitousek. By pulling together a large number of studies from many disciplines, this team reached a provocative conclusion:

> Human alteration of Earth is substantial and growing. Between one-third and one-half of the land surface has been transformed by human action; the carbon dioxide concentration in the atmosphere has increased by nearly 30 percent since the beginning of the Industrial Revolution; more atmospheric nitrogen is fixed by humanity than by all natural terrestrial sources combined; more than half of all accessible surface fresh water is put to use by humanity; and about one-quarter of the bird species on Earth have been driven to extinction. By these and other standards, it is clear that we live on a human-dominated planet.[15]

Each bar in Figure 6.5 (on page 152), taken from the paper by Vitousek and colleagues, measures something different, such as the total percentage of accessible surface freshwater that is used by people, or the percentage of major marine

BOX 6.3

A WARNING? THE ARAL SEA

The degradation of commons usually shows up unevenly. Part of a pasture goes bare first, and if the animals there cannot find new grass, the shepherds in that part of the range will be in trouble. But the rest of the shepherds in the area may not know or may not care.

The Aral Sea is a large lake in central Asia in the former Soviet Union. In Soviet times, the streams flowing into the sea were diverted to grow cotton in the surrounding valleys. This thirsty crop took up the water in that arid region and transpired a good deal of it into the atmosphere. The amount of water reaching the Aral Sea dwindled. The sea ultimately shrank by half its area. Islands became a peninsula, and the sea itself eventually divided into separate bodies of water (see series of satellite images). Seaports and their fishing fleets were stranded miles away from the receding shoreline.

The exposed land, which was saturated with minerals, dried out, and the resulting dust storms blew mineral salts over farm fields, suppressing plant growth (see image). The agricultural industry shriveled and people had no work. Those who remained now lived in a hellish desert. Their ecosystem and its vital services had been destroyed, taking the economy with it.

Although the Aral Sea is a dramatic example, similar problems of soil poisoning and lowering of the water table due to irrigation have been seen in a number of

Shrinkage of the Aral Sea from 1973 to 2000, observed in NASA Landsat satellite images.

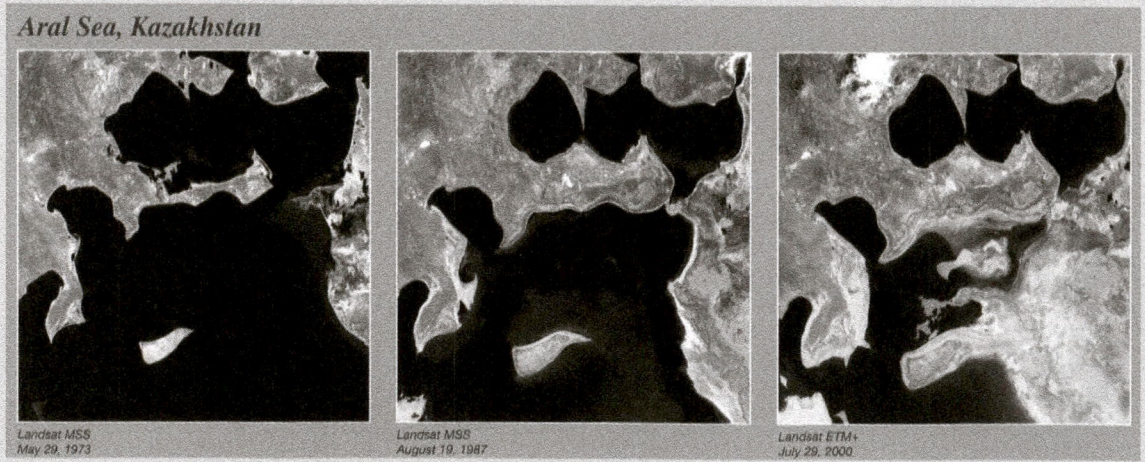

Aral Sea, Kazakhstan

Landsat MSS
May 29, 1973

Landsat MSS
August 19, 1987

Landsat ETM+
July 29, 2000

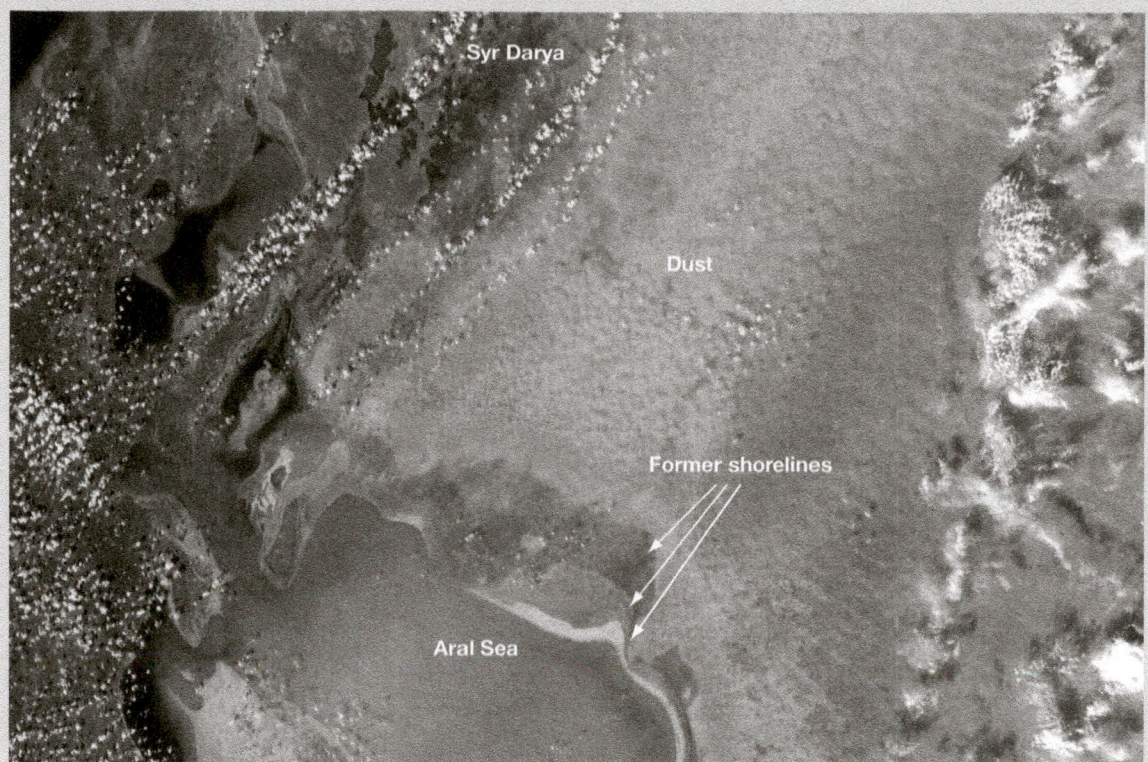

Syr Darya

Dust

Former shorelines

Aral Sea

areas—in Mesopotamia (the land we now know as Iraq), in the Imperial Valley of California, and in western China, to mention a few of the well-known cases. One might wonder if these patches of the global commons going bare may be a warning of worse to come.

fisheries exploited. Perhaps the most notable of these measures is the bar on the left, which shows how much of the land surface of the planet is being used for direct and indirect human benefit. The estimate from which this bar is drawn counts farmland, urban areas, and mines as land under human domination, of course, but it also counts forests and pastureland. If you think back to the discussion of meat eating as a status symbol, you will recall that using land is really a way to capture the products of photosynthesis, because that is what is useful to people. So the left-hand bar means that humans are benefiting from about half the sunbeams captured by plants.

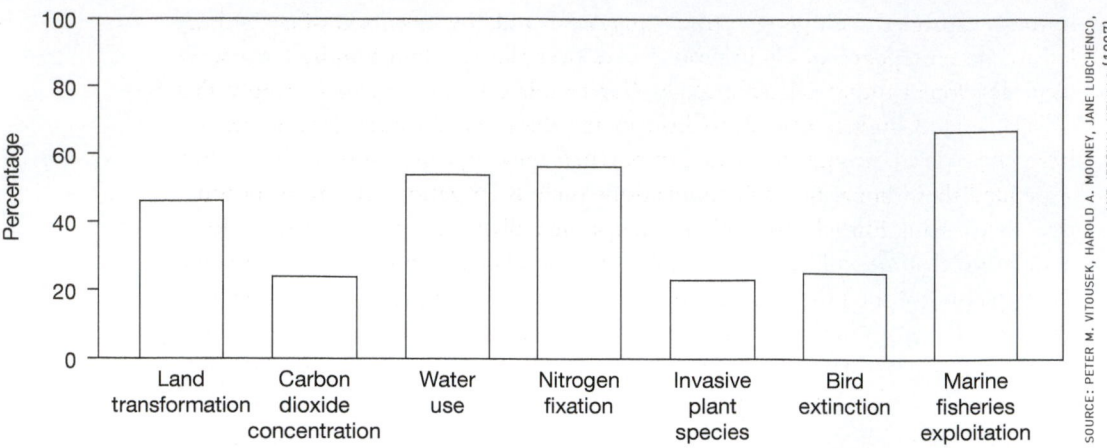

SOURCE: PETER M. VITOUSEK, HAROLD A. MOONEY, JANE LUBCHENCO, AND JERRY M. MELILLO (1997).

FIGURE 6.5

Human dominance or alteration of several major components of the Earth system. Aspects of the human footprint in nature are expressed as (from left to right) the percentage of the land surface transformed; percentage of the current atmospheric carbon dioxide concentration that results from human action; percentage of accessible surface freshwater used; percentage of terrestrial nitrogen fixation that is human-caused; percentage of plant species in Canada that humans have introduced from elsewhere; percentage of bird species on Earth that have become extinct in the past two millennia, almost all of them as a consequence of human activity; and percentage of major marine fisheries that are fully exploited, overexploited, or depleted.

The percentages assigned in Figure 6.5 are likely to be uncertain by considerable margins in several cases. In short, this is no more than an initial attempt to summarize many different estimates of different kinds of ecosystem services used by people (such as water use) or exploitation of ecosystem goods (such as fisheries depletion). What is significant, despite the crude nature of the estimates, is that these measures are all significantly different from zero. Rather, they are around 20 percent of some theoretical maximum or higher. Humans are having an effect that can be measured in the global metabolism. The human world without edges is a significant presence in the world at large. If human populations and economic activity continue to grow as they have done over the past century, and if environmental impacts grow with them, then 20 percent may grow to 80 percent or toward global limits. We might still be able to double the water available to human populations according to Figure 6.5, but water use increased tenfold in places like southern Arizona during the twentieth century. Growth at such a pace is not possible for long anywhere.

What does this discussion have to do with Earth's carrying capacity? Humans have been able to change the apparent carrying capacity dramatically. First, prehistoric humans expanded their geographical ranges. Second, the invention of agriculture led to the emergence of civilization. Third, over the past two hundred years, we have developed industrial technology. Garrett Hardin was wrong to imply that a well-defined limit is placed on human numbers and resource use. So far, we have been clever enough to evade limits when they appeared, and we have often expanded these limits through innovations such as irrigation and more productive crops. Paul Ehrlich and other environmentalists have been warning for a generation that we will need to change our ways because the earth *is* finite and the trajectory of the human economy has risen very steeply. Yet much of our society, including many who call themselves environmentalists, behave as if the limits were not a concern because the prices that should rise if scarcity is looming have not risen in comparison to our incomes.

Yet in our reliance on ecosystem services, we often place demands on ecosystems that are inevitably local. For example, because freshwater is so heavy to carry, it is not feasible or cost-effective to transport it in quantity over long distances. In this respect, our world does have edges. In our impacts on ecosystems, we are cumulatively beginning to have large-scale effects, both regionally and globally in some ways. When markets exist, as Julian Simon shrewdly pointed out in his bet with Paul Ehrlich, it is sometimes possible to sense rising shortages and find substitutes. So we may not need new fisheries if what we care about is protein. We may be able to switch to soybeans if we continue to overharvest the seas. But what about ecosystem services that aren't provided by markets, such as flood control or clean air? These can lead to tragedies of the commons when people run up against limits, as they did in the Aral Sea ecosystem.

In sum, the environmental problems we find in the world today may be both local and global. They come to our attention through the decline of nonhuman species or in the suffering of human populations in specific places. But they may also reflect global limitations in the ability of nature to supply the ecosystem goods and services desired by people. These environmental challenges, remember, are the product of apparent human success—of our ability to expand the carrying capacity to a degree unmatched by any other species. We are the abundant omnivores.

In the next four chapters, you will see how the successes of our species give rise to the grand challenges of sustainability that we introduced in Chapter 1: energy and climate change; population and urbanization; conserving biodiversity; and creating the institutional structures to guide a global economy toward sustainability.

FURTHER READING

Cronon, William. Chap. 7 in *Changes in the Land: Indians, Colonists, and the Ecology of New England*. New York: Hill and Wang, 1983.

Diamond, Jared. *Guns, Germs, and Steel: The Fates of Human Societies.* New York: Norton, 1997.

McNeill, William H. *The Human Condition: An Ecological and Historical View.* Princeton, NJ: Princeton University Press, 1980.

Millennium Ecosystem Assessment. An overview with links to additional information is available at www.millenniumassessment.org.

Vitousek, Peter M., Harold A. Mooney, Jane Lubchenco, and Jerry M. Melillo. "Human Domination of Earth's Ecosystems." *Science* 277 (1997):494–99.

KEY TERMS

agriculture	carnivore	food web	omnivore
appropriation	cartogram	herbivore	permanent
bureaucratic	division of labor	macroparasite	settlement
organization	ecosystem services	metabolism	specialized roles

154   CHAPTER 6: THE MOST SUCCESSFUL SPECIES?

Chapter Seven

CLIMATE CHANGE

THE DILEMMA OF FOSSIL FUELS IN AN INDUSTRIALIZED ECONOMY

One environmental issue that nearly everyone has heard of is **global warming**. In this chapter we first describe the basic science of climate change. Then we discuss the industrialized economies that generate greenhouse gases, primarily through the burning of **fossil fuels**. Few people are aware of how their everyday choices—to eat beef, fly in an airplane, or buy a new car—affect energy consumption and therefore the global climate. The response of Earth's climate system to the addition of greenhouse gases is delayed and complicated. In the language of Chapter 2, climate change is hidden in the invisible present. Moreover, the atmosphere is the largest commons: greenhouse gases emitted anywhere on the planet affect the global climate. As Chapter 3 explains, it has been very difficult to reach international agreements to govern the atmospheric commons. The deep, pervasive link between people's everyday choices and climate change makes this environmental problem a grand challenge of sustainability.

The basic science of global climate change has been known for more than a century. Since 1988, the Intergovernmental Panel on Climate Change, an international body linking the worldwide community of climate scientists to national governments and the international community, has assembled a firm scientific consensus to warn the world that humans are changing the global climate—a warning recognized by the Nobel Peace Prize in 2007. As we'll see, some action has been taken, and there is much talk about doing still more. But little is being done yet that would bring the driving forces of human-caused climate change to a standstill.

In that sense, the classical model of environmental reform has not yet worked. That is, the threat identified by science is not being met with technological and institutional changes that could credibly manage global warming. The environmental problem is still worsening, and its driving forces are not yet being controlled.

WHY CLIMATE MATTERS

As we saw in Chapter 5, climate forms the basic architecture of the natural world (see Fig. 5.6). What makes one place different from another is shaped by the general pattern of weather it experiences—that is, its climate: the annual cycle of the seasons, the amounts of precipitation a place receives, and the frequency of extreme events such as droughts and storms. As we saw in Chapter 5, the climate also shapes the local ecosystem and determines which plants and animals can survive in that place. Even landscapes created by humans, such as parks and backyards, look different in Maryland than they do in Utah. The crops and forests of tropical Indonesia (see Fig. 5.12) do not resemble those of New England (see Fig. 2.2).

Climate is the long-term average of weather. Over the history of the planet, climate has changed. In Canada and much of the United States, a trained eye can see the signs of glaciers that covered the land about ten thousand years ago, so Earth's climate has varied dramatically while humans have inhabited the planet. What is happening now, as humans cause the global climate to change, is that long-term

Learning Objectives
When you have finished studying this chapter, you should be able to

↘ explain the greenhouse effect to an elementary school student or grandparent;

↘ discuss at least two different changes in the global climate system, such as warming, retreat of polar ice caps and mountain glaciers, rise in sea level, and changes in the frequency and intensity of storms and droughts;

↘ identify several activities in your daily life that are directly linked to the burning of fossil fuels and suggest some that are dependent on fossil energy indirectly;

↘ look into energy conservation opportunities in your life and at your college, such as using public transit or installing insulated windows, to find out firsthand why these measures are not being adopted;

↘ explain the implications of filling your car's gas tank on national security, the U.S. economy, and global climate;

↘ articulate your own and others' opinions about an international climate agreement by using ideas from this chapter and Chapter 3.

averages are changing. Indeed, the weather we are seeing today is increasingly departing from expectations based on historical experience.

It is important to bear in mind that the weather varies a lot. So a change in *climate* is subtle and hard to see. Warming—a rise in average global temperatures—is the change that has received the most attention. Perhaps more significant to human societies, though, are changes in the frequency of extreme events such as severe storms, droughts, and heat waves.

Are such changes occurring? It does seem that news reports of tornadoes, crop failures, wildfires, and floods have been increasing in recent years. But that is partly because humans have been settling in places where their vulnerability to extreme events is higher. In the developed world, scenery draws people to coastlines exposed to hurricanes, housing subdivisions are built next to forests, and cities like Las Vegas are built in deserts. In many poor countries, expanding populations have pushed into areas where agriculture is easily disrupted by bad weather, increasing the risks of storm damage and famine. Scientists who study the atmosphere agree that these changes in human vulnerability have increased much more rapidly than the weather itself has changed. But as we will see, the weather *is* changing.

As the climate changes, ecosystems on land and sea are being altered, and the ecologists who study them project larger disruptions in the decades to come. Watersheds, forests, croplands, and oyster beds that yield ecosystem services used by humans are also being changed. Some of the severe droughts that have brought hunger to eastern Africa and economic loss to the U.S. Southwest during the first decade of this century may be a reflection of climate change. These are, in any event, representative of the stresses that climate change will bring.

For the most part, though, residents of a rich nation like the United States may barely notice the disruptions. Coffee or orange juice may cost more. There may be news of starving people in a distant country as reductions in crop yields in places like Russia lead to higher grain prices in markets everywhere. But the world without edges gives people with means the ability to shift sources of supply, often without consumers noticing. That is, climate change may not affect daily life in a rich country very much, allowing skeptics to dispute whether climate change is even under way or whether it is caused by humans.

THE GREENHOUSE EFFECT

The principles of climate change are not seriously in question, although the field science is complicated enough that some scientific questions still need to be resolved. Energy from the Sun reaches Earth as light—technically, electromagnetic radiation in the visible part of the spectrum. This energy heats the earth, and Earth re-radiates a portion of that energy as **infrared radiation**—a form of light with wavelengths that are too long to be visible to the human eye, although you can *feel* infrared rays that come from the heat lamps used to keep food warm in some cafeterias. A portion of that heat is retained in the earth and the planet's atmosphere.

The way the light from the Sun heats the Earth involves the **greenhouse effect**, named after the structures gardeners build to protect their plants from frost and encourage growth. You experience a similar effect when you get into a car that has been parked in a sunlit space. The interior of the car is noticeably warmer than the outdoor temperature because the car windows turn the vehicle into a greenhouse. Energy comes in as sunlight and warms the seats, carpets, and dashboard. This is why the seats can be uncomfortably hot if you're wearing shorts.

The high temperature inside the car shows up as an excess of infrared radiation, but infrared radiation is blocked from escaping by the window glass. Glass is transparent in the visible wavelengths used by our eyes, but it is opaque to infrared light. As a result, one of the main mechanisms of heat transfer is blocked by the windows.

There is some leakage of heat to the outdoors, though, so in a few minutes a **steady state** is reached, in which the temperature inside the car stops rising but is higher than the air outside. The elevated temperature inside is a result of the greenhouse effect (Fig. 7.1).

Because we cannot see temperature or heat, it may be easier to understand the process with an analogy. Consider a bathtub with its drain open and the faucet running. As water flows through the faucet, filling the tub, the flow through the drain increases with the water pressure until a steady state is reached and the water level stabilizes. Now, imagine the drain clogs slightly but does not stop the outflow of water. The water flowing out slows, until the water pressure in the tub rises high enough for the outflow to equal the amount flowing in.

That higher level is the analogue of global warming: the temperature of Earth rises, like the level of water in the slightly clogged bathtub, until a steady state is reached, with energy leaving Earth at the same rate as the sunlight brings it in. The gases that block infrared wavelengths of light, which make up the bulk of heat radiation, are called **greenhouse gases**. If the concentration of those gases is increased, more infrared light is blocked and the net effect is a rise in global temperature.

By far the most significant greenhouse gas is water vapor. If Earth had no atmosphere and thus no greenhouse effect, the planet's average temperature would be about 33°C colder, below the freezing point of water. Looked at that way, the greenhouse effect produced by the atmosphere and the water in the oceans is an integral aspect of the architecture of the natural world.

The science of the greenhouse effect was worked out by French physicist and mathematician Joseph Fourier in an 1827 paper in which he remarked that humans might change the conditions on Earth sufficiently to alter the heat balance of the planet. Others confirmed Fourier's theoretical analysis, and by the end of the nineteenth century, Swedish chemist Svante Arrhenius estimated that a doubling of

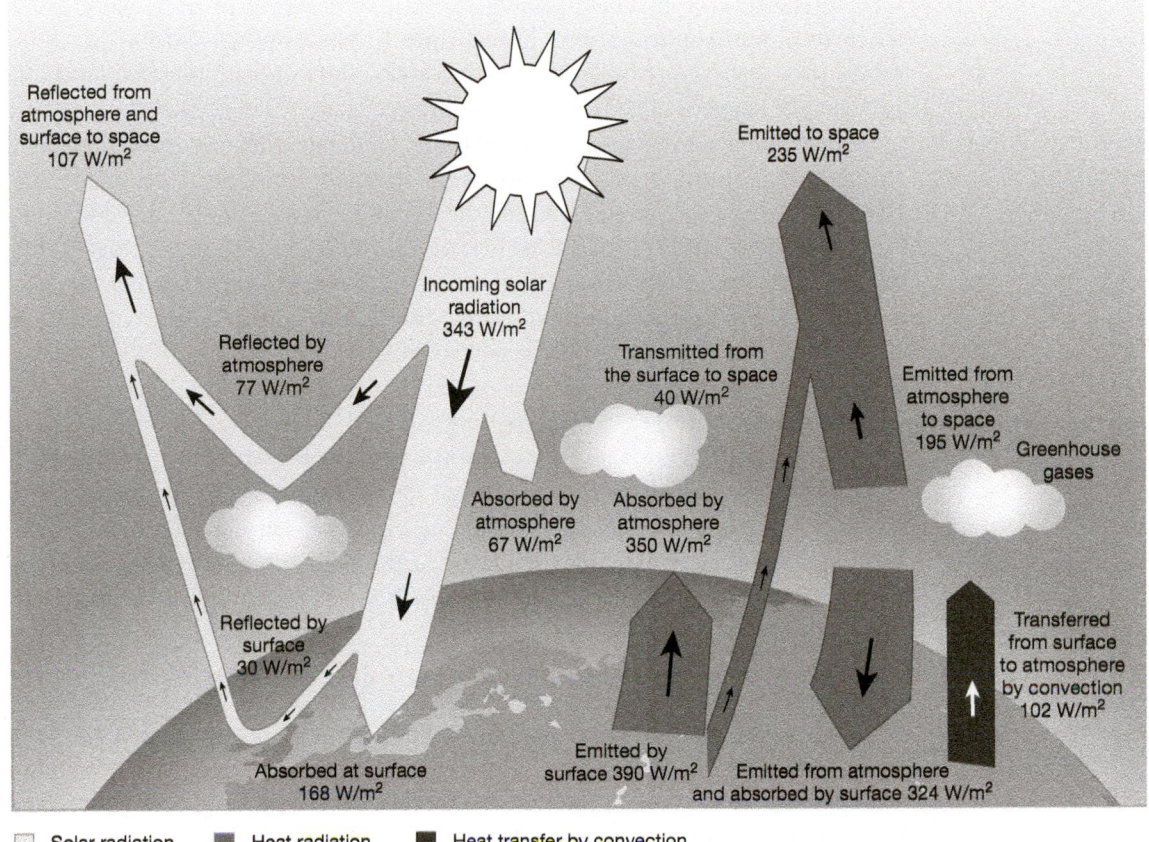

SOURCE: GEOENGINEERING THE CLIMATE: SCIENCE, GOVERNANCE AND UNCERTAINTY, POLICY DOCUMENT (2009).

FIGURE 7.1

Earth's energy budget, showing input of solar energy, loss of energy from Earth to space, and transfers of energy from the surface of the planet into the atmosphere. Changes in the composition of the atmosphere can alter the balance between solar inputs and losses to space, changing the average temperature of Earth.

the **carbon dioxide** (CO_2) in the atmosphere would raise Earth's average temperature by about 5°C or 6°C. It is an estimate that still stands more than a century later.

Humans have been turning Arrhenius' projection into fact, adding to the greenhouse effect by changing the composition of the atmosphere through emissions from agriculture, industry, and transportation. Climate scientists call these additions **anthropogenic radiative forcing**—changes in the heat balance shown in Figure 7.1 caused by people. As of 2005, 63 percent of these changes were being caused by CO_2, the most prominent greenhouse gas. The other major contributions came from methane (19 percent), nitrous oxide (6 percent), and other industrial gases that also affect the stratospheric ozone layer (12 percent).[1] The gases in the last category are regulated by an international treaty agreed to in 1989 called the Montreal Protocol, and these gases are slowly decreasing as emissions are phased out. Methane and nitrous oxide are removed from the atmosphere more rapidly than

CO_2. If emissions from human activities ceased, the impact of those gases would lessen over time compared with carbon dioxide. The time needed for the atmosphere to return to conditions that prevailed before the Industrial Revolution would be centuries, however. The human footprint on climate is substantial and durable.

Methane and nitrous oxide are products of agriculture and industry. Methane is the primary constituent of natural gas, and it is generated in many ways, such as in rice paddies and other wetlands and in the digestive tracts of mammals. Cattle generate enough methane that their manure can be tapped as a source of fuel. Methane is a more powerful warming agent than CO_2, pound for pound, but methane is much lower in concentration, so its net greenhouse effect is about one-third as large as that of CO_2. Nitrous oxide is formed in the combustion of fossil fuels as oxygen and nitrogen in air are exposed to high heat. In addition, nitrous oxide is formed when nitrogen-containing fertilizer is digested by microorganisms. So the fertilizer that is not taken up by plants can contribute to global warming. As with CO_2, both methane and nitrous oxide are by-products of processes valued by humans—from growing crops, to producing meat and dairy products, to burning fossil fuels. Altering emissions of all three of these greenhouse gases will require changes to patterns of economic growth that have considerable momentum.

GREENHOUSE GASES AND ACCOUNTING FOR CARBON

What has been happening to the concentration of greenhouse gases? Figure 7.2 shows data on the concentration of CO_2 that have been collected since 1958 on Mauna Loa, a volcanic mountain in Hawaii. First, notice the sawtooth pattern: the concentration of CO_2 goes up and down each year. This is Earth "breathing"—the CO_2 level drops in spring and summer as plants take up more CO_2 than they release through respiration, then the CO_2 level increases in autumn and winter as the leaves drop and decay. (The data from Hawaii reflect the behavior of the atmosphere over the Northern Hemisphere.)

Next, look at the numbers on the vertical scale: the total concentration of CO_2 is tiny, about 390 parts per million (equivalent to 1 teaspoon in about 3.6 gallons). That's the concentration of peanut butter if you were to spread one tablespoon of peanut butter on twenty loaves of bread—hard to call that a peanut butter sandwich. The change in total CO_2, as you can see, has been about 70 parts per million over the past half-century, or about 20 percent.

The other important greenhouse gases, methane and nitrous oxide, have been growing in concentration faster than CO_2 as meat production and fertilizer use have been increasing to meet demand in China, Brazil, and other rapidly growing economies. In addition, the large quantities of methane trapped in the permafrost in polar regions may be released as the climate warms and the permafrost melts. This would be an example of a positive feedback effect: the more the planet warms, the more of this potent greenhouse gas will be released, warming the planet further.

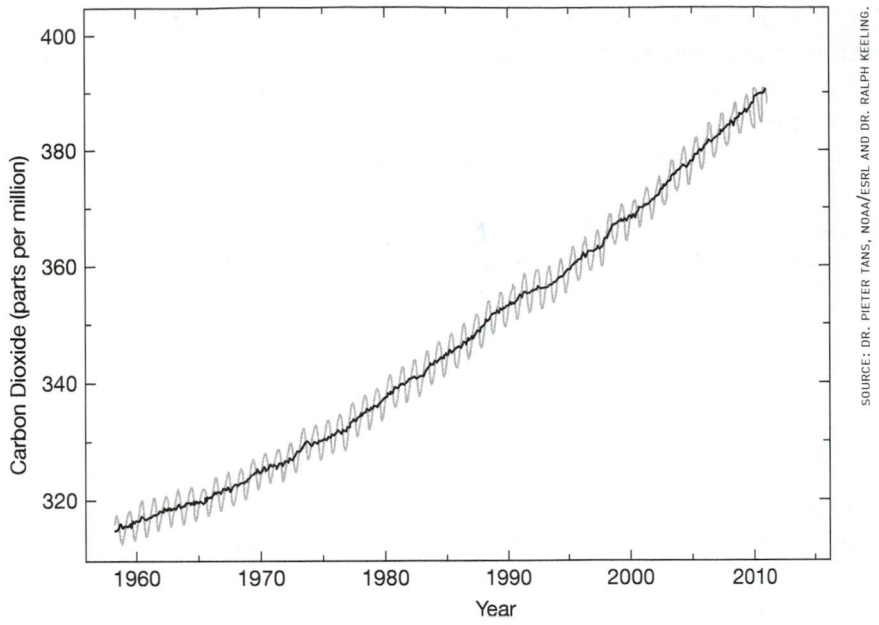

SOURCE: DR. PIETER TANS, NOAA/ESRL AND DR. RALPH KEELING.

FIGURE 7.2
Atmospheric concentration of carbon dioxide recorded at Mauna Loa in Hawaii, 1958–2011.

Even though the concentration of all the greenhouse gases is very small, bear in mind that a slight change in the clogging of the drain can cause a perceptible change in the water level in the tub. Earth's average temperature is the result of an equilibrium between the heating effect of the sunlight that falls on the planet and the loss of heat through the atmosphere. The equilibrium temperature shifts in response to changes that seem very small by the standards of our everyday experience. This is one reason the invisible present is invisible: we cannot perceive the shift in greenhouse gas concentrations except through changes in climate, and these take decades to become apparent.

Carbon dioxide flows in a global carbon cycle (Fig. 7.3). The carbon cycle is an example of a **biogeochemical cycle**, an accounting mechanism used by scientists to track the flow of basic constituents of the natural world, including carbon, oxygen, nitrogen, and water, as they move through the principal components or "compartments" of the biosphere. Carbon shows up in the atmosphere as CO_2, and it moves into plants through photosynthesis, becoming part of the wood, leaves, roots, and a wide array of organic chemical compounds. The transfer of carbon into living things is called net primary production, a technical term measuring the net capture of sunlight by plants. The carbon dioxide also is taken up (absorbed) by seawater, forming carbonic acid. In all of these transformations, carbon moves from compartment to compartment, playing different roles in the chemistry of life and nature.

Earth's atmosphere holds roughly 800 gigatonnes of carbon (GtC). Rather than trying to imagine how many tons of coal contains 1 GtC, it's better simply to think

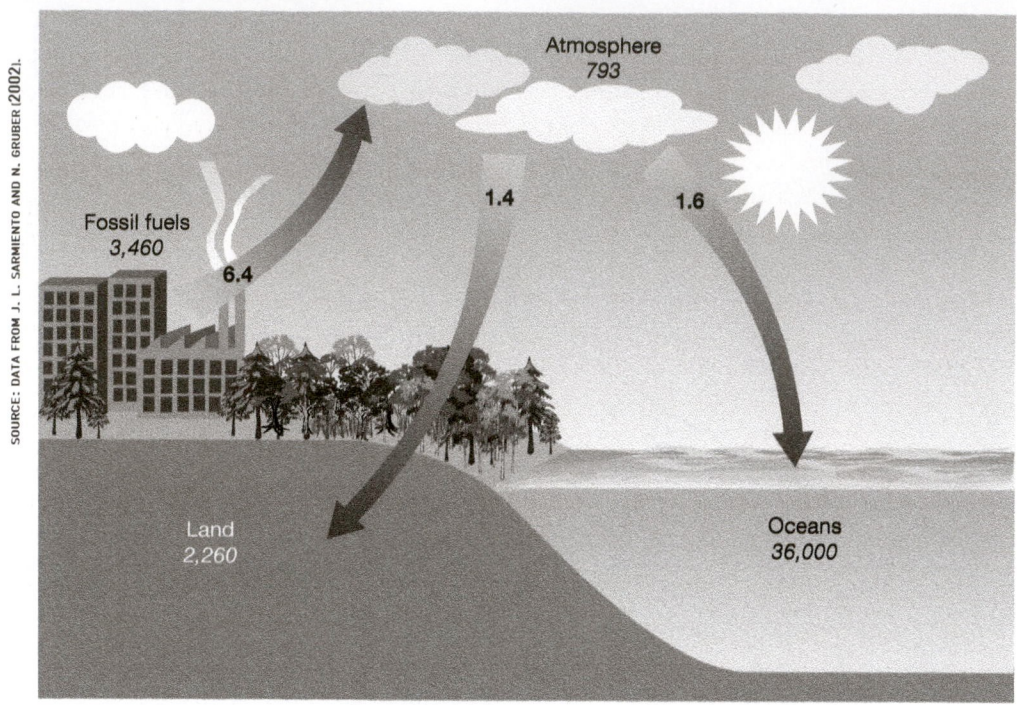

SOURCE: DATA FROM J. L. SARMIENTO AND N. GRUBER (2002).

Atmosphere
793

Fossil fuels
3,460

6.4

1.4

1.6

Land
2,260

Oceans
36,000

FIGURE 7.3
Earth's carbon cycle. Each of the major compartments of the planet holds carbon (italic numbers, in gigatonnes). Each year, because of both natural and human-caused changes, there are small net flows (numbers inside arrows, in gigatonnes per year) between the compartments.

of this reservoir of atmospheric carbon as a large number. Earth's rocks and ocean waters hold a great deal more. Each year, as the seasons unfold and living things process carbon, about 90 GtC flows back and forth between the atmosphere and ocean compartments. A similar flow of about 70 GtC takes place between the atmosphere and land. These are flows, not net transfers, which are much smaller (see Fig. 7.3). The flows can be seen in the sawtooth pattern of the atmospheric CO_2 data in Figure 7.2. The burning of fossil fuels contributes about 6 GtC each year. This is a net addition, which can be seen in the rising trend in Figure 7.2. Fossil fuels are burned, but because no human process soaks up this carbon, it stays in the environment.

GLOBAL WARMING

As many skeptics allege, global warming *is* a "theory." Global warming is a prediction that Earth's atmosphere will continue to warm, as it began to do in the twentieth century, to levels not seen on the planet for hundreds of thousands of years. This prediction comes from the laws of physics as applied in about fifteen computer models at universities and government laboratories around the world. These models employ the same physics as Fourier's 1827 paper but break up the atmosphere into thousands of individual cells so that the physics and chemistry of the gases within each cell can be tracked as sunlight interacts with water vapor, pollutants, and the rest of the atmosphere. As noted above, the broad outlines of these model analyses affirm the numerical results arrived at by Arrhenius in the nineteenth century using far simpler scientific means. In that sense, the basic magnitude of the greenhouse effect is well settled.

Evidence continues to mount that the model predictions are correct—that global warming is not *only* a theory. This has significant implications for the world, both natural and human. One summary of a large and diverse set of data from different parts of the world is shown in Figure 7.4. The dark wavy line shows the average temperature of the Northern Hemisphere over the past thousand years. The shaded area indicates the uncertainties in the underlying data. Over the course of the twentieth century, the trend in temperature turned up sharply. This is consistent with the idea that the globe is warming.

Why would Earth's temperature change so abruptly? Because humans are burning coal, oil, and natural gas at a rapid pace, altering the heat balance of the planet and its biogeochemistry. This assertion has been the subject of a long political debate, but it is not seriously disputed within the scientific community.

FIGURE 7.4
Average temperature of the Northern Hemisphere over the past thousand years. The baseline temperature (located at 0.0 in the figure) is the average temperature over the period from 1961 to 1990. The numbers on the vertical axis are degrees Celsius.

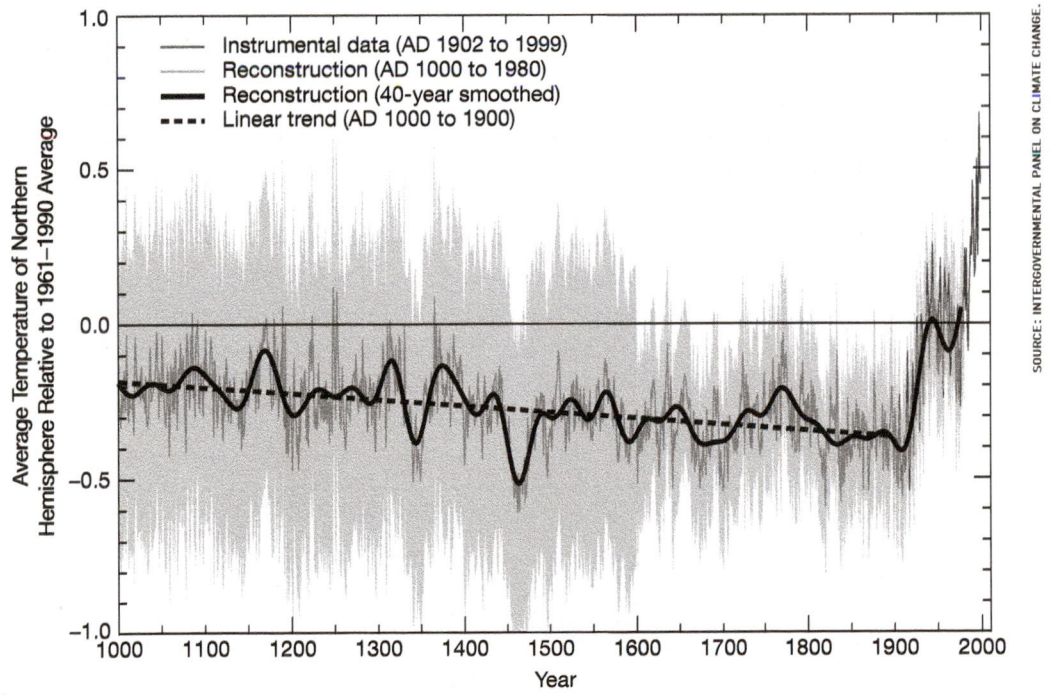

SOURCE: INTERGOVERNMENTAL PANEL ON CLIMATE CHANGE.

A WARMING WORLD

Even a small change in CO_2 (and other greenhouse gases) implies observable and worrying changes in the natural world. The basic notion is simple and stark: if climate establishes the architecture of the living world, then a change in climate will shift the architecture of ecosystems and the species within them. Climate changes are occurring on a time scale far shorter than the evolutionary process in long-lived species such as trees or mammals. This means that the shifting of the architecture of life is less like erosion and more like an earthquake: the whole structure shakes and the parts that are unable to adapt go extinct. The collapse of some species, in turn, sets off a ripple effect of changes throughout the landscapes they inhabit.

One important example of a dramatic but invisible environmental shift is the change in the chemical composition of the oceans. Carbon dioxide in the atmosphere mixes with water, forming carbonic acid. (The CO_2 in a carbonated soft drink turns the water into a dilute solution of carbonic acid. When the soft drink can is opened, the hiss you hear is pressure being released and some of the carbonic acid turning back into water and CO_2. When the bubbles of CO_2 gas rise to the surface, the CO_2 escapes.) As the CO_2 concentration in the atmosphere rises, the oceans are absorbing more CO_2, which makes them more acidic. This process of ocean acidification has been clearly documented by oceanographers in all the world's seas.

The rising acidity of the ocean has a profound consequence: it is becoming harder for creatures with shells to form their shells. The shells of clams and many other species are primarily calcium carbonate. This compound dissolves in acid, so as the ocean water becomes more acidic, it is causing problems for species with shells, just as we would if something caused our bones to weaken. Projections indicate that some species will be driven extinct in the coming decades as the oceans' chemistry changes. This would disrupt food chains. For example, a major food of some species of Pacific salmon is the pteropod, a small shelled creature. If the pteropods die out, so might some varieties of wild salmon.

As the climate changes, species that can move in response to changing weather conditions seem likely to do so. This means that tropical diseases such as malaria may reappear in temperate zones, when these zones warm enough to become favorable for malarial mosquitoes and other pests that are now limited by climatic conditions such as a hard freeze every winter. Species that cannot move, such as plants and animals already adapted to mountain peaks, will face extinction. In the mountains of southern Arizona, one finds life forms not seen at lower altitudes, but they do not have higher, cooler environments to migrate to. These "sky island" ecosystems seem to be getting more fragile. Already, the frequency of forest fires appears to be at unusually high levels.

A warming climate has other effects in addition to a rise in temperature. In some interior areas of continents, longer droughts are likely. In the oceans, warmer water gives storms more energy, so that more severe storms can be expected. Consensus has not been reached about whether these effects might be appearing already. What's clear is that storms such as Hurricane Katrina (2005) and Hurricane Irene (2011) imposed catastrophic damage in places such as New Orleans and Vermont of a kind projected to occur more often in a future, warmer world. In some places, such as Florida and the Gulf Coast, insurance companies have declined to offer coverage to homeowners whose property is vulnerable to hurricanes and storm surges.

Existing climate models do not have enough power to predict which areas will be affected by droughts or specific storms. Scientific teams are collecting more data in the field and building more complex models in an attempt to clarify these issues. A warmer climate implies, for example, that less snow will be stored on high mountain slopes in places like the Sierra Nevada of California. This means that rivers fed by snow and ice will change their flow patterns, with large implications for agriculture and for the rivers' ecology and hydrology. So there is great interest in more precise and accurate predictions.

Sea levels will rise as the warmer ocean water expands and as the total amount of water in the oceans increases with the melting of glaciers. The magnitude of **sea-level rise** to be expected by a specific date is not known precisely, but in 2007 the Intergovernmental Panel on Climate Change estimated a rise of as much as 6 feet by the end of the twenty-first century.[2] This would have a bigger impact than one might think at first because low-lying ecosystems, including wetlands, coral reefs, and fertile soils in river deltas all over the world, will be affected. Significant economic impacts will be felt, particularly in poor countries such as Bangladesh, where millions of people live in low-lying areas. Box 7.1: That Sinking Feeling (page 166) gives an account of what this looks like to a citizen of a country that is literally disappearing. Environmental pressure is likely to be felt in extreme events such as storms, in which a higher sea level may allow surging waters to flow much farther inland. It will be possible to build sea walls and dikes, as the Dutch have done, to preserve high-value lands, and this seems likely to be done in some places. But consider the Mississippi River delta, a low-lying coastal ecosystem that stretches for hundreds of miles. As Hurricane Katrina revealed, a sizable part of New Orleans is already lower than sea level, protected only by fragile levees. Clearly, all defensive strategies have practical limits, even in a country as rich as the United States.

We have controlled our indoor climates for a long time, so we are likelier to remain clueless about climate change, which is hidden in the invisible present. The changes in temperature, storm intensity, streamflow, and ecosystems are slow, and it seems unlikely that they will stay newsworthy. So we are faced with the problem

BOX 7.1

THAT SINKING FEELING
by Fathimath Musthaq

Fathimath Musthaq is from the Indian Ocean nation of the Maldives. One of her grandfathers was a Maldivian island chief. Fathimath left home when she was sixteen to study abroad, graduating from the United World College of the Atlantic in Wales in 2005. She wrote this essay in 2006 at Williams College, where she was studying political science, biology, and environmental studies—the latter because her homeland faces inundation as global climate change raises sea levels in the coming decades. This essay appeared in the online science magazine Earth & Sky (*www.earthsky.org*).

I still remember the time I looked out the window of the Monarch airplane and marveled at the majesty and beauty of what I saw spread beneath me. Even 72 hours of waiting in London's busy airport could not dampen my excitement of seeing those perfectly round living "organisms." These "organisms" protect

Alidhoo Island, Maldives.

my people and sustain their livelihood—a livelihood that is becoming harder to maintain as the temperature of the seas increase and this lethal thing that we call climate change grows stronger by the day.

The organisms that I talk of are the coral reefs, a splendid creation of nature, yet vulnerable to the slightest change in average temperature. These coral reefs protect the small low-lying islands where my great-great-grandfather lived his peaceful life of catching tuna and climbing coconut palms. Today, they protect the island where my father earns his income by luring the world to marvel at the splendor of our little islands.

Once upon a time, these creatures used to protect my childhood playground— the white sandy beach with dozens of coconut palms and the tall trees we call *Hirundhu ga*. I remember waking up every morning and running into the ocean to catch the colorful playful little fish that hid in the corals. But those days and that playground exist only in my memory and in the memories of those who were fortunate enough to be born at least 20 years ago. In 1998, massive coral bleaching (due to increased temperatures) caused severe damage to our reefs. The corals are now recovering slowly and, even today, many of the best dive spots and the whitest beaches are found in the Maldives. However, because of our vulnerability to climate change, I doubt my children will grow up, as my father would proudly say, "children of the sea."

We have lived on these little islands in the Indian Ocean for thousands of years. Our way of life is based on the unique geography of our archipelago. We have 1,192 islands of which people live on 199 islands and 86 islands are used exclusively as hotels for tourists. The population of the Maldives is about 350,000 with approximately one-third of the population living in the capital city Malé. Every year we have approximately 450,000 tourists. Although we import much of what we eat nowadays, the main source of nutrition remains different fishes, mainly tuna. The reasons for our sound economy are the tourism industry and the fisheries industry, both of which depend on our natural environment.

My friends from school in Wales and the U.S. tell me how lucky I am. It is true. Maldivians are lucky to be living in a country where the majestic beauty of the islands and the diversity surrounding the coral reefs sustain a simple and yet harmonious way of life.

The dark side that nobody wants to acknowledge is that when the temperature of the seas increase, our corals will die, our tuna will migrate, and our seas will grow and eat away the little islands. This is dreadful news for a country where the majority of people know only one way of living, a way of living rarely glimpsed in the modernized and busy world that we live in today.

It comes as no surprise that Maldivians would be reluctant to think about migrating to another land. What? Leave home?

I spoke to the director of the Environmental Assessment project in the national government, and he told me what I already knew. Maldivians are going to stay in their fragile little islands and defy nature until the big, heaving seas come and engulf the last remaining house. Even the angry 2004 "tsunami that showed the Maldivians in five minutes what would happen to their home over the next 50 or 70 years" has not prompted conversation about moving. Neither did it change people's minds drastically.

I don't blame them. To dig out the roots of ancestry and history and plant them somewhere where the soil will not nourish them is bound to erase a culture, a tradition, and a way of life.

Today, the government is busy engaging itself in "adapting" not "moving." The hesitance, no, the outright defiance of the Maldivians to moving leaves the government no choice but to invest millions in building sea walls and reclaiming land. While my people are trying their best to defy nature, sometimes tripping and falling, across the world in the land known as America, people are driving hundreds of miles in their carbon dioxide emitting automobiles. We know. We know that some things such as sinking islands are inevitable in a world that may continue to heat up. One of my classmates put it very bluntly to me once in a seminar. Life goes on.

Life will go on, for Maldivians and for the rest of the world.

While we are not ready to acknowledge the devastating impacts of climate change, Americans are not ready to change their lifestyle either. In fact, most Americans, I dare say, are unaware of the existence of the Maldives. However, it is my hope that the conflict of interest would be resolved before it is too late. It is time for Americans and the rest of the world to wake up and become educated in a global sense and realize the global tragedy we face today. There are things to be done: energy conservation, funding research for more efficient energy sources, planting trees, and so on.

While the global attention is focused on the wars in the Middle East, another war is being fought in the little islands of the Maldives. A war against nature, the most powerful enemy there is. Who will help us? Who can help us? Who is going to switch off their lights and drive a hybrid? Who is willing to do research and find new ways of adapting? Or even ways of relocating?

In a world without edges it is the responsibility of the international community to aid those forced to be in the frontline of the battle against nature, especially when we, the Maldivians, are among the least responsible for causing climate change.

We may be the first to go down in history as people who lost an identity to nature. But, we may not be the last.

Source: "That Sinking Feeling," by Fathimath Musthaq. *Earth & Sky*. Courtesy of Earthsky.org.

we encountered with the Maya in Chapter 4: when the people who are able to control things do not get feedback, the whole system can go off the rails. As we will develop further in Part III, people *can* adjust quite rapidly in some circumstances, as we did when energy prices rose dramatically in the 1970s.

THAW

As global climate responds to increased concentrations of greenhouse gases, the climate models predict that temperatures in polar regions will rise faster and by a greater amount than in tropical and temperate zones. Evidence supporting this prediction can be seen in changes to the physical geography of polar regions. These changes are leading to the retreat of the ice caps and the retreat of ice in the Arctic Ocean. This has dramatic implications for rising sea levels and for international rivalries over ocean areas that are increasingly open to travel, fishing, and oil exploitation.

An impressive example of changes in the polar environment came on an Antarctic ice shelf known as Larsen B, which was located on the eastern shore of the Antarctic Peninsula, the piece of Antarctica that extends northward toward Chile. Figure 7.5a (on page 170) is a satellite image of Larsen B at the end of January 2002. The map overlay shows that the area is a bay covered by solid ice that is about the size of the state of Connecticut. (On fast roads, one can drive across Connecticut, about 120 miles, in about two hours.)

Thirty-five days later, on March 5, 2002, a new satellite image showed that Larsen B was gone (Fig. 7.5b, page 170). It had been there for thousands of years. What remained was a collection of icebergs, some about as large as Hartford, the capital of Connecticut. The breakup of the ice shelf opened an area of more than 1,200 square miles, nearly the size of Rhode Island. It was the largest ever seen over such a short time. It is one of the most dramatic manifestations of climate change observed so far.

On the other side of the world, the North Pole has seen a rapid retreat of the ice-covered area of the sea (Fig. 7.6, on page 171)—far larger in extent than Larsen B—in the last decades of the twentieth century. This retreat is unfolding over decades instead of a month. The loss of sea ice affects polar bears, which use the ice cap as hunting habitat. In the United States, polar bears have been listed as threatened under the Endangered Species Act, although it is not yet clear what that status will mean in practical terms. A study of a range of climate models projects that polar sea ice will vanish entirely in the summer by 2100. There is now sometimes open water at the pole. The last time open water appeared there regularly was before the last Ice Age, tens of thousands of years ago.

With such dramatic effects stemming from a well-understood response of nature to human activities, one would think the case for action is clear. It is. The problem is that the kind of action needed sounds drastic, because the world economy as it is currently structured is premised on access to cheap fossil fuels.

FIGURE 7.5
(a) The Larsen B ice shelf on January 31, 2002, and (b) on March 5, 2002.

(a)

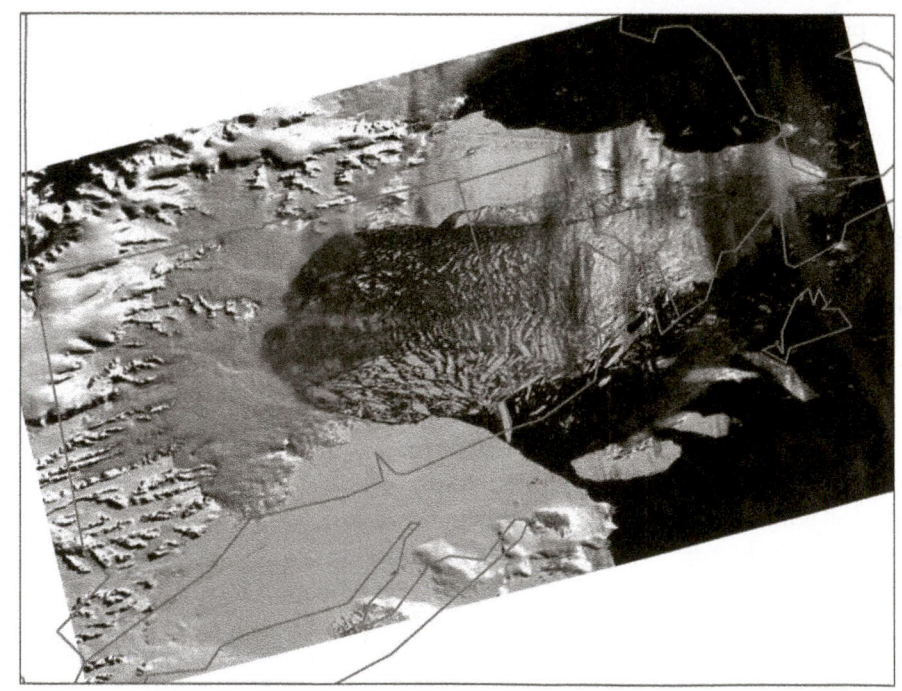

(b)

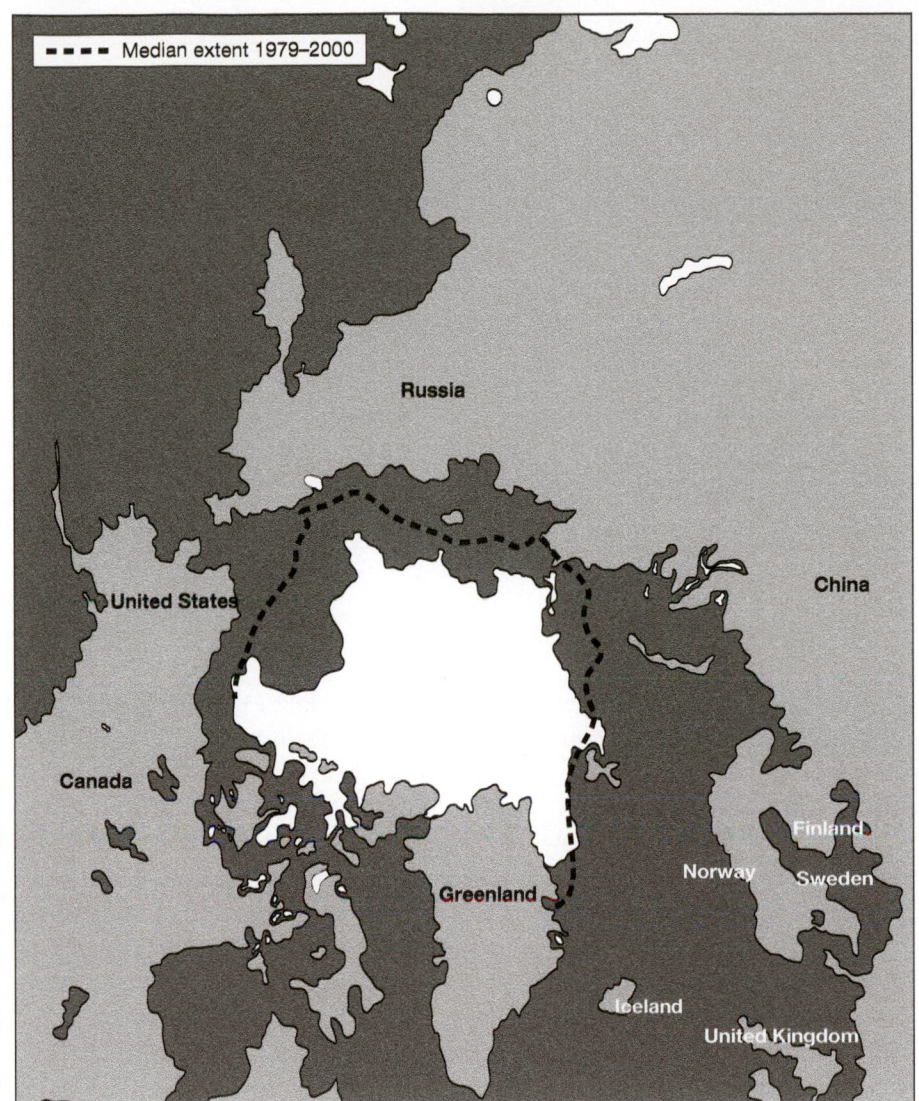

Median extent 1979–2000

Russia

United States

China

Canada

Finland

Norway

Sweden

Greenland

Iceland

United Kingdom

FIGURE 7.6
The extent of polar sea ice at its minimum in 2009. The dotted line shows the median extent of the sea-ice minimum for the last twenty years of the twentieth century. Note that Norway and Sweden would fit into the new area of open water.

ENERGY AND THE CONTROL OF NATURE

The most successful species learned to capture and control the energy of the Sun with the invention of agriculture ten thousand years ago. Sunlight is captured by crop plants and some of that energy is used by humans, either directly or indirectly, when plants are fed to animals consumed by people. More than two hundred years ago, we learned how to efficiently mine and utilize the energy in coal

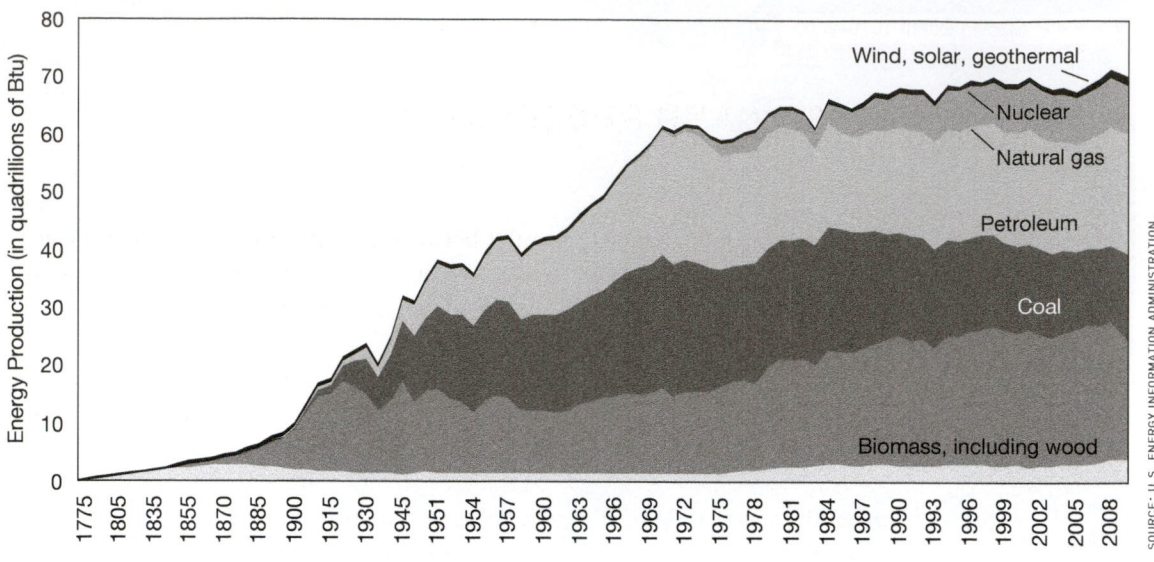

SOURCE: U.S. ENERGY INFORMATION ADMINISTRATION.

FIGURE 7.7
U.S. energy production, 1775–2009, by source. The different sources are added to one another, so that the bottom line shows biomass (wood, up until 1945), while the next line adds coal production, and so forth.

and launched the Industrial Revolution. Figure 7.7 shows the striking increase in energy consumption, beginning in the early nineteenth century, as humans learned to harness energy from sources beyond their own bodies and those of domesticated animals.

The industrial economy draws its energy mainly from fossil fuels, though nuclear energy and hydropower are quite important in some nations. Fossil fuels are mineral resources that were once living things: coal is formed from plant materials such as ferns; oil and natural gas are also formed largely from the remains of living things. The major deposits of fossil fuels being exploited today formed more than 100 million years ago, many times longer than humans have walked the Earth. Like all of today's living things, fossil fuels are the product of the food webs that begin with photosynthesis, the transformation of sunlight into chemicals useful for life. Fossil fuels are sunbeams from long ago.

The harnessing of fossil fuels is linked to the development of **thermodynamics**, the science of heat. Thermodynamics provided a conceptual framework for turning heat into work. The basic principle is a simple one: things expand when heated, and the burning of a fuel turns a solid, such as coal, or a fluid, such as natural gas or heating oil, into a hot gas. The hot gas pushes things, such as the blades of a fan (called a turbine when it is designed to be spun by hot gas) or the piston in a car engine. In this way, flame becomes force, and heat can be transformed, in part, into other, more useful forms of energy, such as the ones we find today in engines, computers, and air-conditioners. Turning burning oil into cool air from

BOX 7.2

ENERGY AND PROSPERITY

One indicator of the basic connection between energy consumption and the economic circumstances of people may be seen in the cartogram below, which distorts the area of each of the world's countries in proportion to its fuel use.

In this map, the rich nations, including the United States, Japan, and the countries of the European Union, have areas much larger than they would have on a normal map. The poor countries, including most of sub-Saharan Africa and much of Latin America, appear shrunken.

A similar point is made in the graph, which compares energy use to **gross domestic product (GDP) per capita**. As one might expect, a clear association is evident between income (measured by the average economic output per person) and energy use.

The biological machinery of the human body runs at about 100 watts. The average American's energy use—more than 10,000 watts—is about 100 times that of a person, whereas a resident of Argentina uses energy at a rate equal to about twenty people. One might think of this in terms of having (imaginary) "energy

Fuel use cartogram.

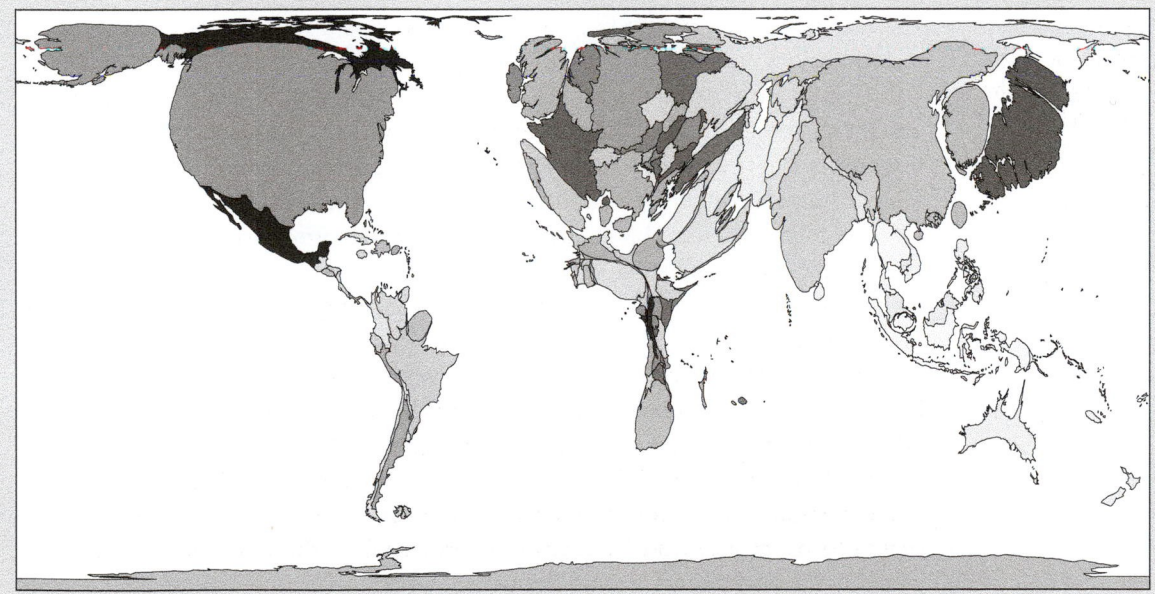

SOURCE: SASI GROUP (UNIVERSITY OF SHEFFIELD).

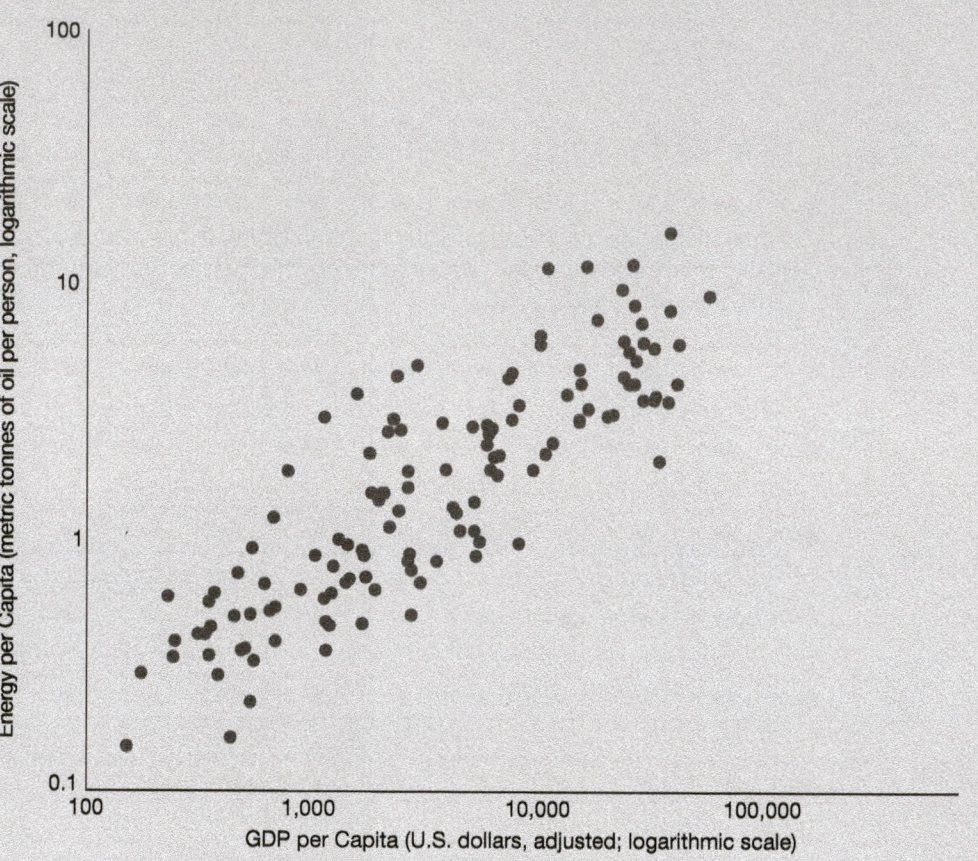

SOURCE: WORLD BANK.

GDP per Capita (U.S. dollars, adjusted; logarithmic scale)

Relationship between GDP and energy use, 2004.

servants." Americans have 5 times as many energy servants to do their bidding as Argentinians. Yet even in poverty-stricken Africa, energy from nonanimal sources is significant. Today, worldwide, 90 percent of that energy comes from fossil fuels.

The table also reminds us of something every visitor to a poor nation has experienced: someone in a rich country takes for granted many things, such as operating a washing machine, that are far from the experience of most people in a poor one.

Yet although a clear correlation exists between energy use and economic welfare, the association is not a rigid one. Japan has a GDP per person close to that of the United States, but a Japanese person uses energy at less than half the rate that an American does. The difference reflects in part the much larger size of the United States and our consequent use of more energy for transportation of goods and people. Another significant factor is differences in life-style. For example, most American residences are much larger than Japanese ones, and few Japanese houses are centrally heated. The differences among nations also suggest significant opportunities for lowering energy use that would not require decreases in health and well-being, though they would mean changes in life-style.

HUMAN-ENERGY EQUIVALENT OF TOTAL ENERGY CONSUMPTION, BASED ON AN AVERAGE HUMAN METABOLIC RATE OF 100 WATTS (APPROXIMATELY 2,000 CALORIES PER DAY, OR 3 MILLION BTU PER YEAR).

	Primary energy consumption, 2006 (million Btu/person)	"Energy servants" (1 human = 100 watts)
World	72.4	25
Africa	15.9	5
United States	334.6	115

Source: U.S. Energy Information Administration, International Energy Annual 2006, Table E.1c (World Per Capita Total Primary Energy Consumption [Million Btu], 1980-2006), www.eia.doe.gov/iea/wecbtu.html.

an air-conditioner sounds like magic, and if you think about it, so does the transformation of a lump of coal into megabytes of a music video flowing over the Internet. These examples, multiplied dozens of times in each person's day, are the result of the engineering knowledge built upon thermodynamics during the past two centuries.

Although the concepts of thermodynamics are used to control heat from all sources, the highest concentrations that could be tapped economically came from fossil fuels. It was only in the last third of the twentieth century that the even higher concentration of energy in uranium could be harnessed economically in commercial nuclear power. By then, coal, petroleum, and natural gas—burned in relatively simple furnaces or chambers such as the ones in the engine of a car—supplied more than 90 percent of the energy used in the world's richest nation.

This enormous array of technological capabilities made possible by harnessing energy led to the industrialization that dramatically increased material wealth in many economies (see Box 7.2: Energy and Prosperity, page 173). This connection can be directly visible. We can turn up the thermostat dial to warm our house in winter, enabling us to live in a cold climate through the consumption of energy.

Commuters feel the pinch of rising gasoline prices, reminding them how much they rely on affordable energy. A neglected power mower won't start, so a homeowner perspires behind a push mower to cut the grass—a reminder in another way of the gains we derive from energy.

These connections are also indirect: food in supermarkets has traveled, on average, more than a thousand miles from farm to shopping cart; nearly all plastics are manufactured from petroleum; and the more than one billion personal computers in the world depend on a reliable supply of electricity, including those that are portable. Some of these connections have significant environmental consequences: contaminated water in urban Africa can't be pumped or treated because of a lack of machines and reliable electric power; the mining of coal allows rainwater to percolate into rock crevices, forming acids and other pollutants that leach into streams, killing fish and sometimes whole ecosystems; the burning of fossil fuels releases gases into the atmosphere that are causing glaciers to melt and altering the chemistry of the oceans.

These examples could be multiplied many times over. Life in an industrialized economy such as Spain is obviously different in material terms from life in a developing country such as El Salvador. The differences can be traced in nearly every instance to the harnessing of energy.

ENERGY SOURCES

The energy we use comes from a variety of sources. As you can see in Figure 7.8, fossil fuel still provides about 85 percent of the energy used in the United States. Notice that the estimates of energy use in residential and commercial buildings, as well as in the industrial sector, exclude the consumption of electricity, which is reported separately. Of course, electricity is used almost entirely in industrial, residential, and commercial applications. The opportunities for generating and using electricity more efficiently are rather different, however, from those in the other three large end uses. All of those involve the direct harnessing of combustion, whereas electric energy is produced in turbines driven by hot gases—or by wind or falling water in a dam—and then distributed to end users.

Renewable energy sources, such as wind and hydropower, also draw energy indirectly from the Sun. As we discussed in Chapter 5, the heat of the Sun drives the weather, propelling wind and lifting water from the oceans up into the mountains through the precipitation cycle. Dams have been built on many streams to intercept the water flowing downhill in order to capture the water's energy to turn turbines.

Another renewable source is plant material. Wood has long been a major source of energy for humans, and today, crops are being grown and new crops are being developed to provide fuels to burn in engines and furnaces. To date, efforts to

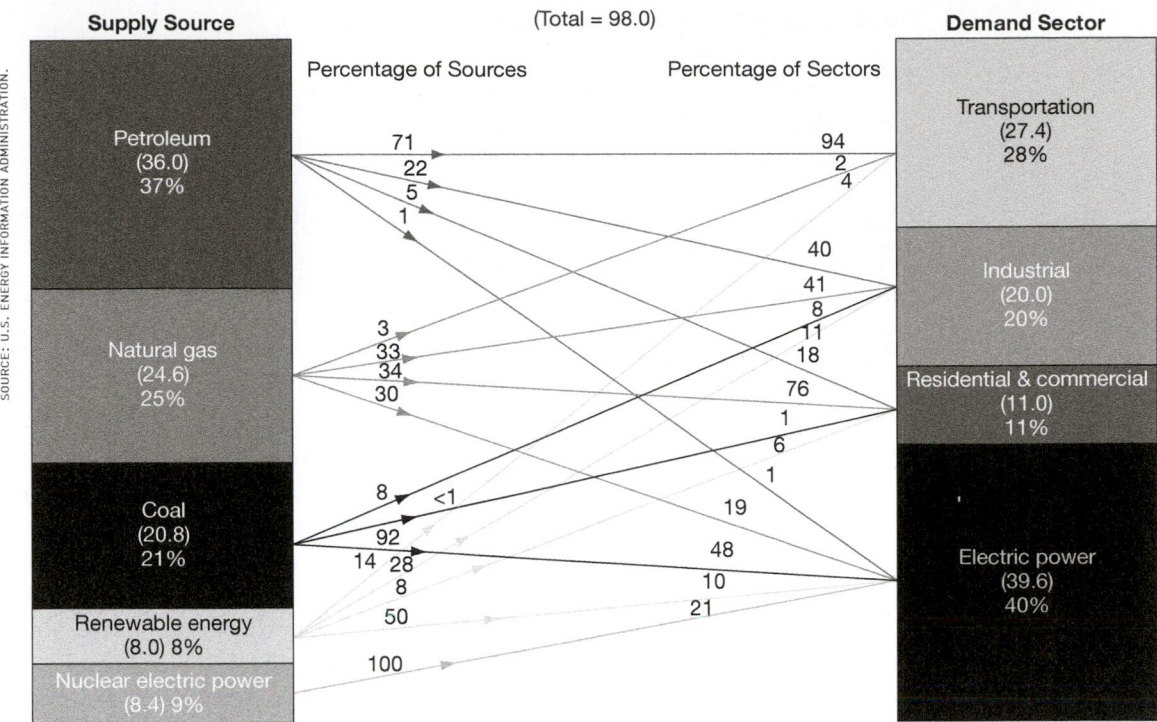

develop fuels from crops have suffered from significant problems, including the fact that growing crops can require substantial inputs of fossil fuels, but in the future, this may be a promising approach to developing renewable sources of liquid fuels for vehicles and aircraft. When plant materials are burned, they release carbon dioxide, but an equivalent amount was taken out of the atmosphere when the plant was growing, so the whole cycle should be roughly neutral in greenhouse gas emissions in favorable circumstances.

Sunlight can also be captured in photovoltaic cells—semiconductors somewhat like those in computer chips and flat-panel monitors—which convert light directly into electricity. Photovoltaic cells are expensive to manufacture, but the cost of production is being lowered enough that this approach may become competitive with more conventional approaches. The quantity of solar energy falling on Earth is roughly 10,000 times as large as total energy use by humans. The most efficient plants capture 3 to 10 percent of the sunlight falling on them, and the efficiency of photovoltaic cells can reach the low end of that range. From these numbers, one can see that, in principle, it would be possible to meet the world's energy needs not from ancient sunlight but from recent sunlight, using only renewable sources. One of the authors of this book, for example, installed photovoltaic panels on his

FIGURE 7.8
U.S. primary energy consumption, 2010. (Numbers in parentheses are in quadrillion Btu.)

roof, cutting his monthly electric bills by 99 percent; the initial cost was high, but with tax credits, the cost will be paid off in full in about twelve years. Still, the leap from a technology that can work in principle to one that becomes standard practice could be a long one for an economy built around inexpensive fossil fuels.

By far the most important fuel from an economic standpoint is petroleum. Changes in oil prices have tended to ripple through the markets for other fuels because oil can be used in residential, commercial, industrial, and transportation applications, and thus it competes against all other energy forms. As oil prices rise, other sources, including renewables, become more competitive economically. As you can see in Figure 7.9, the price of gasoline, the form of oil that people are most familiar with, has risen steeply over the past century. But if one takes out the effect of inflation, the cost of gasoline (and other forms of oil) actually fell over the course of that century. This has meant that alternatives such as solar energy have faced ever stiffer competition. This is the main reason fossil fuels remain dominant in the world's economies.

The most environmentally problematic fossil fuel is coal. Coal is mostly carbon, and when it is burned, it releases the most carbon dioxide of any fuel per unit of useful energy. So coal contributes more to climate change, proportionally, than other fuels. In addition, the impurities in coal are harmful to people and the environment. Coal often contains significant quantities of sulfur. When burned, the result is sulfur dioxide, a major contributor to health-threatening air pollution.

FIGURE 7.9
U.S. gasoline prices since cars became popular. The nominal price is the price charged at the pump; the real price is adjusted for inflation using the consumer price index through January 2012.

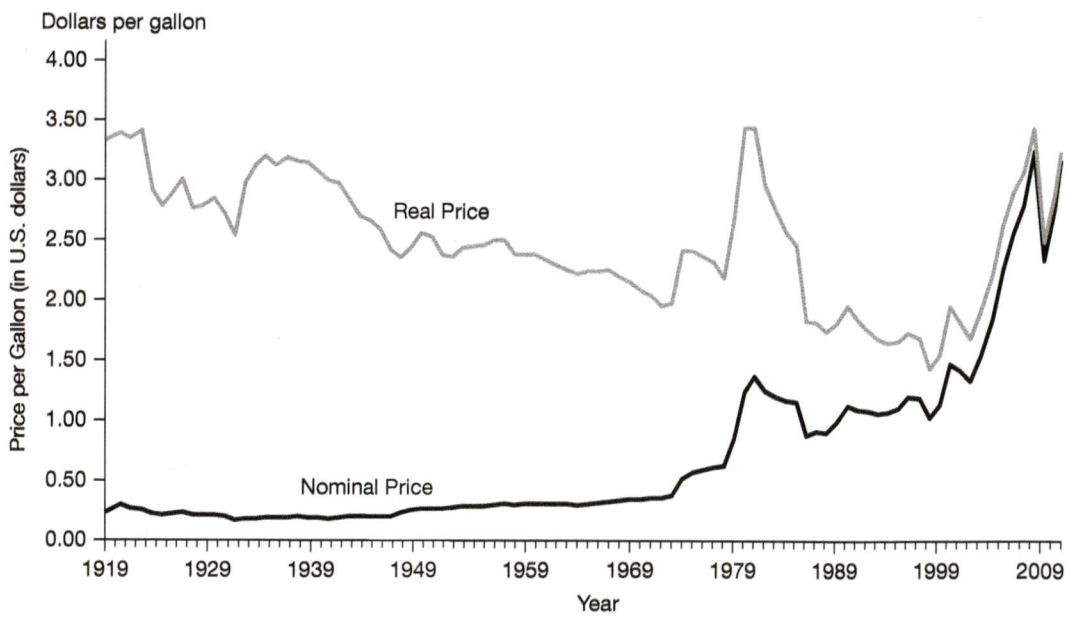

Coal also contains trace impurities such as mercury, and coal-fired electric power plants are a major source of the mercury generated by human activities. Mercury is a potent neurotoxin that can affect the development of children's brains. But coal is more abundant than any other fossil fuel. The world's largest reserves are in China, where electric power plant construction proceeds at a rapid pace to meet the demands of an industrializing economy.

There is one major alternative to fossil fuels that many environmentalists don't like to talk about—nuclear power. In routine operations, nuclear power plants are cleaner than fossil-fired ones per unit of energy produced. Nuclear plants emit no CO_2 because the steam they use to spin turbines comes from splitting uranium or plutonium nuclei to boil water. Atomic power also emits very little air pollution. Indeed, because of the impurities in coal, a properly operating nuclear power plant actually emits less radiation than an average coal-fired plant. Today, more than one hundred nuclear plants supply about one-fifth of the electricity used in the United States. France, the world leader in nuclear energy development, generates about four-fifths of its electricity from atomic energy.

As with every form of energy, however, nuclear power has risks, which were dramatically illustrated during the 2011 disaster in Japan. Although the Fukushima nuclear complex meltdowns were caused by a magnitude 9.0 earthquake and tsunami that was larger than had occurred in the recent past, the accident revealed frailties in the design of the whole system. Engineers had warned that an earthquake of this magnitude was reasonably likely to occur and that the reactors and spent-fuel storage systems were not designed to withstand such an event. Those particular problems will now be taken into account, but it is likely that other severe nuclear accidents—due to issues that are as yet unknown—will occur, just as plane crashes persist despite increasing improvements in safety and training. Worry about these risks led the German government to embark on a phaseout of nuclear energy in that country in 2011. This is having the ironic effect of increasing Europe's emissions of greenhouse gases, even though some European countries, including Germany, have been converting to renewable sources such as wind power far more rapidly than the United States.

An unresolved environmental concern of nuclear energy is waste disposal. When the nuclei of uranium or plutonium undergo fission, highly radioactive nuclear fragments are produced. The fission process also generates substantial quantities of other material that becomes mixed with and contaminated by radioactive elements. Although most of the radioactive elements decay rapidly, some long-lived radioactive constituents pose substantial biological hazards. These must be isolated for long periods—in some cases, for thousands of years. The geoscientists and engineers who have studied nuclear waste disposal think this is not an insoluble problem, and a number of designs with layers of safety features have been developed in prototype. But assuring the survival of human institutions to take care of these waste-isolation facilities for thousands of years is impossible, and controversy continues over how to seal and abandon a waste repository safely.

In addition to waste, making the fuel for nuclear power plants is a process that overlaps with the creation of ingredients needed for nuclear weapons. Trying to slow the proliferation of nuclear weapons while more nations clamor for nuclear power has proved to be a thorny political problem.

Asian economies are now leading the way with a new generation of nuclear power plants, so today's students will get to see how reliable they are. There may well be additional pressure to develop nuclear power as a way to generate electricity with fewer greenhouse gas emissions. But the cost of building and operating nuclear plants is high, and this factor may determine how many are built. In any case, the choices in coming decades will likely not be controlled by the critics of nuclear power, who have, for better or worse, played a key role in stalling the nuclear option for a generation.

The largest and least costly "supply" of clean energy is conservation. Technical improvements, such as better-insulated buildings, combined with smarter use of energy will give us more of what we want with less generation of greenhouse gases and other pollutants. **Energy efficiency** is a large but fuzzy resource, because estimating the technological potential for more efficient energy use is only a crude measure of what can be achieved practically; most estimates of technological potential suggest that 10 to 50 percent of U.S. energy use could be rechanneled into more efficient patterns.

Efficiency is not free, and installing and using more efficient methods can require some behavioral adaptations, such as learning how to install weather stripping around the edge of a drafty window. This is not hard and the friendly staff at your local hardware store will be glad to show you how. The problem lies in the number of windows and doors that need to be weatherstripped for the full potential of that particular improvement to be captured. There are literally millions of them, and most are in private dwellings whose occupants don't like the idea of energy inspectors dropping by. This is a social barrier, not a technological one, but it is not easily hurdled.

One way around the barrier is higher energy prices. If the cost of heating oil skyrockets, homeowners will have reason to economize on fuel. Greater energy efficiency is one way to do that while maintaining the warmth of a home during a cold winter. Higher prices, together with programs by utility companies and government agencies, have made significant contributions over the past generation. Another approach is to require more efficient technology in new buildings and in renovated ones. This can be done by provisions of building codes administered by local governments and followed by builders. Tighter windows, more insulation in walls, and other measures have gradually made their way into new and upgraded structures. Mileage standards for new cars provide a similar approach to raising energy efficiency. There is a long way yet to go, however, in the sense that energy users can still reduce their consumption substantially while saving money, even taking into account the cost of more efficient technology. We return to this question below, when we discuss international action to respond to climate change.

ENERGY AND THE WORLD WITHOUT EDGES

Energy resources are not distributed evenly throughout the world. Figures 7.10 and 7.11 provide a graphic demonstration of disproportionality in the ownership of petroleum. This fact has profound political and economic implications (see Box 7.3: Geology, Economics, and Politics, page 182).

Oil is a finite resource. As with all fossil fuels, oil is not being produced at a significant rate in the natural world. All the fossil fuels we pump, mine, and burn now may eventually be resupplied through natural processes, but these will take millions of years. Wood, by contrast, regrows fast enough that forests in the United States actually increased in area during the twentieth century. Sunlight pours in each day. It is accordingly relevant to ask how much of the different fossil fuels is available and how long we might be able to use them, particularly as they form the basis of the material comforts and the economy that we now have.

FIGURE 7.10
World oil reserves by country (cartogram). The total reserves for the top fourteen oil-producing nations are given in billions of barrels (Bbbl) and as a percentage of the world's reserves.

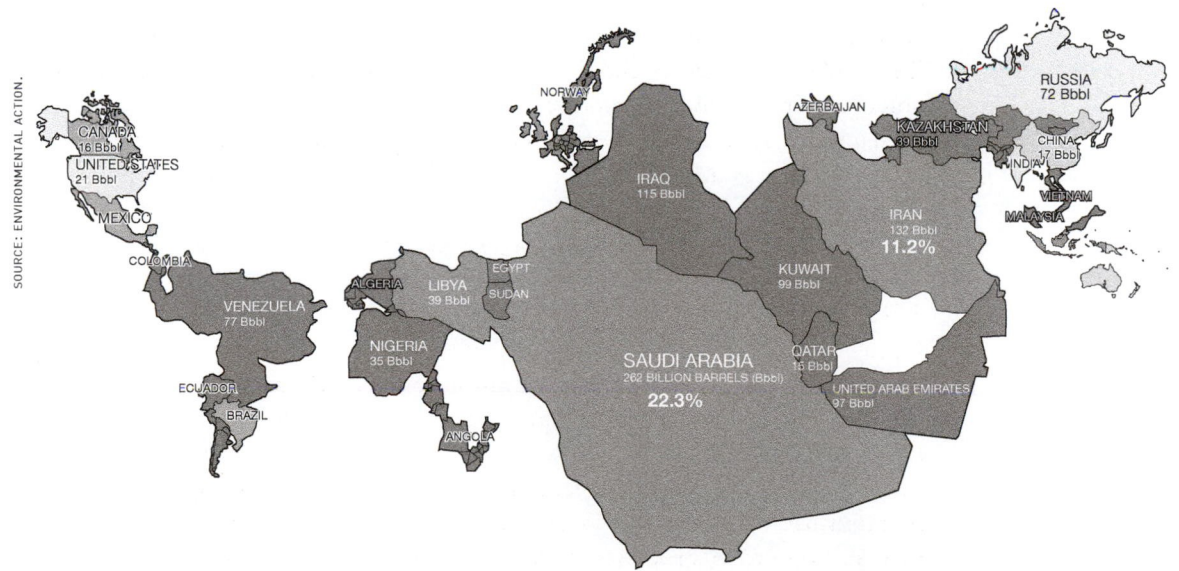

BOX 7.3

GEOLOGY, ECONOMICS, AND POLITICS

The cartograms in the "Energy and Prosperity" box and in Figure 7.10 show energy use and the distribution of energy resources, respectively, and they are strikingly different in appearance. That is, the places where the most energy is used are not the same as where it is produced. The figure below shows where oil imported into the United States comes from. Although well more than a third comes from Canada, Mexico, and western Europe, almost two-thirds comes from South America, western Africa, and the Middle East. America's principal allies in Europe and Japan are even more heavily dependent on imports of fossil fuels, with Russia supplying nearly all of the natural gas burned in Europe, as well as a large fraction of the oil.

U.S. oil imports, monthly average, June 2005 to November 2006.

The other major oil-producing countries include Venezuela; Nigeria and Angola in Africa; and Saudi Arabia, Iraq, and Iran in the Middle East. Each of these countries is a focus of intense concern. Several have been politically unstable, including Nigeria, Angola, and Iraq. Iran and Venezuela have been highly critical of America and its allies, and Russia and the United States have had a nervous and ambivalent political relationship since the Cold War ended in the 1990s. U.S.

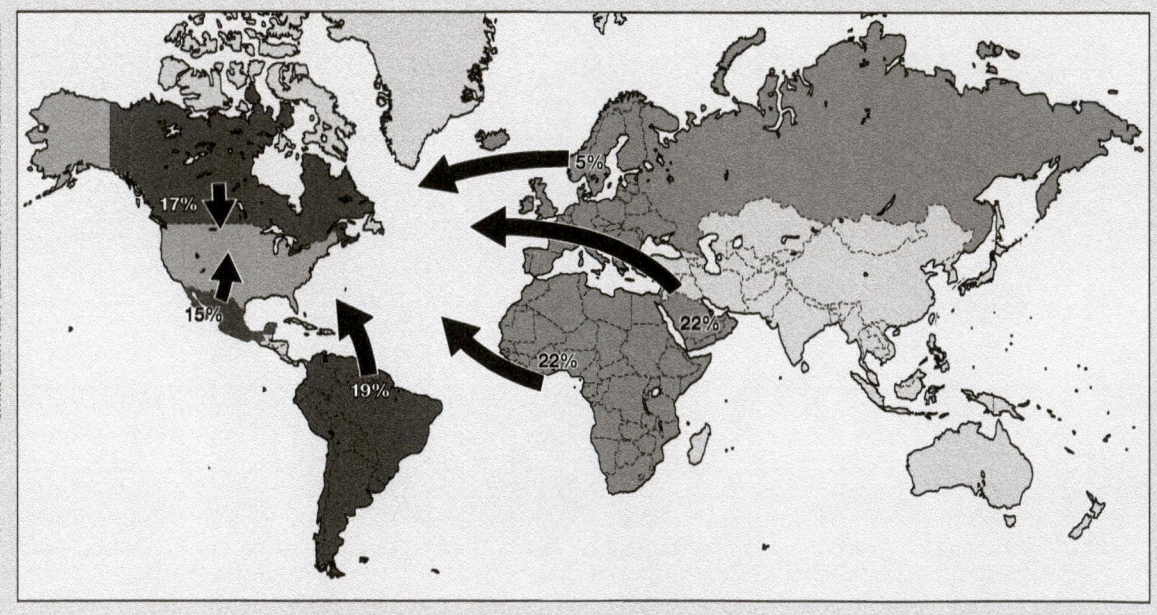

SOURCE: LUGAR ENERGY INITIATIVE, CITING DATA FROM U.S. ENERGY INFORMATION ADMINISTRATION.

national interests are tied economically to politically risky producers of fossil fuels, most visibly in the Middle East.

The relationship between energy and national security is one of the most important in the world without edges. In 1941, as the world slid into World War II, the United States instituted an embargo against Japan, cutting off that nation's access to petroleum. In response to this threat, the Japanese government launched a surprise attack on Pearl Harbor, which brought America into the war. What is perhaps surprising in light of that history is that few Americans today think about which countries supply the oil they pump into their vehicles. Some of those countries are not friends of the United States.

The question of how much fossil fuel remains turns out to be harder to answer than you might think. Figures 7.10 and 7.11 show **oil reserves**, rather than the total amount of petroleum contained below ground. The distinction is important. Reserves represent oil resources that have been physically discovered and made ready for production: the oil is in fields that have been drilled and people have a good estimate of how much oil can be produced economically from those fields.

FIGURE 7.11
Oil reserves by region.

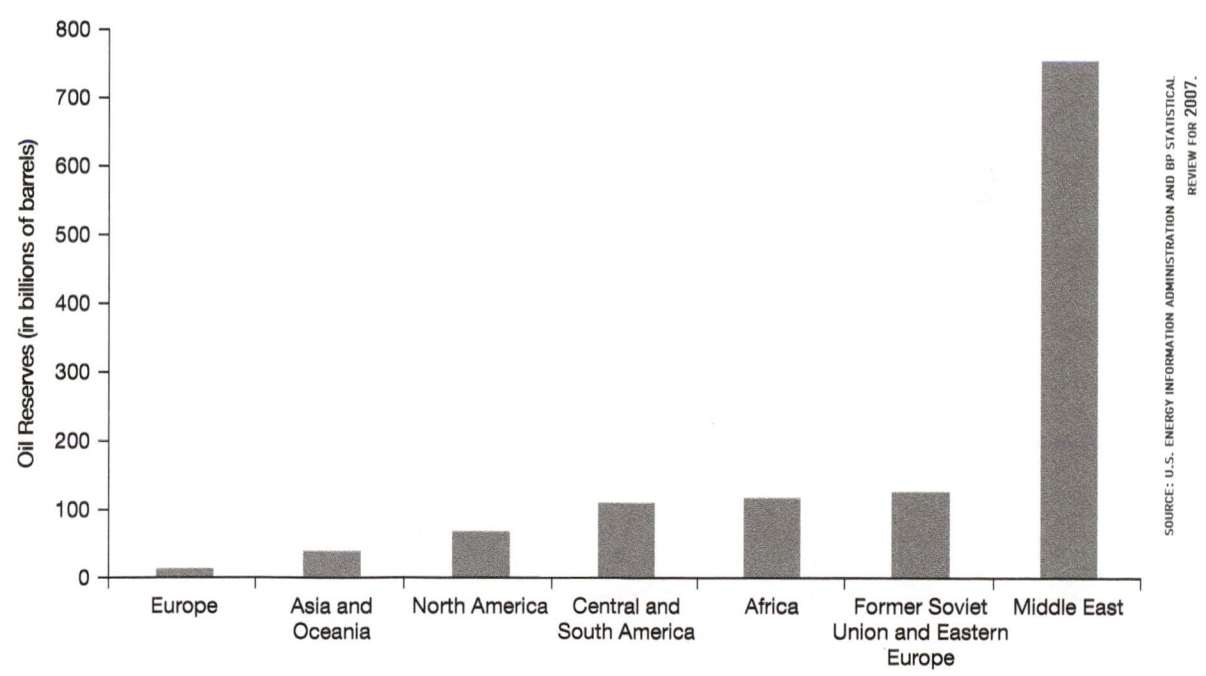

SOURCE: U.S. ENERGY INFORMATION ADMINISTRATION AND BP STATISTICAL REVIEW FOR 2007.

How much more is there? No one knows precisely, because it is likely that more will be discovered in the future. Energy firms and governments are spending hundreds of millions of dollars each year trying to find it; the *Deepwater Horizon* well that caused the largest oil spill in American history in 2010 was part of such an exploration program. Still, it is clear from studying the production data from a large number of countries that we are now in a period in which oil production is likely to begin to decline, as the reserves accessible to us at low cost are being depleted.

Figure 7.12 plots data from American oil fields, but it shows a pattern seen elsewhere, too. Oil production in the United States reached a peak in 1970 and has declined ever since. Because U.S. oil consumption has continued to rise, oil imports have climbed, though increased U.S. production temporarily halted this rise in the first decade of the twenty-first century. In 1973, a coalition of oil-producing countries called the Organization of Petroleum Exporting Countries (OPEC) temporarily interrupted the flow of oil to force price increases. The effect was dramatic: the world economy tipped into recession, and the resultant social

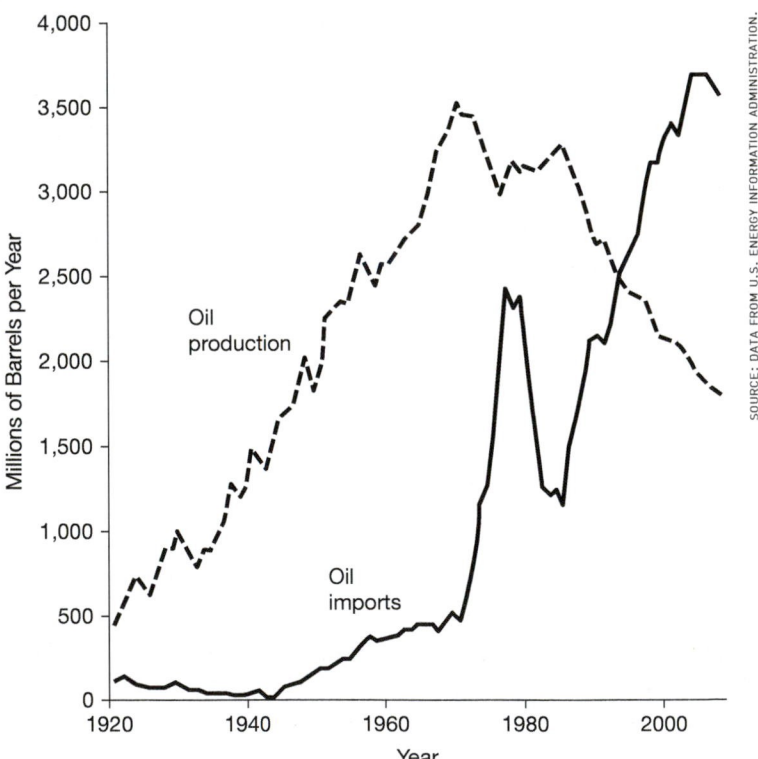

FIGURE 7.12
U.S. oil production and imports, 1920–2008.

crises stimulated both economic and political action. Eight years later, the discipline of the OPEC coalition cracked and oil prices sank. They did not rise to the heights achieved in the 1970s until early in the twenty-first century (see Fig. 7.9).

Throughout most of this period of turmoil and adjustment in world oil markets, U.S. production has dwindled. Kenneth Deffeyes, a geologist at Princeton University, is one of many experts who have concluded that the reason for this decline is that no more cheap oil can be obtained here.[3] The United States was the first nation to produce petroleum commercially, in 1859, and America has been the object of intense exploration activity ever since. So it may not be surprising that production has begun declining here first, too.

What is unsettling is that oil production has also peaked elsewhere. Deffeyes and many other geologists and energy analysts argue that the all-time peak in oil production may have occurred in the first decade of the twenty-first century. Remember the bet between Paul Ehrlich and Julian Simon that we discussed in Chapter 6, however. It is hard to pin down when we are running out of a finite resource, because markets are volatile and people are clever. New technologies and new energy sources such as wind power will enter energy markets. The economic logic is clear. As cheap energy declines, higher prices will make sources and technologies that are now unprofitable worthwhile. Yet there is no doubting our reliance on petroleum today, as shown in Figure 7.8, and it is worth wondering how smoothly energy consumers and governments will make the transitions that lie ahead.

The situation is similar with natural gas, which is found together with oil. But there is a far greater known reserve of coal. In addition, coal is distributed differently in Earth's crust, and the largest reserves are in China and the United States. The problem with coal is a different one, as noted earlier—burning coal releases more carbon dioxide per unit of energy than any other fossil fuel. China and the United States are also the largest contributors to global warming. Moreover, mining and burning coal generates large quantities of other pollutants that cause harm to humans and ecosystems unless controlled effectively; this has proven hard to do in many places.

TACKLING THE GRAND CHALLENGE OF CLIMATE CHANGE

Faced with the emerging evidence of climate change, nations are beginning to respond. This will be a long, painful adjustment—one that will unfold during much of your life and may even involve your career. There is much opportunity, as a result, for constructive change, from individuals to firms to governments and the wider global society.

The emissions of greenhouse gases are the exhalations of the industrial economy of the world: fossil fuels have created the world without edges. Oil, natural gas, and coal account for nearly all of the additions to the atmosphere. So thinking about how to reduce those emissions leads directly to deep changes, not only to the gasoline used in cars—the one form of energy Americans monitor closely because they so frequently fuel their vehicles—but also to manufacturing, transportation, and the heating and cooling of our living, shopping, and work spaces. Nearly all of our energy is generated and handled with machines such as furnaces and electric power plants that are costly to build and that last for decades. Even in a rich nation, we cannot afford to change overnight. In practical terms, the United States has yet to begin the process of change. And in China, a large new coal plant was being opened at the rate of one a week during the first decade of this century.

Not only will it be costly and difficult to reduce greenhouse gas emissions, the attempt to do so must overcome the fact that the atmosphere is a commons, one that may be damaged if the shepherds put too many animals on the pasture, to use the metaphor of Chapter 3.

Greenhouse gases emitted anywhere on Earth go into the atmosphere, where they circulate over the entire planet. This is why measuring CO_2 on a mountaintop in Hawaii (Fig. 7.2) can detect coal-burning electric power plants outside Shanghai, natural gas flared in Kazakhstan, and truck exhaust in Mexico. Among the shepherds with flocks on the pastures that are the world's atmosphere, some have sheep that eat a lot more grass than others. An American emits carbon more than twice as fast as a Japanese (see Box 7.2, page 173). As with other commons, it is not easy to secure agreement on who should go first and how severe the reductions should be.

Yet it is far from impossible for humans to control climate change. Although the world economy is deeply dependent on fossil fuels today, the cost of making large reductions in greenhouse gases is surprisingly affordable. Global consulting firm McKinsey & Company published an analysis in 2009 (see Fig. 7.13) that reaffirms the conclusions of other economic studies:

> Our analysis finds that there is *potential* by 2030 to reduce [greenhouse gas] emissions by 35 percent compared with 1990 levels, or by 70 percent compared with the levels we would see in 2030 if the world collectively made little attempt to curb current and future emissions. This would be sufficient to have a good chance of holding global warming below the 2 degrees Celsius threshold [deemed prudent by scientists].
>
> Capturing enough of this potential to stay below the 2 degrees Celsius threshold will be highly challenging, however. . . . A 10-year delay in taking abatement action would make it virtually impossible to keep global warming below 2 degrees Celsius.

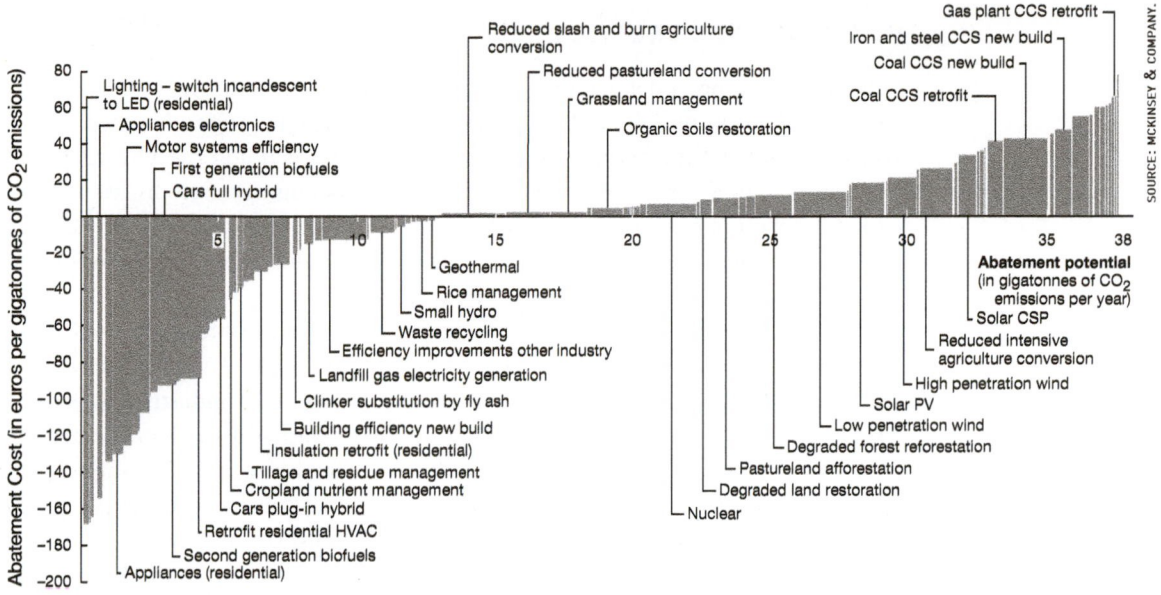

FIGURE 7.13

The estimated costs of different alternatives for reducing greenhouse gas emissions by 2030. This figure provides a large range of technical alternatives, including more efficient residential appliances and the deployment of solar photovoltaic (PV) technology (solar cells). An important point here is that many of the "costs" are below zero—that is, adopting such options would save money. This is true, for example, when a new energy-efficient refrigerator lowers electricity consumption enough to more than pay for its higher cost. (CCS = carbon capture and storage; CSP = carbon sequestration program.)

What would such an effort cost? We find that, if the most economically rational abatement opportunities are pursued to their full potential—clearly an optimistic assumption—the total worldwide cost could be €200 to 350 billion annually by 2030. This is less than 1 percent of forecasted global GDP in 2030.[4]

The McKinsey study is notable, in part, because it represents a statement from a leading voice in the business world. The costs of *failing* to control greenhouse gas emissions, meanwhile, were estimated by the British Chancellor of the Exchequer (roughly equivalent to the U.S. Secretary of the Treasury) to be between 5 and 20 percent of the world economy. That larger number comes close to the economic losses of the Great Depression of the 1930s. There is now significant awareness among large corporations that climate change is not only real but that the responses to climate contain business opportunities as well as threats.

INTERNATIONAL TREATIES

Politicians have been deliberating these matters for some time. In 1992 a Framework Convention on Climate Change was opened by the United Nations. This is a **treaty** that has been ratified by 189 nations so far, including the United States. The word "framework" means that the signatory nations agree to discuss a problem, but have not necessarily adopted any actions to respond to that problem. In effect, the framework convention is an agreement that identifies a commons to be governed and articulates principles to guide governance.

In 1997, a concrete step was taken under the framework—a treaty known as the Kyoto Protocol. The Kyoto Protocol came into force in 2004 with an expiration date of 2012. The Kyoto Protocol is a commitment by most industrialized countries to decrease their emissions of greenhouse gases, just as if some of the herders with larger flocks agreed to limit the number of animals they put into the pasture. These large herders are called Annex B countries. The Annex B countries agreed to commitments to reduce their greenhouse gas emissions below the level they were releasing in 1990. The developing countries that signed on to the Kyoto Protocol were exempt from reductions. Because they are developing nations, China and India were included in this exemption. This so unsettled critics in the United States, it eventually became the world's only developed nation *not* to ratify the Protocol.

In 2009, representatives of many nations gathered in Copenhagen, Denmark, to craft a successor to the Kyoto Protocol. They failed. The logic of the commons had proved too powerful. The search for a practical way to govern the global atmosphere continues in the international community.

The Kyoto Protocol is, so far, the high-water mark of international agreements to respond to climate change. But even this treaty would not have halted climate change, even in the long run. Because CO_2 can stay in the atmosphere for several hundred years, the effects of today's elevated levels of greenhouse gases will take time to work their way through Earth's biological and geological architecture. What the treaty targets would do, if met, is slow down the rate at which humans put additional greenhouse gases into the atmosphere—not halt them. But as the McKinsey analysis (see Fig. 7.13) demonstrates, the economic burden of slowing warming to an environmentally manageable level is affordable. This burden can only be lifted, however, by a world ready to deal with its grandest commons. This has yet to happen.

Of course, no climate policy being considered today would return the atmosphere to its composition before the Industrial Revolution. We are adding more and more animals to the pasture. Climate remains a grand challenge.

This is a very large opportunity for readers of this book, however. It is an opportunity in every professional field, from law to medicine to engineering. It is an opportunity in government and business—including businesses that do not yet exist. And it is an opportunity at every scale—from the individual consumer to the international system where agreements are now being negotiated. It is an opportunity that has tangible links to virtually every great public issue, from the politics of the Middle East, to the health of the economy, to how we should meet our obligations to one another in a world built around fossil fuels that cannot stay that way.

If enough of the people now in college or entering the job market seize these opportunities, climate and energy will indeed be the defining issue of a generation. The question is how many will do so, and whether good middle-class jobs await those who do. An opportunity to work on the solution of a vast commons dilemma such as climate change does not automatically mean a job that pays a reasonable salary and has health insurance. Part of a grand challenge is the challenge of human ingenuity. It must be feasible for a lot of talented people to get involved in the response.

During the Industrial Revolution, humans drove a hard bargain with nature. Through the harnessing of fossil fuels, we gained understanding and mastery that changed the human condition. The labor of an individual person could be multiplied by many times; the scientific discoveries that emerged in industrial societies transformed health, economic management, manufacturing, and the handling of information; and the spread of democracy altered the relationships between followers and leaders.

But the burning of fossil fuels is now altering the natural world in fundamental and pervasive ways. We do not fully grasp the trajectory that we have fashioned for ourselves and the world we inhabit. The shape of that path will be mapped, and then carved, as we face the grand challenge of climate change.

FURTHER READING

Archer, David. *The Long Thaw: How Humans Are Changing the Next 100,000 Years of Earth's Climate.* Princeton, NJ: Princeton University Press, 2009.

Kolbert, Elizabeth. *Field Notes from a Catastrophe: Man, Nature, and Climate Change.* New York: Bloomsbury, 2006.

National Research Council. *Understanding and Responding to Climate Change. Highlights of National Academies Reports.* Washington, DC: National Academies, 2006.

Oreskes, Naomi. "The Scientific Consensus on Climate Change." *Science* 306 (2004): 1686.

Stern, Nicholas. *Stern Review on the Economics of Climate Change.* London: Her Majesty's Treasury, 2007. Available at www.hm-treasury.gov.uk/independent_reviews/stern_review_economics_climate_change/stern_review_report.cfm.

KEY TERMS

anthropogenic radiative forcing

biogeochemical cycle

carbon dioxide

energy efficiency

fossil fuel

global warming

greenhouse effect

greenhouse gas

gross domestic product (GDP)

per capita

infrared radiation

oil reserves

sea-level rise

steady state

thermodynamics

treaty

Chapter Eight

HUMANS AND THEIR HABITATS

POPULATION GROWTH AND URBANIZATION

In Chapters 6 and 7 we began to see how humans modify their habitats, creating farms and cities that transform natural landscapes into human civilization, bringing forests and fishing grounds and mineral-rich terrain into the human economy, powered by fossil fuels to a large extent. In this chapter, we examine the growing numbers of our "most successful" species.

Since the rise of environmentalism about half a century ago, growth in human population has been a central concern of environmentalists. The reasoning is hard to argue with: people cause environmental problems; more people will cause more problems. In this chapter we examine this proposition and come to two conclusions that may surprise you. The first is that the rise in human numbers is a challenge, but not a grand challenge. This is one environmental problem that has a reasonably clear solution in view, though the solution does not look environmental. The second is that urbanization—the rapid rise in the number of people living in cities—is a grand challenge, a linked set of problems to which we have no clear solutions, even though the problems pose stark and basic threats to both humans and nature.

One way in which humans have been the most successful species is through the rapid growth in human population since the seventeenth century. As this book was being written, the human population crossed the 7 billion mark. The Census Bureau, the agency responsible for counting heads in the United States, has a population "clock" that provides an estimate of the number of people on Earth.[1] The clock is updated once a minute. In the time it takes to deliver a one-hour lecture, the number of people in the world increases by about four thousand. This means that, after a day of classes, an entire university's worth of potential students has been added to the human race.

This happens every day, with a net population increase of about 75 million people per year—the equivalent of five cities the size of Los Angeles. Figures like these have understandably led people to think human population growth is a mighty environmental challenge. Indeed, Garrett Hardin presented his model of the tragedy of the commons (see Chapter 3) in a discussion of population in which he articulated the need to regulate reproductive choices. Yet **demography**, the science of population, shows that the growth in human numbers has a distinctive history, one that suggests a rather more complex—and unexpectedly optimistic—appraisal of population growth in the generations ahead. In fact, today's students may live to see human population growth come to a halt, and their children are very likely to see that watershed event.

A look at Table 8.1 suggests why Americans may be surprised by world population growth. We live in a place that is not like other parts of the world. As you can see, the major concentrations of our species are in Asia, a region unfamiliar to most Americans. The density of population in Asia is about 8 times higher than in North America. Africa is where the world's most rapid population growth is occurring. That region is also one few Americans have visited.

Learning Objectives
When you have finished studying this chapter, you should be able to

---

↘ look for evidence of the demographic transition in the history of your own family;

↘ notice differences in the age structure when you travel to other countries;

↘ evaluate arguments that population growth by itself constitutes an environmental threat;

↘ recognize the way that infrastructure, such as water supply pipes, provides ecosystem services for the community in which you live;

↘ look for the influence of rapid urbanization in international news of business developments, civil unrest, natural disasters, or famine.

TABLE 8.1 WORLD POPULATION ESTIMATES, 2010. TOTAL FERTILITY RATE IS THE AVERAGE NUMBER OF CHILDREN BORN TO EACH WOMAN, IF CURRENT TRENDS CONTINUE.

	Population estimate, 2011 (millions)	Growth rate, 1980–2011 (%/year)	Total fertility rate, 2005–10
World	6,974	1.45	2.54
Africa	1,046	2.49	4.64
Asia	4,207	1.51	2.28
Europe	739	0.21	1.53
Latin America and the Caribbean	597	1.61	2.30
North America	348	1.01	2.03
Oceania	37	1.55	2.49

Source: UN Department of Economic and Social Affairs, Population Division, World Population Prospects: The 2010 Revision.

THE DEMOGRAPHIC TRANSITION

Until about 1650, human populations did not grow steadily, but experienced major setbacks in times of epidemic disease, warfare, and ecological disasters such as crop failures due to drought. Beginning in Europe in the mid-seventeenth century, and then gradually throughout the world, a new pattern took hold. Human populations grew, almost without interruption. By 1800, people began to worry that surging human numbers would outstrip the ability of agriculture to feed them. Yet now, two hundred years later, what we are seeing is a striking slowdown in population growth—one that is unfolding, with painful exceptions, without the catastrophes of hunger, plague, or war. The historical pattern that has emerged is called the **demographic transition**. This transition describes humans' ability to slow population growth, a surprising and perhaps essential element of what it takes to be the most successful species.

The demographic transition is a simplified story, or **scenario**, of the way human populations have grown over the past four centuries (Fig. 8.1). Before the transition began, populations were low, with high birth rates and high death rates. Many infants and children died before they could reproduce themselves. Then the death rate decreased as people gained access to better food and clean water, and the population rose rapidly. Later, the birth rate decreased, too, as preferences for

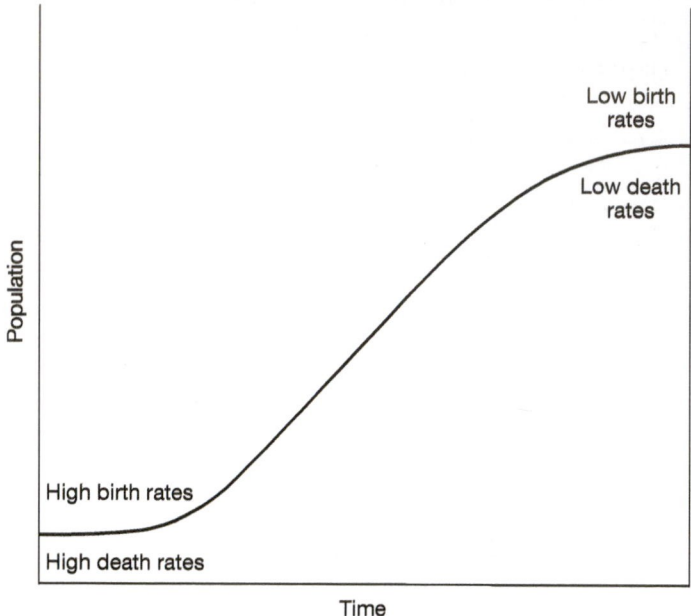

large family size decreased, so that the growth of population slowed and headed toward stability. At the end of the transition, birth rates and death rates are low, and the population is roughly stable; however, the total population is much larger, having grown through the transition.

Figure 8.1 shows a simple relationship: the population is increased by births and decreased by deaths. When the birth rate is higher than the death rate, the population grows, and when they are roughly equal, the population is roughly constant. Within any community or nation, however, there is another way for population to change—by migration of people into or out of the territory. With some exceptions, however, migration has been a relatively small factor numerically, so we will not discuss migrations here, even though migration has sometimes been prominent in environmental thinking about population growth.

The demographic transition is a creation of theorizing, like the tragedy of the commons. As with other influential models, the demographic transition organizes a lot of history simply and insightfully. But it is not a law of nature, and the scenario does not fit historical evidence perfectly. Bear that in mind as we go through the transition scenario.

Before the transition begins, population is growing slowly or not at all. Large declines occasionally happen, such as during the Black Death pandemic in medieval Europe. Figure 8.2 shows the population of Egypt, a country where historical records reach far back into ancient times. From tax records, for example, it is possible to make educated guesses about the population size, and these can be checked against travelers' accounts or other estimates. Drawing together a number of accounts and studies, the demographer T. H. Hollingsworth inferred that the population of Egypt had experienced large changes over time, many tied to war, disease, and climatic events such as crop failures.[2] The fluctuations are large, but with no clear long-term trend of increase until the beginning of modernization in the early nineteenth century.

Egypt before 1800 typifies the situation before the demographic transition began. Birth and death rates were both high. Average life expectancy was roughly forty years, but most of this was due to high mortality among infants and small children:

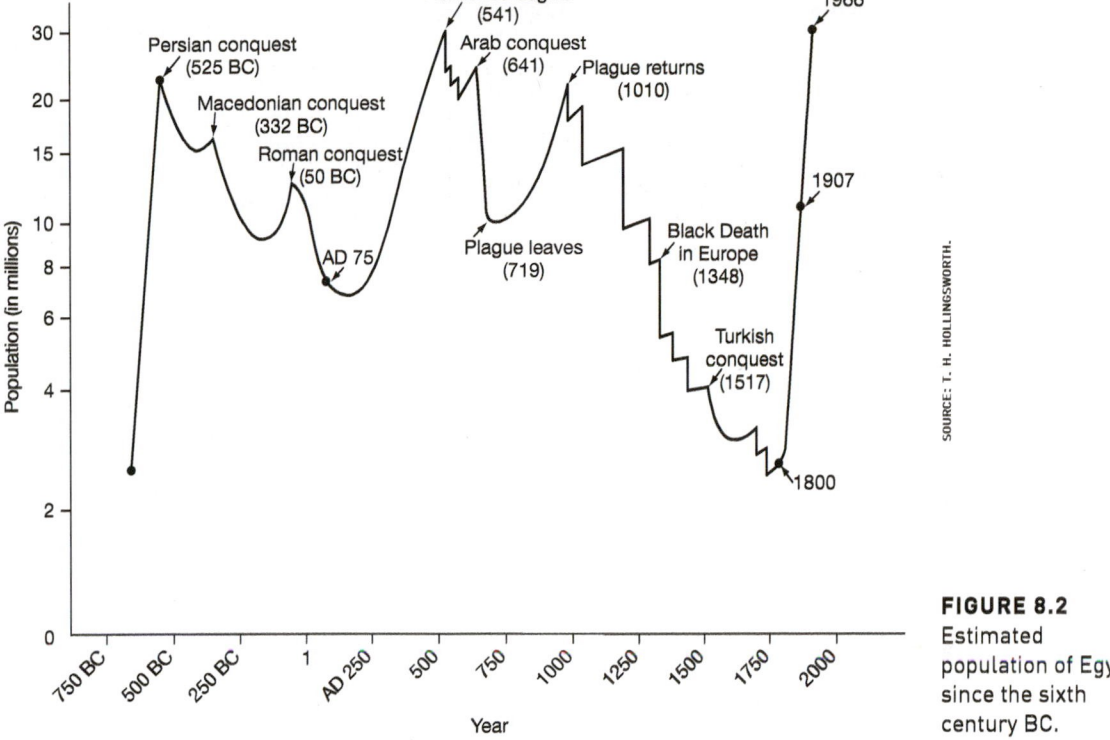

FIGURE 8.2 Estimated population of Egypt since the sixth century BC.

SOURCE: T. H. HOLLINGSWORTH.

roughly half of all children died before five years of age. Family sizes were large, with women giving birth to four to seven children. Once past the hazards of childhood, life expectancy commonly ranged up to seventy, a life span recorded in the Bible as threescore and ten. This was a world in which human experience was different in important respects from the world we live in today in rich countries such as the United States. However, in sub-Saharan Africa, many children still die young and families are large. This was true in Asia and Latin America, too, until recently.

FALLING DEATH RATES AND POPULATION GROWTH

Beginning in the seventeenth century, populations in Europe and then North America took off. The reason was a sharp decline in the death rate, especially among children. For some time after death rates fell, birth rates remained high. This meant that populations continued to grow, as births outnumbered deaths decade after decade.

Why did death rates drop? First, children were better fed, which gave them a better chance of surviving illnesses. The introduction of crops from the New World into European agriculture led to rapid increases in agricultural output. Maize (corn) and potatoes could be grown on land that would not support wheat or rice. Larger crops pushed down prices, which made it harder for the more numerous young people to stay to work the land. Thus, rising food production drove people out of agriculture and into the cities to seek work.

The crowded cities were rife with diseases that spread more quickly in dense populations, but clean drinking water and improved control of sewage during the nineteenth century dramatically lowered death rates from epidemic disease.

Finally, scientific medicine greatly expanded humans' ability to deal with infectious diseases. Interestingly, medicine based on science emerged only after the demographic transition was well under way. In the nineteenth century, mass vaccinations were instituted against smallpox and the childhood diseases that had traditionally taken a high toll—diptheria, typhoid, whooping cough, and others. Other infectious illnesses, such as tuberculosis, were driven into the shadows by antibiotics in the twentieth century.

The fear once inspired by tuberculosis, cholera, and other infectious diseases can be seen now in the social response to HIV/AIDS, an infectious disease whose spread can be slowed via education and public health programs, and whose management (not cure) can be improved but at high cost. Moreover, tuberculosis is now reemerging as a threat, as inappropriate use of antibiotics has led to the unintended creation of resistant strains of the bacillus that causes the illness.

FALLING BIRTH RATES AND SLOWING POPULATION GROWTH

The record of rising survival rates due to public health goes back, for Americans, to the time of your great-grandparents—too long, now, for most to have family stories. So let one story stand for many. Elizabeth Pollard (a member of the family of one of this book's authors) was born in 1883 in central Missouri, one of eight children. She married and had four children by 1920, when her husband Ralph Hammond died at the end of a great influenza epidemic, which killed 40 million people around the world. Her daughter Virginia, born in 1910, had three children, one of whom, a daughter born in 1950, had two children, born in 1977 and 1980. In a century, the number of children per woman in this family dwindled from eight to two.

This pattern of smaller numbers of children in successive generations is one seen throughout North America, Europe, and Japan—the countries that are now rich. These countries all experienced sharp declines in birth rate as per capita incomes rose with industrialization. As a result, population growth in the United States as a whole has slowed dramatically over the century since Elizabeth Hammond was born. Consider your own family's history over the past four generations in this light.

This pattern is intertwined with changes in the status of women. Women began to have fewer children, motivated in part by the improvement in public health. Mothers could count on their children surviving to adulthood, so they did not need to bear so many to ensure having descendants who would care for them in old age or carry on their family's legacies.

The changes in population growth came in the wake of urbanization. On a farm, children could do chores and help with planting, harvesting, and caring for animals. In cities, having large numbers of children was not an economic advantage, however. Competition in urban societies came to center on education—so it has made sense to invest in the schooling and nurture of a small number of children. It was increasingly advantageous for women to gain an education, which led to a delay in marriage and the birth of the first child, further decreasing family size on average. Moreover, with social insurance programs like Social Security providing a safety net for retired people, it has become less important for people to rely on their children in old age.

All these factors have in turn dramatically widened opportunities for women. In rich societies, women now commonly work when their children are young, increasing family income. With greater opportunities, family size has fallen further, so that in many European nations and Japan, the population is projected to decrease substantially during this century.

The demographic transition (Fig. 8.3 on page 198) is a change from a situation of high birth rates and high death rates, to one of low death rates followed by low birth rates. At the outset, one sees societies like that of Gilbert White, the "stationary" man whom we met in Chapter 2. People made their living from farming and were rooted in traditional, place-bound social patterns. The early phase of rapid population growth is associated with technological changes, in nutrition and health, associated with the initial driving down of death rates. Technological changes in the form of industrialization and improvements in overseas transportation, in turn, led to urbanization and, in Europe, to the creation of international colonial systems. The values of traditional societies changed, as both economic and social arrangements gave way to the often disorderly onslaught of industrialization and city life. This was the time in America of unions, of great industrial fortunes like that of the Rockefellers and Fords, of jazz, and of race riots and ethnic politics in cities. Today we are seeing social turbulence in Brazil and China, as those nations go through significant changes of their own.

And everywhere, even in poor countries, families are choosing to have fewer children. This is a product of public policy in China, which has had a one-child

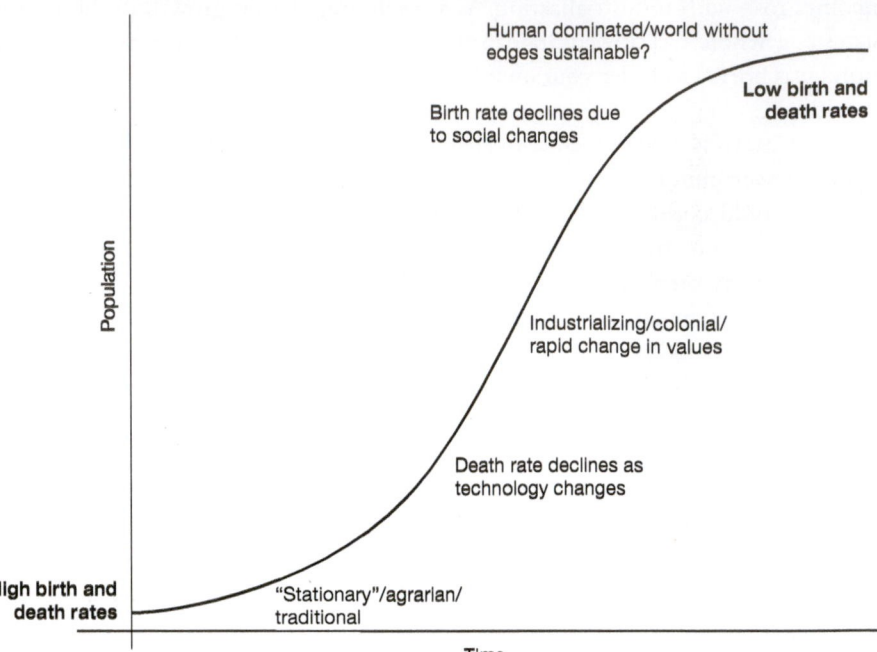

Human dominated/world without edges sustainable?

Low birth and death rates

Birth rate declines due to social changes

Population

Industrializing/colonial/ rapid change in values

Death rate declines as technology changes

High birth and death rates

"Stationary"/agrarian/ traditional

Time

FIGURE 8.3
The demographic transition, with phases related to broad phases in human history (see Chapters 2, 4, and 6).

limit on family size for a generation. Elsewhere, the decline in family size has been voluntary, but it has been marked and rapid. The later phase of slowing is associated with social changes, including the change in the status of women, in the lowering of birth rates. The rapid rise of population in the early phase means that the total population at the end of the demographic transition is much larger than it was at the beginning.

WHERE IS POPULATION HEADED?

The transition scenario captures the history of human populations reasonably well over the last 350 years. The trajectory of rapid growth due to falling death rates, followed by slowing growth from falling birth rates, was first observed in Europe and then in North America and Japan. During the second half of the twentieth century, the demographic transition also has emerged in most of the rest of the world. Figure 8.4 shows how this has happened quantitatively over the past sixty years. The projections from about 2000 forward show the patterns projected by demographers, the scientists who study human population.

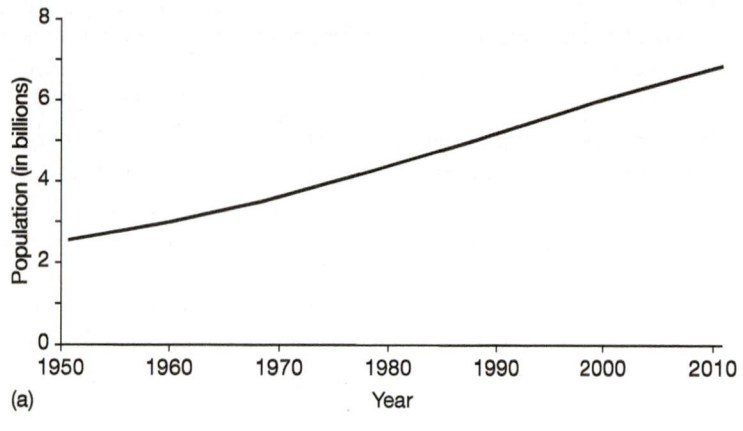

(a)

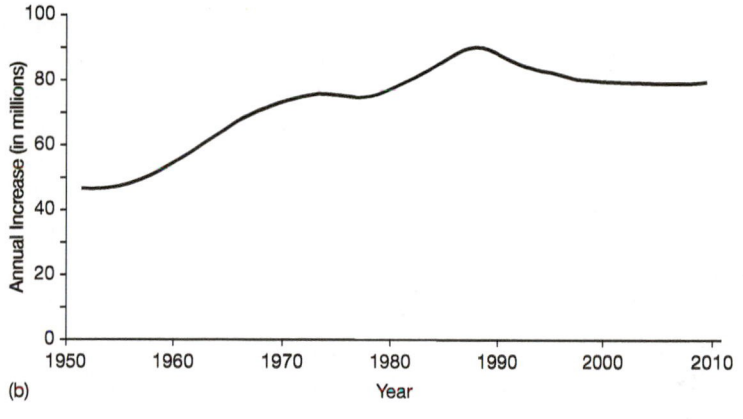

(b)

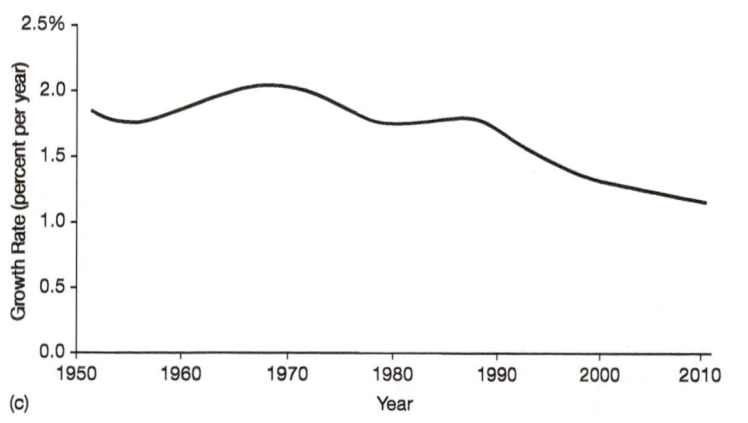

(c)

SOURCE: UN DEPARTMENT OF ECONOMIC AND SOCIAL AFFAIRS, POPULATION DIVISION.

FIGURE 8.4
World population
since 1950.

World population is still rising (Fig. 8.4a), and the projections indicate substantial growth—to about 9 billion by 2050. But although population continues to rise, a trend is clearly visible: globally, human population is leveling off (Fig. 8.4b, c). "By 1995," according to the UN Population Division, "44 per cent of the world population lived in countries where fertility was at or below replacement level (2.1 children per woman). Forty-nine countries, including China, were in that group and many of those countries had been experiencing below-replacement fertility for at least a decade or two."[3]

One can see the leveling off by looking at how fast population is changing year by year in Figure 8.4b. This figure plots the net change in human numbers each year. It shows the speed with which population is growing. If you think of human population as a car going down a road, the total population (Fig. 8.4a) is the total distance the car has traveled, and annual population growth (Fig. 8.4b) is the speed of the car. Notice that the annual growth line shows substantial fluctuations due to the effect of famines, epidemics, and other disasters that together have a large enough impact to show up in a graph with units in the tens of millions.

Demographic projections are made by looking at the **age structure** of the population—the numbers of people of different ages in a group (see Box 8.1: Age Structure and Social Change, page 202). Women bear children during a limited period of their lives, from about age fifteen to about forty. Demographers use historical data to make informed judgments about how many children, on average, each woman will have. That is, they are estimating how large families will be for the women of childbearing age in the population. This varies from culture to culture, and it can be affected substantially by events. For example, the severe economic decline in Russia as the Cold War ended in the early 1990s led to a sharp decrease in birth rates. Those rates have now partly rebounded as the Russian economy has strengthened. To take into account future uncertainties, demographers prepare different population projections based on variations in the average number of children per woman. The projections in Figure 8.5 lie in the middle range of the projections considered plausible by scientists today.

Figure 8.4c divides the numbers in Figure 8.4a by those in Figure 8.4b to determine the annual percentage of growth of the human population. What we see is that, since its peak in the mid-1960s, the rate of growth has been slowing for more than a generation. Demographers have projected that the growth rate will cross zero later in this century. What this means is that today's students were born after what may well have been the all-time peak in the annual growth rate of the human race. This is rarely acknowledged in environmentalists' statements of concern about population growth.

The rapid fall in family size emerged with little government intervention in the nations that are rich today. Since the middle of the twentieth century, public policies, augmented by the efforts of private organizations such as the Planned

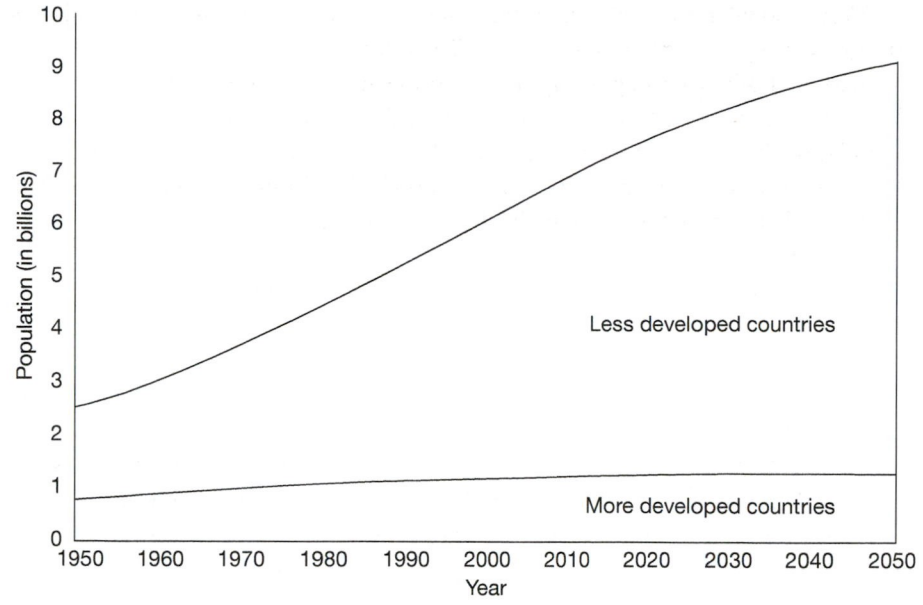

SOURCE: UN DEPARTMENT OF ECONOMIC AND SOCIAL AFFAIRS, POPULATION DIVISION.

FIGURE 8.5
World population projected to 2050.

Parenthood Federation, have been adopted to encourage smaller families in the developing world. The best known is China's one-child policy, instituted in the late 1970s. Effective enforcement has sharply lowered the rate of growth of China's population, although concerns have been raised about the ways such a policy may violate human rights. In other countries, the focus is on family planning—educating parents and giving them the capability to choose the number of children they will have and when to have them.

The overall result has been substantial disproportionality. As indicated in Table 8.1 and Figure 8.5, population growth is concentrated in the poor countries of Africa and Asia (outside of China), much of it in the Middle East. Growth rates are generally declining even in these parts of the world, but family planning is likely to be needed in the next forty years if the human population is to reach stability.

The observation that population growth rates decrease with modernization suggests that steps such as educating girls to increase their economic opportunities, and economic development in general, may help to bring down population growth. The idea that development might be a contraceptive is one that runs against the assumption by many environmentalists that economic growth and population growth are both to be opposed. What is needed are ways to channel economic development in more environmentally responsible directions. Population programs targeted at developing countries also need to provide sustained support for family planning,

BOX 8.1

AGE STRUCTURE AND SOCIAL CHANGE

The slowing rate of population growth has an important implication: in places where the birth rate falls, the population grows older on average. In the United States, the most rapidly growing segment of the population, in percentage terms, is the group of people eighty-five and older. In much of Africa and Asia, where the average number of children born to a woman is more than four, half the population is under the age of thirty (see figure). These are aspects of the age structure—the way members of a population are distributed by age. A population with a rapid birthrate has a more pyramid-shaped age structure, with relatively more young people, than a population with a low birthrate. Nations where people live longer have more older people, and their age structures reflect this fact with a more even distribution across age classes. When visiting a foreign country, a tourist can often observe differences in age structure when attending religious ceremonies, using public transportation, or any time he or she is in a large crowd of locals.

The age structure of a society has major implications for public policy and the way a society spends its money. The rising number of older Americans means that the number of people who rely on Social Security for their income is going up, and therefore the financial stability of the government's Social Security trust fund and the taxes used to replenish it have become significant political issues.

In countries with large numbers of children, by comparison, education and health care are vital, but are often scarce or of poor quality. Inadequate education, in turn, hobbles the economy when large proportions of the population seek work with limited skills.

The media and politicians usually do not treat these social challenges as ones arising from the structure of the population. Perhaps this is because the government can do little about age structure in the short term, even when an explicit population policy is in place, as in China. Yet the composition of the population is a major element of the human environment, often shaping the economy and the life chances of many people.

The demographic transition is unfolding over the same centuries-long period as industrialization (discussed in Chapter 7). Industrialization required raw materials from afar and markets for mass-produced goods, giving rise to greatly expanded world trade and travel—the phenomena we call the world without edges in this book. There is now a world culture in which cooking, language, and ways of life encounter each other with a frequency and intensity that was unknown in

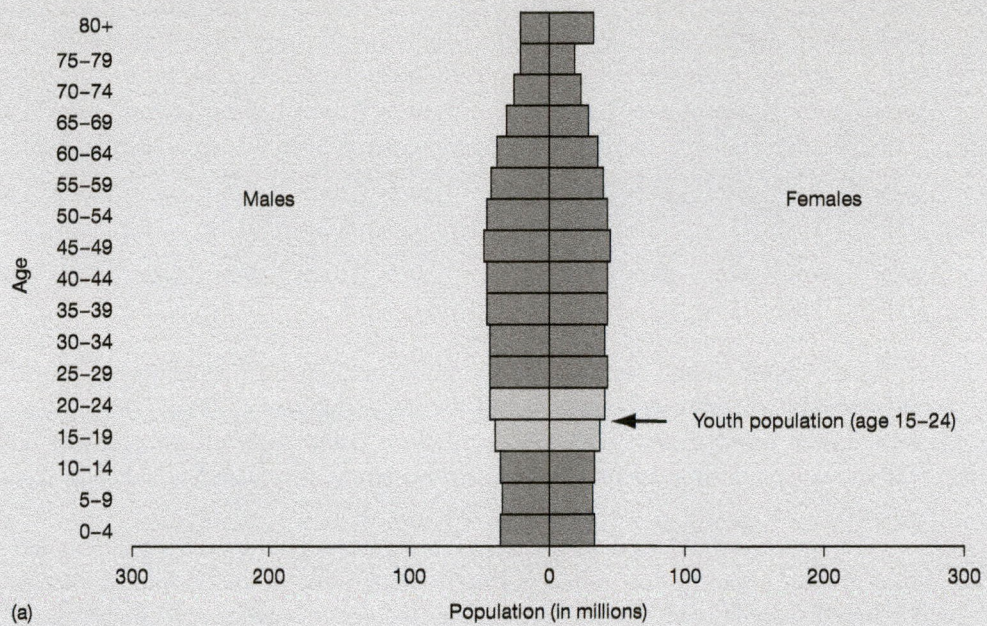

(a)

Population (in millions)

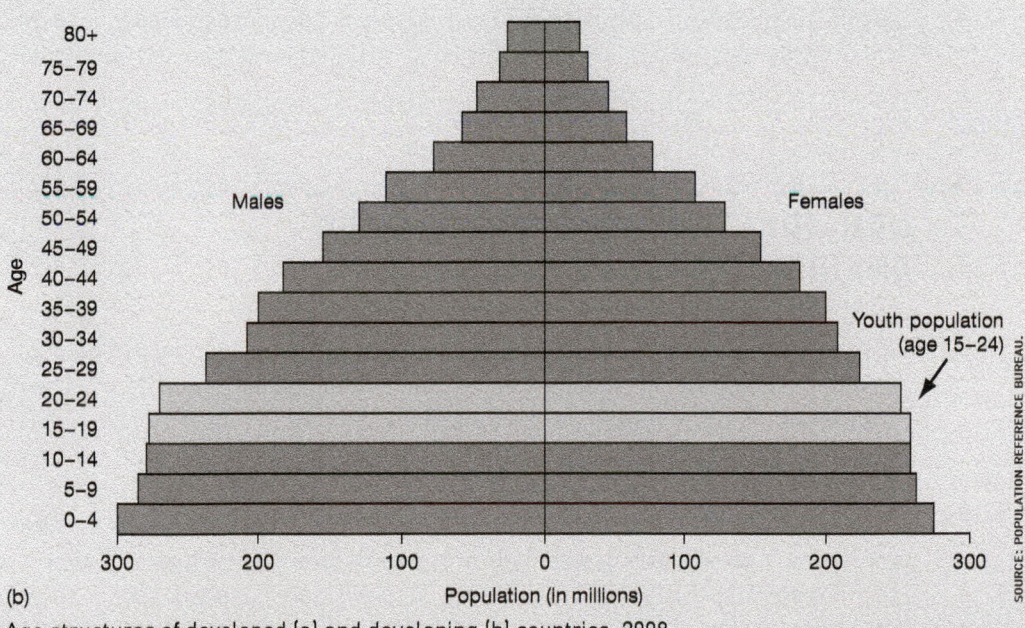

(b)

Population (in millions)

SOURCE: POPULATION REFERENCE BUREAU.

Age structures of developed (a) and developing (b) countries, 2008.

the preindustrial world. This cross-cultural mixing also produces conflicts—from racial discrimination to religious strife to the rise of political Islam.

In thinking about sustainable development and its call to pay attention to the needs of future generations, the size of those future generations, their age structure, and the way they interact with one another in the world without edges all play significant roles.

women's education, and raising the age at which women begin childbearing. These steps would accelerate the conclusion of the demographic transition. The demographer John Bongaarts estimated that such efforts could lower the ultimate size of the human population by more than 2 billion below projected levels during this century.[4]

The spontaneous decline in growth rate is historically unprecedented. Population growth began to slow in places experiencing none of the traditional checks on human numbers, such as famine, war, or epidemic disease. Instead, populations slowed and then largely stopped in the most prosperous nations. The demographic transition is associated with striking social changes, some of which can be clearly seen in the history of both rich and poor societies, and some of which are now emerging in them today. Many of the most striking social and environmental changes can be seen in cities.

URBANIZATION, AN UNRECOGNIZED ENVIRONMENTAL CHALLENGE

This brings us to a grand challenge of sustainability that is just coming to be recognized by environmentalists and those working in environmental policy: the challenge of urbanization. Urbanization provides a striking contrast to climate, a grand challenge whose importance has been clear for some time. Urbanization has mostly been ignored, even by the institutions focused on human and economic development.

Urbanization is also strikingly unlike population, a challenge that, as we have seen, is not a grand challenge, though it requires continued support of a demographic transition that has been under way for well over a century. Urbanization raises questions for which we have no answers yet. Densely settled cities such as New York are often highly efficient, so that the environmental impact per person is much lower than in suburbs where people drive a lot and have lawns to water. So cities hold a key to sustainability. Yet urban growth is fastest in poor countries, where environmental conditions are already bad and where the economic, technological, and political conditions that led to improvements in wealthy cities like New York have yet to emerge.

In 2008, the number of humans living in cities exceeded those living in rural areas for the first time, a trend that seems likely to continue to gather momentum during the rest of the twenty-first century. Why does that matter from an environmental perspective? It matters because urbanization is transforming developing countries where more than two-thirds of all people live (see Fig. 8.5). The environmental threats to people include infectious diseases from contaminated water and poorly controlled pests such as mosquitoes; air so polluted that it induces asthma and shortens life spans; and the wide range of environmental, social, and economic perils that come from living in slums.

These problems all reflect a basic fact: a city is a physical and social mechanism to deliver ecosystem services needed by a concentrated human population. This statement may not seem obvious at first. Remember that ecosystem services (Chapter 6) are essential for human life and well-being and that many of them, such as clean water, are obtained locally and in a situation in which the logic of use can unfold into a tragedy of the commons.

Because ecosystem services are unnoticed much of the time, both rich and poor contribute to and must face the environmental challenges of cities. Inhabitants of poor cities face environmental problems in daily life, including life-threatening problems such as cholera and severe flooding. At the same time, inhabitants of rich cities, like most Americans, cause environmental problems of which we are unaware. We remain unaware of these for two reasons. First, because the negative consequences of our actions are distant—as with overfishing to meet demand in rich cities. Second, some of the environmental impacts of high consumption are so slow to emerge that they disappear into the invisible present—as with global warming. As a result, the choices of consumption and investment made by the rich are transforming both the natural world and the societies in which all humans live. In Chapter 14 we will return to the question of consumption and the world without edges that those of us who live in rich nations shape through our lifestyle choices. Here, it is worth reflecting on the fact that if city dwellers cannot manage these challenges, sustainable development is unlikely to be achieved.

URBAN GROWTH

Urban habitats now house a majority of people (see Box 8.2: What Is a City? on page 206, which discusses the complications involved in defining the word "city"). This is an unprecedented development. There are more than 3 billion urban residents now,

BOX 8.2

WHAT IS A *CITY*?

A neat division of human populations into rural and urban is not quite correct, because we don't have a precise definition of "urban." From an airplane, we can see the obvious difference between a built-up city and a rural area, but the exact boundaries are hard to define and usually do not correspond to the jurisdictional lines separating a city from its surrounding area.

Consider the area around Mexico City shown in the figure. One can see the official city, the Distrito Federal, in the dark shading. This corresponds to the District of Columbia, where the city of Washington is located. But as residents of Washington know, a wider metropolitan area surrounds the official city, and the same is true of the area around Mexico's capital (gray shading). An even larger "urban subsystem" links urbanized areas into a megalopolis that is similar to the one that stretches along Interstate 95 from Washington, D.C., to New York City. The spread of settlements beyond the formal jurisdiction of a city that one sees in Mexico City is also typical in cities around the world, including Phoenix, Boston, and Atlanta, to mention only a few of many possible U.S. examples.

Each Mexico City is a different size, and there are valid reasons to count each one as a "city." The federal district encompasses the formal government, which in this case is the capital of the nation. The metropolitan area includes the major economic activities of Mexico City. The megalopolis includes the transportation network that feeds the city's population and brings in workers whose jobs are found in the megalopolis. Someone considering investing in or legislating or planning for environmental needs would have to take into account all three of these cities, even if one of them was the central focus.

An important study of urbanization by the National Research Council commented that "cities such as Buenos Aires, Mexico City, London, and Tokyo can correctly be said to be declining or expanding in population, depending on how their boundaries are defined."[1] This isn't just a problem for policy makers and investors. Decisions about where public transportation will be located, for example, affect ordinary citizens and families. Will the transit system serve only the dense core, or should it be used to draw development along major corridors into the countryside—as the American interstate highway system has done?

Many families pool the income earned from a branch of the family working in the city with the crops and other income earned by another branch of the family in the countryside. When the economy suffers a major downturn, the people who live this way can retreat to their agricultural relatives' homes. When there is

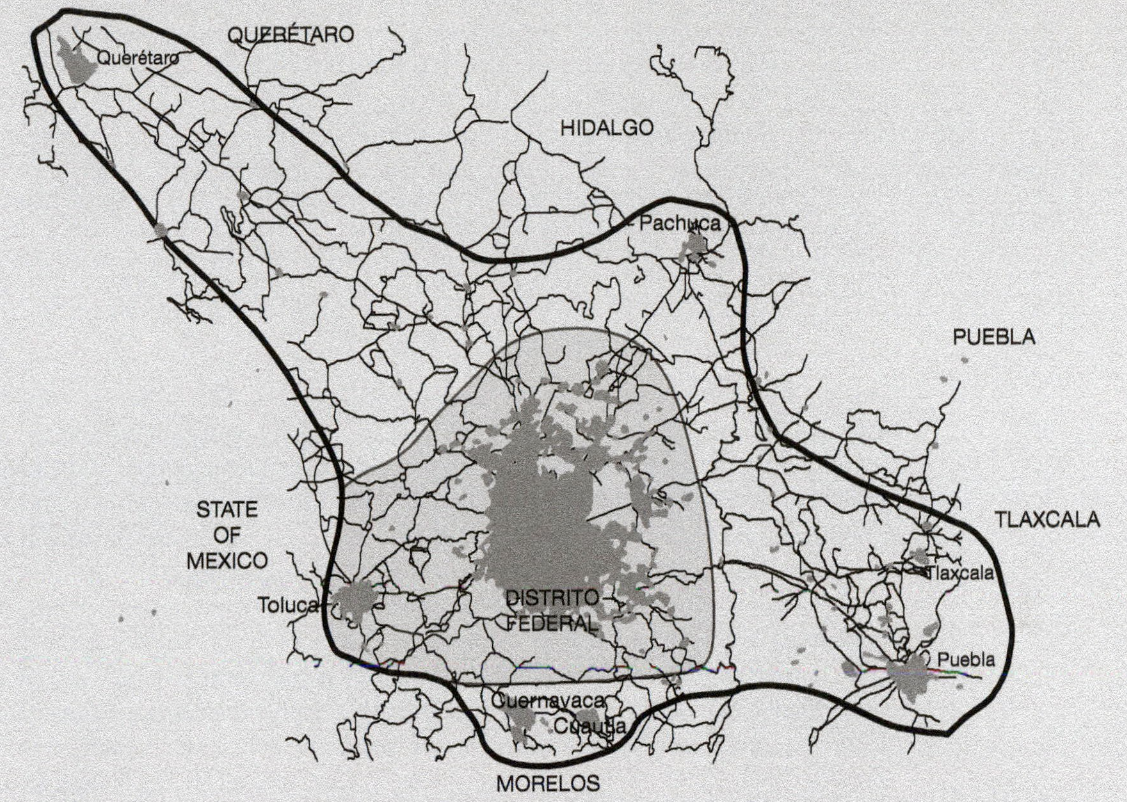

	Federal district	8.6 million
	Metropolitan area	17.9 million
	Megalopolis	23.2 million (24% of national population; 35% of urban population)

SOURCE: NATIONAL RESEARCH COUNCIL.

Three definitions of Mexico City.

a drought or other natural disaster, the country cousins can come to work in the cities. How should these factors be taken into account in planning education or health care? Will investors have the workers they need if they create businesses in the central core, where congestion is high and housing is expensive? All these questions depend on an understanding of urban populations that is incomplete or simply lacking in many of the world's cities, and each is linked, in turn, to environmental conditions in the city and its surrounding countryside.

What we *can* say is that an urban sense of place is complicated and perplexing. Although governmental jurisdictions and property lines are sharply drawn on a map, the underlying environmental and social patterns are fuzzy and broad. The water that serves a neighborhood may fall as rain in a watershed far beyond the city limits. Some of the food that feeds an urban resident comes from far away. Air pollution from an industrial facility outside the reach of a city's government may increase asthma among urban children. Meshing these fuzzy edges with the sharp lines of jurisdiction and property is a persistent challenge of urban governance, including attempts to handle environmental problems.

1. National Research Council, Panel on Urban Population Dynamics, *Cities Transformed: Demographic Change and Its Implications in the Developing World*, eds. Mark R. Montgomery, Richard Stren, Barney Cohen, and Holly E. Reed (Washington, DC: National Academies Press, 2003), 136.

and the number is projected to rise to more than 6 billion later in the century. As Figure 8.6 shows, all of the future net increase in human population is projected to go into cities, the great majority in developing countries. Demographer Joel Cohen called urbanization a population implosion.[5] In the last two hundred years, urban population has gone from 2 percent (1 out of 50 people) to 50 percent (1 out of 2 people). The doubling of urban populations means that, during the lifetimes of today's students, humans will create, somewhere in the world, urban settlements containing as many people as in all the cities that now exist. This is, potentially, a great opportunity: to build a sustainable urban habitat that works economically, environmentally, and socially for its residents. And it is a daunting challenge because this habitat is already being built, willy-nilly, and much of it is locking in a dependence on automobiles and other technologies that will be difficult to implement in an environmentally sustainable fashion.

Urbanization is now spreading rapidly to the portion of the world that has only begun to industrialize recently (Table 8.2 on page 210). Which of the twenty-one largest cities in 2010 have you visited? How many can you find on a map? Few Americans have been to many of these places, and we know little about them. It is no wonder that the environmental challenges of these cities are relatively unknown.

The most rapid rise of urban populations in the past twenty years has occurred in Asia and Africa. Consider the fourth and fifth bars in Figure 8.7 on page 212: Africa now has more urban residents than the United States and Canada combined.

Most popular attention has focused on the mega-cities with populations of more than 10 million (Table 8.2), but as you can see in Figure 8.8 on page 212, about half of the urban population is found in settlements with populations of less than half a million, and cities of between 1 and 10 million are gaining in the share of overall population housed.

By the early twenty-first century, the rapid urbanization of the world's population was unfolding in distinctive ways in different parts of the world. Only one generation ago, when the parents of today's students were in school, the poor nations were predominantly agricultural, with low levels of urbanization. Since

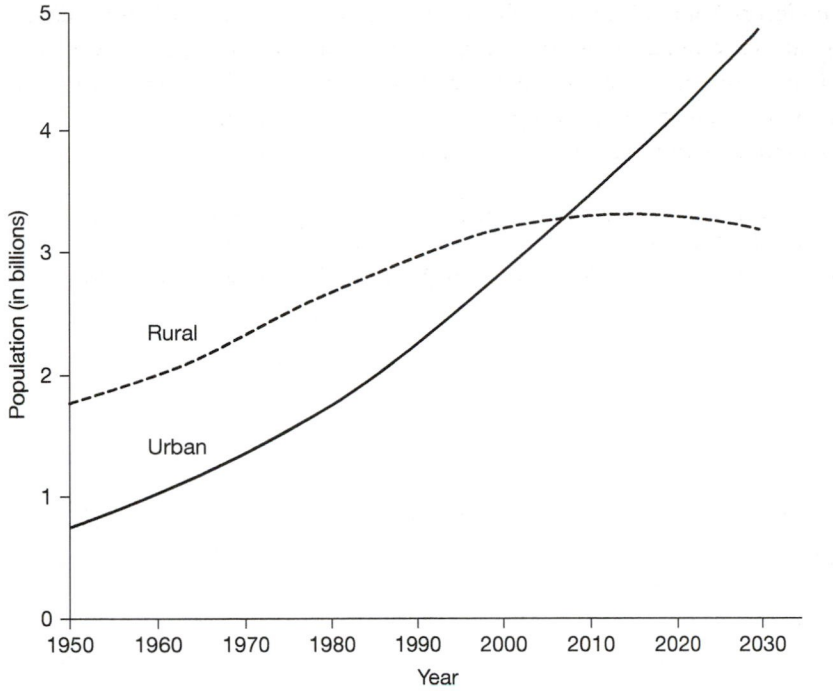

SOURCE: NATIONAL RESEARCH COUNCIL.

FIGURE 8.6
Global urban and rural populations, 1950–2030.

1970, China, India, and smaller nations such as Singapore have become richer and more urban.

Latin America, at 77 percent urban, has already gone through an urban transition like those of North America and Europe, with national population growth rates declining since the 1960s. Growth in the region's mega-cities has slowed considerably as the costs of congestion have made smaller urban areas more attractive. Yet, with the world's highest levels of economic and social inequality, Latin American cities have large slum populations that continue to grow.

In Africa, where some 38 percent of the population lives in urban areas, urbanization is more recent and more rapid in proportional terms because of higher population growth rates, rural poverty due to low agricultural productivity, and wars that drive people into cities as refugees. The spatial and economic structure of African cities reflects choices made by Europeans in the colonial era, when trading centers for agricultural products and natural resources produced for international export replaced an older network of market settlements serving agrarian populations. The colonial cities were designed by Europeans with small enclaves for themselves, and adjoining indigenous districts built with little attention to water and sanitation, roads, transportation, and energy supplies. The lack of infrastructure

TABLE 8.2   CITIES WITH MORE THAN 10 MILLION INHABITANTS IN 1950, 1990, 2010, AND 2025 (PROJECTED) (MILLIONS)

1950		1990		2010		2025	
City	Population	City	Population	City	Population	City	Population
1. New York–Newark, NJ	12.34	1. Tokyo, Japan	32.53	1. Tokyo, Japan	36.67	1. Tokyo, Japan	37.09
2. Tokyo, Japan	11.27	2. New York–Newark, NJ	16.09	2. Delhi, India	22.16	2. Delhi, India	28.57
		3. Mexico City, Mexico	15.31	3. São Paulo, Brazil	20.26	3. Mumbai (Bombay), India	25.81
		4. São Paulo, Brazil	14.78	4. Mumbai (Bombay), India	20.04	4. São Paulo, Brazil	21.65
		5. Mumbai (Bombay), India	12.31	5. Mexico City, Mexico	19.46	5. Dhaka, Bangladesh	20.94
		6. Osaka–Kobe, Japan	11.04	6. New York–Newark, NJ	19.43	6. Mexico City, Mexico	20.71
		7. Kolkata (Calcutta), India	10.89	7. Shanghai, China	16.58	7. New York–Newark, NJ	20.64
		8. Los Angeles–Long Beach–Santa Ana, CA	10.88	8. Kolkata (Calcutta), India	15.55	8. Kolkata (Calcutta), India	20.11
		9. Seoul, Korea	10.54	9. Dhaka, Bangladesh	14.65	9. Shanghai, China	20.02
		10. Buenos Aires, Argentina	10.51	10. Karachi, Pakistan	13.12	10. Karachi, Pakistan	18.73
				11. Buenos Aires, Argentina	13.07	11. Lagos, Nigeria	15.81
				12. Los Angeles–Long Beach–Santa Ana, CA	12.76	12. Kinshasa, Democratic Republic of the Congo	15.04

13. Beijing, China	12.39	13. Beijing, China	15.02
14. Rio de Janeiro, Brazil	11.95	14. Manila, Philippines	14.92
15. Manila, Philippines	11.63	15. Buenos Aires, Argentina	13.71
16. Osaka–Kobe, Japan	11.34	16. Los Angeles–Long Beach–Santa Ana, CA	13.68
17. Cairo, Egypt	11.00	17 Cairo, Egypt	13.53
18. Lagos, Nigeria	10.58	18. Rio de Janeiro, Brazil	12.65
19. Moscow, Russia	10.55	19. Istanbul, Turkey	12.11
20. Istanbul, Turkey	10.52	20. Osaka–Kobe, Japan	11.37
21. Paris, France	10.49	21. Shenzhen, China	11.15
		22. Chongqing, China	11.07
		23. Guangzhou, China	10.96
		24. Paris, France	10.88
		25. Jakarta, Indonesia	10.85
		26. Moscow, Russia	10.66
		27. Bogotá, Colombia	10.54
		28. Lima, Peru	10.53
		29. Lahore, Pakistan	10.31

Source: "Table 6: Population of Cities with 10 Million Inhabitants or More" by the Department of Economic and Social Affairs, *World Urbanization Prospects: The 2001 Revision*. United Nations Publications.

FIGURE 8.7
Urban population by region, 1950, 1990, and 2005.

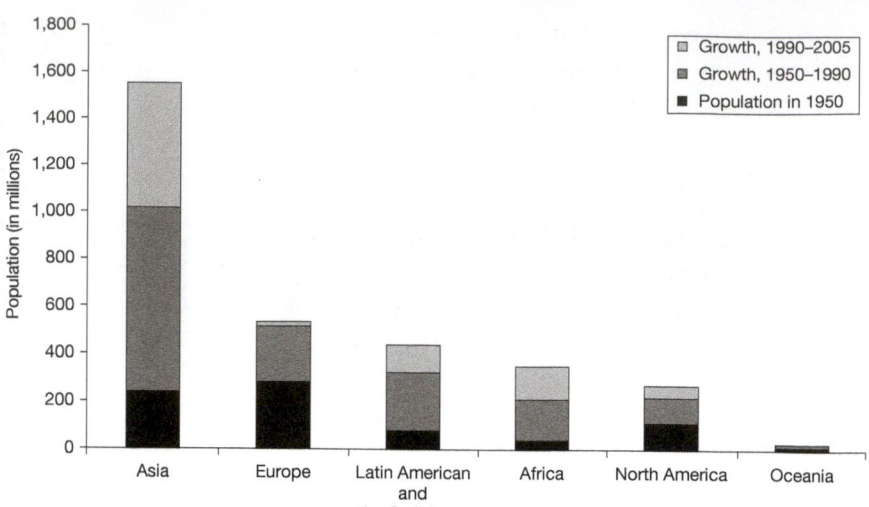

FIGURE 8.8
Urban population by size of settlement.

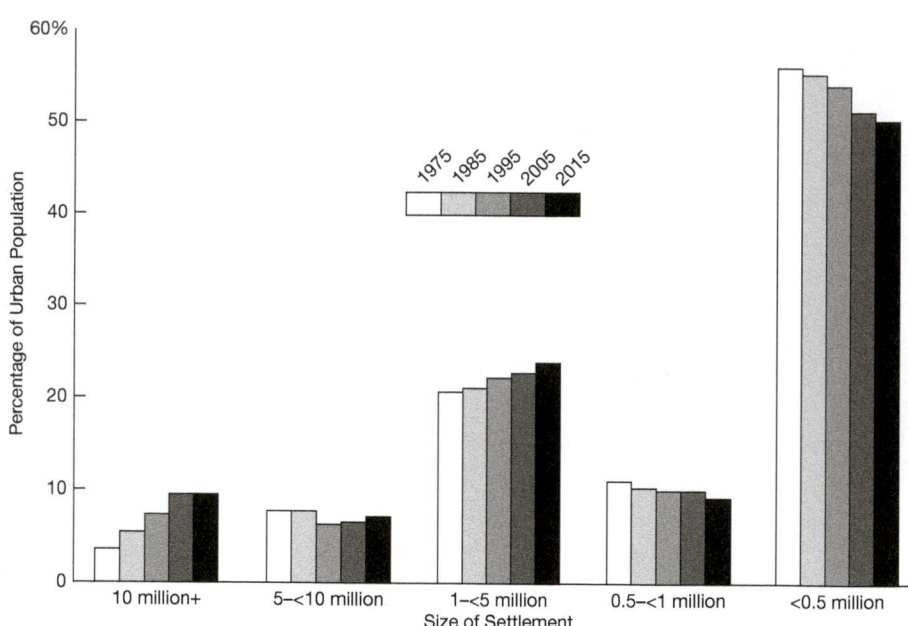

for the poor, followed by rapid urban growth, has produced large slum populations living at high levels of risk from disease and environmental hazards such as flooding.

Poor economic performance in sub-Saharan Africa since independence, half a century ago, has led to urban economies dominated by **informal employment**. That is, most people are working jobs such as doing laundry (Fig. 8.9), or are in small-scale commerce. Their jobs are not within businesses recognized by governments. This means, in turn, that workers have no legal protection, such as minimum wage requirements or antidiscrimination enforcement, and have no benefits, such as vacations or health insurance. The businesses— many of which consist of the owner and family members only—do not pay taxes, so a large portion of the economy is "off the books." Urban workers nearly all fall into the statistical category called "nonagricultural workers." More than three-quarters of nonagricultural employment in sub-Saharan Africa is in the informal sector, nearly all in low-wage and low-profit activities, and informal enterprises account for only 41 percent of the economic output of these countries. This means that informal workers account for a smaller proportion of the total output than their numbers. That is, they are poorer than average. There is little industrial employment or the sort of jobs Americans seek. African economies are not well integrated into the global economy, and they still depend on the export of natural resources and agricultural products to pay for imported manufactured products, just as in colonial times.

FIGURE 8.9
Doing the laundry in Nima, a suburb of Accra, the capital of Ghana, January 2006.

Asia, the world's most populous region, is roughly 40 percent urban, with a varied urban landscape. Pacific Asia, the coastal regions from Japan to Southeast Asia, has undergone a remarkable economic transformation over the past generation, as China and the newly industrializing countries of eastern Asia have rapidly increased income and levels of urbanization. China is now home to sixteen of the world's twenty most polluted cities, as rapid economic changes have pressed the ability of governments to protect and to improve public health. In western China and southern and interior Asia, urbanization is also rapid but economic growth has not been so meteoric, and poverty burdens nearly a third of India's urban population. Population growth rates remain high in Bangladesh and Pakistan, although they are declining. Urban populations shrank or grew slowly in central Asia in the severe economic and political disruptions following the collapse of the Soviet Union.

Beyond these regional generalizations, each city has a history and a population that will lead the city in its own direction. Cities attract settlers and retain residents because they offer opportunities for employment, for meeting and being with people, and to better one's situation. As migrants to every slum will affirm, they are there because they want to be. Seizing the opportunities and taking the risks of city life, some will fail and others succeed. Often, people will do things they could not have done in rural settings, and sometimes they will push the urban community and economy in a new direction altogether—whether it is in opening up a new kind of business, such as an organic food market; or making new links to distant communities by sending earnings back home; or disrupting the community by committing a crime or introducing a previously unknown disease, such as avian flu. The dynamism of cities makes each urban area a unique place, a distinctive social and environmental setting around which loyalties and antipathies can form.

As with population as a whole, urban population growth is heavily concentrated in the cities of low- and middle-income nations (Fig. 8.10). The United Nations estimates that by 2025–30 (when today's college students will be in their forties and have maximum influence in the world), there will be 71 million new urban residents in developing countries *each year*, living both in new urban settlements and at greater density in existing cities. That number is roughly 5 times the size of Los Angeles. Growth of this magnitude is the reason it makes sense to talk

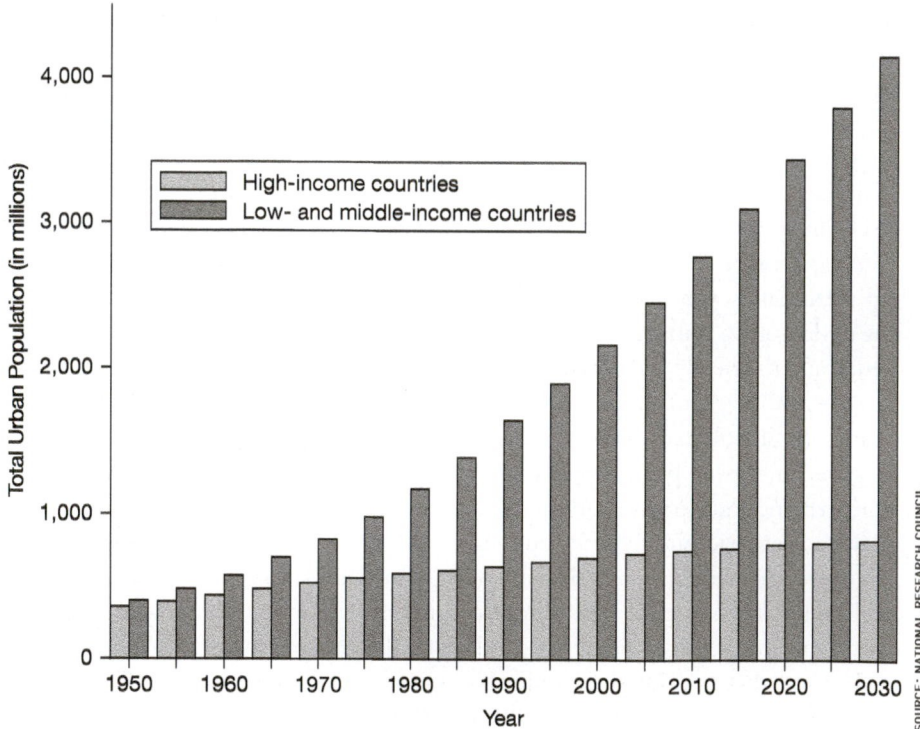

FIGURE 8.10
Growth of urban populations in high-income and low- and middle-income countries, 1950–2030.

SOURCE: NATIONAL RESEARCH COUNCIL.

about a doubling of the urban habitat—and why it is important to build this new urban world in a way that is environmentally sound and sustainable in both human and environmental terms.

URBANIZATION, POVERTY, AND THE ENVIRONMENT

One must understand this opportunity realistically. The great urbanization now under way will not come as suburban sprawl as we understand it in America, but through the expansion of slums. Environmental quality within cities faces immense challenges as a result.

As people move into cities to find a place in the urban, cash-driven economy, they usually face a shortage of affordable housing. One consequence is the rise of **informal housing**—residential spaces that are not planned or legal or served by municipal services such as electricity or piped water. Often, the construction of informal housing occurs on marginal lands, such as the steep slopes in Tijuana, Mexico, a rapidly growing city just south of San Diego (Fig. 8.11). Steep slopes are hazardous to build on, and when the vegetation cover is removed, there is nothing to hold the soil in place. Heavy rains then trigger landslides, destroying homes.

In other ways, the conditions of life are unfamiliar to people living in the United States. Water is heavy to carry, and it is needed in significant volumes for

FIGURE 8.11
An informal hillside settlement in Tijuana, Mexico, 2005.

FIGURE 8.12
Carrying water in
Accra, Ghana, 2006.

washing, cooking, drinking, or cleaning. Figure 8.12 shows a woman in Accra, the capital city of Ghana, who only has to walk about 200 yards up a gentle slope to go from the community tap to her house. Only! Think about carrying 30 pounds (less weight than she is carrying in that plastic jug) across a college campus before you could cook a meal for your family and wash the dishes. A generation ago, the women of this neighborhood had to go a mile to get their water, so in that sense, 200 yards is an improvement.

Keeping the streets of the neighborhood clean is a formidable task when there are no street cleaners (not to mention machines). Trash collection is infrequent in poor neighborhoods. The fact that the street in Figure 8.12 is so clean is good news: the local municipal assemblyman had organized a workforce of poor, under-

employed men to collect the trash that businesses and homeowners swept into the street. Yet, the overflowing dumpster located at the gateway into the neighborhood of Nima in Accra is an impressive sight to a visitor (Fig. 8.13).

WHY ARE SLUMS GROWING SO FAST?

More than a billion people are now living in slums. Most city dwellers were born in cities, but many, even in slums, have moved there. Why do they go? Some are driven to cities by war, or they want to be reunited with family members separated during the colonial era. But the primary draw is to find a job and to gain better living conditions. That is, life in a slum can be an improvement. If one talks to

TABLE 8.3 COMPARISONS BETWEEN URBAN AND RURAL ENVIRONMENTS IN INDICATORS REFLECTING ACCESS TO BOTH ECOSYSTEM AND SOCIAL SERVICES

Comparison (and units of measure)	Urban	Rural
Population growth: total fertility rates (number of children born per woman over her lifetime)	4.16	5.55
Infant mortality (probability of death in first year after birth)	5.6–7.5%	8.6%
Education (percentage of adults who have completed secondary school)	26%	12%
Water: dwellings with inside supply (percentage with piped water or well)	55%	19%
dwellings without inside supply (time to fetch from nearest supply)	18 min	26 min
Sanitation (percentage with flush toilet)	42%	8%
Electricity (percentage with service to dwelling)	65%	25%

Source: National Research Council, Panel on Urban Population Dynamics, Cities Transformed: Demographic Change and Its Implications in the Developing World, eds. Mark R. Montgomery, Richard Stren, Barney Cohen, and Holly E. Reed (Washington, DC: National Academies Press, 2003), which analyzed data in the Demographic and Health Surveys (DHS) of developing countries.

people in an urban slum, those who were born there tend to complain about their position in society and their living conditions; those who have come to the slum from the countryside, by contrast, say they wouldn't go back.

Consider Table 8.3, which compares the urban and rural environmental settings of poor countries where urbanization is expanding rapidly. "Environmental" here includes access to social services as well as ecosystem services. On average, life chances in cities are better than in rural areas. People have a rational basis for wanting to settle in cities, and these reasons are environmental as well as economic.

This is a recent development, however. Throughout most of human history there has been an **urban penalty**—people died faster in cities. Until the 1890s, European cities had higher death rates than rural areas, and cities grew only when they could draw more migrants to replace those who perished. In Prussia (now a part of Germany), where public health innovations were undertaken early, the infant mortality rate was 240 per 1,000 in 1875, but only 190 in rural areas. As shown in Table 8.3, urban areas in developing countries are now safer for babies than rural ones, a trend that emerged about a hundred years ago with the construction of sewers that collected human wastes and controlled the spread of infectious diseases such as cholera and dysentery, which are major killers of young children. Concern remains that an urban penalty may emerge once again, as transmissible diseases such as HIV/AIDS or bird flu make high-density habitats riskier, particularly in poor countries that have poor public health and environmental management capabilities. This is only a worry, so far, but it is one more reason why the doubling of the urban habitat is a significant challenge.

Where do the jobs that draw urban migrants come from? Cities are the stomach and liver of an economy, places where nutrients are taken in, broken down, and redistributed to make the whole body work. Cities do this by providing a high density of interaction opportunities, creating rapid commercial flows (such as banking and retail shopping) and intense cultural interactions (such as libraries, dance companies, or universities). Cities are places where business firms can concentrate materials for manufacturing (e.g., automobiles, assembled from components made in separate facilities often far away) and deliver specialized services (e.g., hospitals, stock markets, law firms). As a result of all this interaction, services such as transportation and communications themselves become major industries (e.g., telephone networks, Internet access, bus routes, shipping ports).

All these economic activities create jobs. Some of them require high levels of skills unknown in rural societies, such as website design, but others can be easily learned, such as assembly-line work or cleaning hotel rooms. Where schooling is available, children learn to read and to calculate, and they can integrate into an urban economy in ways their parents born in the country may have struggled to do.

TRANSFORMING NATURE WHILE RELYING ON IT

When humans build cities, they change nature, both within the city and in the countryside around them. The ecosystems managed for human use, including cities and farmlands, are generally *simpler* than the ecosystems of untended countryside nearby, but the managed systems are not usually less productive—indeed, the point of agriculture is to raise the abundance of species valued by people.

Urban places are meant to be good habitats for people, but they serve other species as well, often inadvertently. Urban habitats are populated by species such as rats, pigeons, cockroaches, and seagulls, which are considered nuisances by people. All of these creatures subsist well on organic materials discarded by humans—that is, garbage. The high density of human settlement provides good conditions for the spread of micro-parasites (germs, viruses, and parasites like head lice), and humans then build defenses against them, such as pipes to deliver clean water, different pipes to carry away water fouled with human wastes, as well as hospitals, clinics, pharmacies, and health departments to respond to epidemic illnesses. Skyscrapers provide

BOX 8.3

SECOND NATURE

An Excerpt from *Nature's Metropolis*

City, town, and country . . . worked together as a system, joining to become the single most powerful environmental force reshaping the American landscape since the glaciers.

Bison and pine trees had once been members of ecosystems defined mainly by flows of energy and nutrients and by relations among neighboring organisms. Rearrayed within the second nature of the market, they became commodities: things priced, bought, and sold within a system of human exchange. From that change flowed many others. Sudden new imperatives revalued the organisms that lived upon the land. Some, like the bison, bluestem, and pine tree, were priced so low that people consumed them in the most profligate ways and they disappeared as significant elements of the regional landscape. Others, like wheat, corn, cattle, and pigs, became the new dominant species of their carefully tended ecosystems. Increasingly, the abundance of a species depended on its utility to the human economy: species thrived more by price than by direct ecological adaptation. New systems of value, radically different from their Indian predecessors, determined the fate of entire ecosystems.

Differential pricing of species produced dramatic shifts in far-flung regional landscapes. The ecology of first nature had been more local than not: climate aside, species succeeded and failed mainly because of circumstances they encountered in their immediate habitats. Quite the opposite was true of second nature. Chicago, and the economic demand it represented, put new pressures on species hundreds of miles away. Its markets allowed people to look farther and farther afield for the goods they consumed, vastly extending the distance between points of ecological production and points of economic consumption. Now food and other resources made ever longer journeys to reach the places where people consumed them. The cattle that grazed on a Wyoming hillside, the corn that grew in an Iowa field, and the white pine that flourished in a Wisconsin forest would never ordinarily have shared the same landscape. All nonetheless came together in Chicago. There they were valued according to the demands and desires of people who for the most part had never even seen the landscapes from which they came. In an urban market, one could buy goods from hinterlands halfway round the world without understanding much if anything about how the goods had come to be there. Those who bought plants and animals from so far away had little way of knowing the ecological consequences of such purchases, so the separation of production and consumption had moral as well as material implications.

new cliff-like habitats. Quite a few American cities now provide nesting sites for peregrine falcons, as these hunting birds build nests and raise their young on window ledges in tall buildings and feed on the flocks of pigeons in the streets below.

Most of these ecological changes are the result of choices made to provide urban residents with the ecosystem services they need to achieve and maintain well-being—services such as clean air and water, processing of wastes, protection from natural hazards like floods and storms, and food. In a book about the history of Chicago entitled *Nature's Metropolis*, the historian William Cronon calls the adaptation of nature to human needs Second Nature (see Box 8.3: Second Nature, page 220).

In a rich country, we usually take the delivery of these ecosystem services for granted. It is when they are disrupted (for example, by natural disasters) that we begin to realize our dependence on nature and become aware of the relationship between Second Nature and the natural world that is the source of necessary services (First Nature). Significant elements of that dependence are summarized in Table 8.4 (page 222), in which urban problems are related to humans' continuing dependence on ecosystem services. Some services, such as clean air, can be assured only by preventing pollution. In other cases, such as sewage, it is possible to use technological means to substitute for the ecosystem services of streams and wetlands, which provide habitats for organisms that digest and detoxify human wastes.

You can see that Second Nature is shorthand for the engineered structures that people build to manage the natural world for human benefit. A city may be thought of as a physical and social mechanism to acquire and deliver ecosystem services to a concentrated human population. The physical part of this mechanism is often called **infrastructure**, whereas the social part includes markets, government, and community organizations. These institutions, including banking, government, law, and education, reshape human behavior so that the remade world can be managed by urbanized (that is, remade) people. The assembly-line worker and the lady of fashion, the accountant and the drug addict, all are denizens of Second Nature. Yet like all humans, they depend on First Nature.

TABLE 8.4 AN URBAN WAY OF LIFE REPLACES NATURAL ECOSYSTEM SERVICES (FIRST NATURE) WITH TECHNOLOGICAL INFRASTRUCTURE (SECOND NATURE), WHICH CHANNELS FIRST NATURE AND ADDS NEW RISKS.

Problems that appear in Second Nature . . .	reflect continuing dependence on ecosystem services (First Nature).
Wastewater (sewage)	With high population density, sewage treatment plants are needed to serve as artificial streams and wetlands in which microbes can digest toxic or disease-bearing organic matter.
Inadequate or unclean food	City dwellers require food, but food systems can be disrupted by micro-parasites, human disorder, and poverty.
Dirty air	Vehicle and industrial emissions produce sulfur oxides and lead (no natural analog; effects can be delayed and subtle).
Solid and hazardous waste	Industrial materials and chemical hazards are not naturally recycled (no natural analog).

When these engineered systems for delivering ecosystem services break down, human health suffers, death rates rise, and productivity falls due to illness. After Hurricane Katrina struck New Orleans in 2005, large areas of the city were rendered uninhabitable, and the damage was so extensive that large areas remain unoccupied today. Replacing the services of Second Nature is costly. For some resources, such as food or housing, the costs translate into unequal access. For other resources, such as safe drinking water or urban air quality, the costs may mean that everyone suffers the resulting infectious diseases or asthma because of the tragedy of the commons.

WHY IS URBANIZATION A GRAND CHALLENGE?

Ecosystem services have usually been delivered via capital-intensive infrastructure networks: water pipes, sewers, roads, electrical wires, and gas pipelines. The economies of scale captured via network development have been so great that they have shaped urban form. The cost of water delivered by pipe is so much lower than that of water delivered by hand or by truck that the location of water pipes can influence where people settle. Conversely, when pipes are not laid before land is occupied, the pattern of land use can make later installation of sewers or roads much more expensive. Installing infrastructure through settled land is typically disruptive and also contentious, as rights of way have to be cleared through established communities.

Poor cities lack planning and capital. Even if the case for building infrastructure is undeniable on grounds of public health, economic productivity, and political support, the economic constraints and institutional barriers can be daunting. Even when donors can supply capital, the low income of poor people can make it difficult or impossible to pay for maintenance or routine operation. These difficulties can be compounded by the widespread expectation that charges for utility service will be low, making it difficult even to arrange cross-subsidies, in which richer residents pay the costs of poor residents. Under such conditions, of course, repaying capital costs and saving for reinvestment and system extensions are also hard to do.

The problem is not only poverty but the social and political conditions of urban poverty. When a city's population grows through migration of people into the city, migrants may maintain an official residence in rural areas, so that their taxes and votes remain beyond the reach of municipal government. The willingness of all citizens to pay taxes and utility bills can be reduced by evidence of corruption, even when people have the ability to pay. And taxes and fees can be difficult to collect in informal settlements and from an informal economy.

These challenges are evidently great in cities where population is growing but economic output is not. Even in industrializing cities, however, where incomes and employment are rising, it can be hard to overcome institutional barriers. Cities along the Mexico–U.S. border, for example, have grown in income with the rise of an export-oriented manufacturing economy over the past two decades. The residents of these cities clamor for improved water and sanitation services, and their elected politicians see the political advantages of responding to those needs. Yet Mexican cities do not have the legal capability to issue municipal debt, so the financing of public works depends on resources controlled mostly by the national government. Thus, even when the economy can generate resources for infrastructure, there may be institutional and financial barriers.

Because urbanization is proceeding most rapidly in developing countries, the challenge of supplying ecosystem services to their people is especially acute. Conditions are already bad in most rapidly growing cities in poor countries. The financial capital needed to build the sewers, water supplies, and other infrastructure that delivers ecosystem services is lacking. Growing cities replace one landscape, largely natural in its functions, with one in which human control dominates. The creation of a human habitat in cities is expensive and permanent in the way it transforms ecosystems.

The urban environment may be the greatest unrecognized environmental challenge of the twenty-first century. Providing habitable, productive cities in the coming decades is a tremendous challenge, but it is also an opportunity to bring clean water and effective transit to poor urban districts; to channel urbanization so that wetlands and farmlands adjacent to urban areas can continue to provide low-cost ecosystem services; to work with communities to strengthen public health; and to give people access to decent work. Few environmental studies students think of investment banking or municipal bonds as environmental activities. But these are ways of creating a sustainable Second Nature, and they are among the significant opportunities to be grasped in realizing the idea of sustainable development.

The human population has now surpassed 7 billion in the steady march of the "most successful" species. In the twentieth century, a remarkable pattern began to emerge: population growth began to slow in places experiencing none of the traditional checks on human numbers—famine, war, or epidemic disease. Instead, population growth slowed and then stopped in the most prosperous nations—countries where women have become full participants in economic life and where the definition of success for children has become how well they are educated, not how many siblings and cousins they have.

The demographic transition, as this pattern of spontaneous stabilization of population is called, did not mean that environmental problems were solved, of course. Rather, the focus now shifts away from the numerical size of human populations to a closer look at what all of these people are doing. In Chapter 14 we examine consumption, a driving force of the environmental impacts caused by people in rich societies such as the United States. In this chapter, we drew your attention to urbanization, an environmental grand challenge that directly affects human well-being. It is a challenge to which we have no satisfactory answer yet.

FURTHER READING

Cohen, Joel E. *How Many People Can the Earth Support?* New York: Norton, 1995. Chapters 3, 4.

Cronon, William. *Nature's Metropolis: Chicago and the Great West.* New York: Norton, 1991.

McNeill, William H. *The Human Condition: An Ecological and Historical View.* Princeton, NJ: Princeton University Press, 1980.

KEY TERMS

age structure	informal	scenario
demographic	employment	urban penalty
transition	informal housing	
demography	infrastructure	

Chapter Nine

BIOLOGICAL DIVERSITY

SALVAGING THE WEALTH OF NATURE

Earth is increasingly a human-dominated planet. The hope of sustainable development is to live in such a way that two goals can be achieved: meeting the needs of those now living, and enabling future generations to meet their own needs. Because human well-being relies on ecosystem services, meeting our needs requires ecosystems capable of providing essential services such as clean air, food, and the cultural services associated with the sense of place discussed in Chapter 2. As discussed in Chapter 8, even city dwellers depend on ecosystems, although they often are not aware of that dependence. Some services, including the agriculture that provides much of the food we eat, are produced in ecosystems that are managed to serve human purposes, such as farm fields or forest plantations. Other services, such as clean water, rely on wetlands, soils, streams, and other ecosystems that are not actively managed by people. In all of these situations, the species that benefit people, such as corn or chickens, continue to interact with wild species in ecosystems that are not completely under human control, including microbes and pests, pollinators such as bees, plant communities exchanging seeds and pollen with other communities, and the occasional predator, such as the golden eagles of the Channel Islands discussed in Chapter 1. Ecosystem services link humans to the environment in ways that people cannot completely control.

Humans depend on nature, but humans also impose changes on ecosystems. In some cases, the pressure is on individual species, as in the over-harvesting of blue-fin tuna. In some cases, the pressure is a side effect of economic activities, as when humans eliminate the habitats of species to grow crops or build cities; an example is the draining of mangrove swamps in Southeast Asia to make shrimp farms. In some cases, people accidentally or intentionally introduce nonnative species to an ecosystem, which can lead to serious losses among the native species, as with the kudzu vines that have invaded much of the southeastern United States or the zebra mussel in the Great Lakes. Climate change is a further threat to ecosystems. In the "sky islands" at the tops of mountains in southern Arizona, one finds communities of plants and animals adapted to a cooler climate than the future climate that is predicted. Yet they cannot move to higher altitude because they are already at the highest elevations.

Human pressures can drive species extinct. When a species is removed, the ecosystem to which it belonged may reorganize—often in ways that damage the ecosystem services on which humans depend. Eliminating wolves and other

Learning Objectives
When you have finished studying this chapter, you should be able to

↘ explain how biodiversity is related to human well-being;

↘ examine why some familiar places, such as a parking lot, differ in biodiversity from others, such as a well-established garden;

↘ demonstrate the fact that most people living in the world without edges do not understand their dependence on biodiversity;

↘ examine your food supply for signs of its resilience and to consider whether the same can be said of your supply of clean water;

↘ see simple measures of the state of natural and human systems in your everyday life,

such as the health of trees or the frequency of air pollution alerts;

↘ interpret environmental problems in terms of the pressure-state-response framework;

↘ notice how little information is available about the species of plants and animals in your food;

↘ discuss what a slogan like "Save the Whales" might mean for changes in human activities if that slogan were to be taken seriously;

↘ visit a state park or other protected area and notice what is and is not protected—and from what.

predators has enabled deer populations to increase enough, in some places, that the deer have now become pests, carrying Lyme disease–infected ticks, eating ornamental and garden plants, and depressing populations of wildflowers and tree saplings in forests.

In most cases, humans do not intend to drive species extinct. Even people who over-harvest fish or wild plants and animals rarely want to exterminate a species that brings good money. In some cases, most of the damage is done by a small segment of a regional economy, and it may be possible to improve environmental conditions with few economic dislocations. In other cases, humans can be trapped in a tragedy of the commons, as we discussed in Chapter 3, in which their interest lies in capturing whatever they can, even at the cost of extinction. Being trapped in a commons dilemma, however, is not the same as intending to do damage, so even in these cases, solutions may be found: if the commons problems can be managed through effective institutions, people may well find the motive to protect our fellow life-forms. In many places, species once targeted for extermination, such as wolves in the American West and upper Midwest, are now being protected for their own sake by institutions such as the U.S. Endangered Species Act.

The wealth of nature as it now exists is contained in the genetic information in living species and in the ecological relationships between organisms and their biophysical environment. Each species has a set of genes, called a **genome**, that provides the biological instructions to grow a member of that species. Each genome, containing hundreds or thousands of genes or units of genetic information, is unique to the species. The genome that defines humans, from a molecular standpoint, was mapped at the beginning of the twenty-first century. A species is, by definition, a population of living things that can reproduce. When all the members of a species have died or are no longer able to reproduce, the species is extinct. Lost with it is the genetic information that made the species unique. Over the past two hundred years, humans have been accelerating the rate of species loss, perhaps as much as a thousandfold over the rates that prevailed before industrialization. We are eroding the wealth of nature, often without knowing it, and this in turn undermines ecosystem services that we need for our own survival.

To save genes we need to save ecosystems. One cannot preserve living things outside of nature today. For one thing, scientists have not identified the great majority of living things. Second, even for species we have identified, it is extraordinarily costly to preserve plants and animals in conservatories and zoos. In a few isolated cases, breeding of captive animals has worked to raise a species' abundance so that it can be reintroduced into the wild. This is not an approach that can slow the loss of species generally, though, especially when the original habitat has been modified so much that the reintroduced species have difficulty learning to survive and reproduce after they are released.

The genetic heritage of species is irreplaceable and clearly has value for that reason, just as works of art cannot be replaced and may therefore be priceless. Not every species is valuable, of course. Public health workers have now eliminated smallpox in the wild, and the two remaining populations are housed in tightly controlled research laboratories. But mostly, we have treated species as if they were no more than scraps of painted canvas, and we have not stopped to consider that some of them may be treasures like the *Mona Lisa*.

The plight of biological diversity is not obvious. Very few people pay attention to rare plants and elusive animals in distant locations or even in their own backyards. We know of the widespread threat of extinction because of the efforts of a small band of biologists, who established a distinct branch of their science called conservation biology in the 1980s. Conservation of life-forms is an activity with a long history, however. National parks and other biological preserves, for example, began to make their appearance about 150 years ago. As we will see, large tracts of land and coastal waters have been given protected status. The strategy of such protection is to exclude humans from landscapes.

Since the 1990s, however, efforts have also been made to harmonize human uses with conservation of biodiversity, through approaches such as **ecotourism** and the search for medicinally useful chemicals in plants and animals, and through recognizing the rights of indigenous human communities to own and manage their lands and waters. These traditional societies have lived sustainably in their territories for a long time, and some continue to do so even after coming into contact with the world without edges. All three of these newer approaches seek to make biodiversity socially and economically worthwhile, so that the people inhabiting a high-diversity landscape will protect it out of self-interest.

In 2005, an international team of scientists published the Millennium Ecosystem Assessment, an influential drawing-together of knowledge about ecosystems and their relationship to human well-being. This study, sponsored by the United Nations (UN), provided an ecological parallel to the Intergovernmental Panel on Climate Change, which analyzes global warming. The scientists who participated in the Millennium Assessment concluded that "human actions have fundamentally and . . . irreversibly changed the diversity of life on Earth, and most of these changes represent a loss of biodiversity. Changes in important components of biological diversity were more rapid in the past 50 years than at any time in human history."[1]

The wealth of nature is our wealth, too, and biological diversity matters not only because there are rare species in nature but also because there are indispensable species all around us that condition the world by moderating floods, providing shade, and making the world habitable in many other ways (see Fig. 6.4). Some of these species also feed us directly. These are the reasons why the challenge of biodiversity is important, and why it is critical to see that it is a grand challenge.

WHY DOES BIODIVERSITY MATTER?

As we saw in Chapter 6, humans rely on ecosystem services. Ecosystems form the infrastructure of the material world. Without functioning ecosystems, humans would perish for want of clean air and water, food, and much of the built environment that we often take for granted. A fundamental reason that biodiversity matters is that the diversity and abundance of species makes up the living world, and we humans depend on that living world for our survival. We have an interest in the productivity and integrity of the biosphere—the collection of all ecosystems covering the planet. The conservation of biological diversity is part of that interest: it is necessary to our thriving as a species to assure the flourishing of the ecosystem we inhabit.

A second fundamental reason is based in morality rather than material interest. Some have argued that humans have an ethical duty to preserve the species of the world. We are part of a family of living things, and the genetic evidence establishes our family ties (see Box 5.3: Genetics and the Invisible Present in Chapter 5, page 117). If we are in some ways the most successful species, we should exercise our power over the natural world with a sense of obligation—an obligation to look after the natural world as we influence it more and more. The core of the argument is the ethical conviction that, as with the prohibition against murder, we should not take the life of a species by eliminating all of its living members, simply because it is wrong to do so, whatever the material consequences. This is a compelling argument whose force is acknowledged in the Endangered Species Act, a law that has been interpreted to require expensive remedies and the halting of major economic development projects in order to protect species.

So if biodiversity is regarded to be of fundamental value both in a material sense and an ethical one, why have humans sometimes depleted biodiversity in the past, and why are we doing so much more of it now? There are two basic reasons.

The first is that most of human history has been a story of rising domination and control of nature. Agriculture, fishing, and forestry have played central roles in the economic life of most human societies. In each of these activities, humans struggle to subdue ecosystems and bend them to our will. Urban and suburban development replace natural landscapes with artificial ecosystems built around infrastructure. The harnessing of fossil energy has given humans the ability to supplant reliance on sunlight when we wish—although, in the process, we have inadvertently thrown the global climate onto a new and troublesome trajectory.

In some cases, religions have directed humans to extend their reach in a world where nature is perceived to be so vast that humans need fear more for their own survival against the forces of natural disaster than worry about how nature would persevere against people. Even now, environmentalists and others who preach the sanctity of nonhuman life-forms rarely include microbes and viruses among the species deserving protection, and few of us hesitate to employ antibiotics (a word that means "against life") to combat an infection. Moreover, not all the species in nature are equally important to humans. Some, such as the smallpox virus, are ones we have deliberately sought to exterminate. Others, such as rice or wheat, are so valuable that we have designed extensive seed banks to preserve a wide range of genetic variants, so that commercial varieties can be strengthened against disease or changing climate conditions.

As we noted at the beginning of this book, the idea of sustainable development is a human-centered one. Enabling future generations to meet their needs does require that those now alive take care of the biological world. But in meeting the needs of the present, we continue the deeply rooted human tradition of pursuing our interests at the expense of other life-forms. Because all human societies do this, it is essentially impossible for any single person to opt out of the erosion of biodiversity. This does not mean, of course, that efforts to conserve species are hopeless; indeed, as we will see, major advances have been made.

A second reason that humans continue to undermine biodiversity is that we have very little awareness that we are doing so. Humans directly cultivate or exploit several thousand species of plants, animals, and microbes for uses that include the production of food, timber, fiber for paper and cloth, and medicines. But humans rely on a tiny fraction of these species. The UN Food and Agriculture Organization pointed out that more than two hundred thousand edible plant species exist, but fewer than two hundred are used by humans, and three—rice, maize, and wheat—account for 60 percent of the calories consumed by people.[2] Because we simply do not know most of the species living on this planet, it is difficult to take care of them. From a practical standpoint, species can only be saved as components of ecosystems.

The ecosystem services we exploit deliberately connect to people through domesticated populations such as crops and livestock, as well as through a small number of tree and fish species taken from unmanaged ecosystems. We pay attention to the rest of the species mainly if they are pests, such as the wolves that eat wild deer and cultivated sheep, or if we derive cultural benefits from their presence, such as wolves howling in a national park. But the vast majority of species are bystanders that are sometimes affected by humans but lie outside the scope of our attention and understanding. As human domination has risen, more and more of these bystanders have fallen victim to forces of human origin.

PRESSURE, STATE, RESPONSE—AND RESILIENCE

To understand the relationship between humans and biological diversity, it is helpful to use a simple framework devised to guide sustainable development called **pressure-state-response**.[3] Human activities exert pressures on ecosystems through economic activities. These affect the state of ecosystems. Responses, in the form of policies or changes in practice, such as recycling, can then be undertaken in an effort to rein in or reshape pressures.

This framework is helpful in identifying readily available measures that can be used to monitor progress in moving toward sustainability. The acreage used to grow a crop might be used as an indicator of pressure. Farming can alter the state of a nearby stream when fertilizer and pesticides are washed into it by rainfall or irrigation water, lowering the water quality nearby and downstream. In response, farmers can alter the timing and amounts of pesticides and fertilizer applied, to lessen the pressure on the fish and other life-forms in the stream. In each case, data that are routinely collected by wholesale buyers of crops, merchants of agricultural chemicals, and environmental agencies can be put together to guide decision making and to foster more sustainable practices.

As the dotted arrows in Figure 9.1 suggest, however, changes in the state of an ecosystem do not reliably trigger responses, and even the responses that are adopted may not achieve the desired changes in the pressures exerted. Still, pressure-state-response is a useful way to organize learning so that different ways of sensing pressures and changes in states can be linked to responses, making them more effective over time.

Figure 9.1 is an oversimplification, of course. All ecosystems are subject to multiple pressures, both human and otherwise. A stream flowing through an agricultural

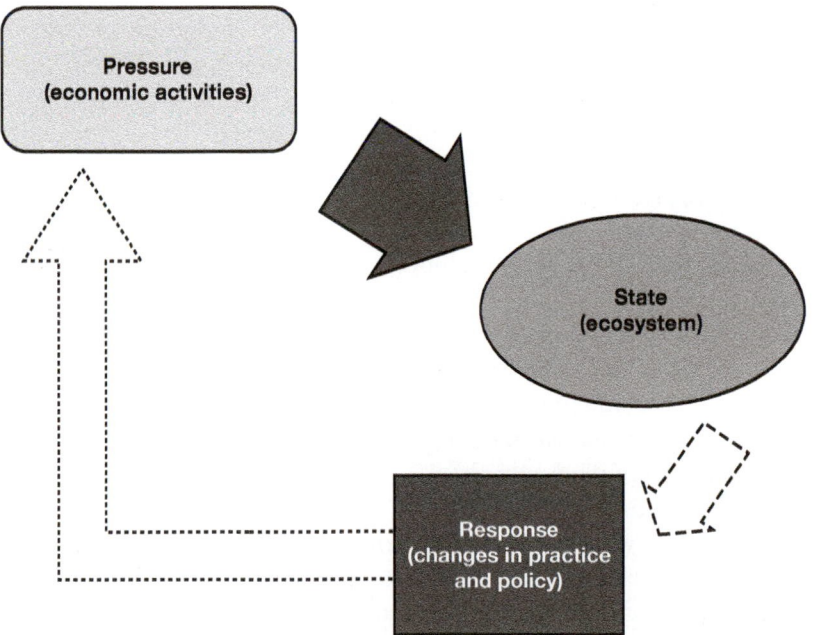

FIGURE 9.1
The pressure-state-response framework is a way to organize progress toward sustainable development.

landscape is affected by drought or heavy rains, so the concentration of pesticides in the water varies with changes in stream volume as well as the practices of the farmers. Human pressures interact with one another and with changes that are not directly caused by humans. Slowly, as the global climate changes due to human activity, every ecosystem will face pressures of human origin.

It is important to understand that ecosystems are adapted to changing profiles of pressures. Periodic, seasonal drought is characteristic of many landscapes, for instance, and the vegetation and animal life in an unmanaged ecosystem reflects the shifting availability of water. This is why a desert looks different from a forest. When species are subjected to human pressures, they may be able to adapt to them in ways that preserve the basic functioning of the ecosystem. Box 9.1: Resilience and Ecosystem Function on page 233 describes a remarkable example: an ecosystem that survives even though all the species visible within it have changed.

The property of resilience discussed in the box on page 233 is critical to an understanding of biodiversity. Each species in an ecosystem plays a role in the resilience of the whole assemblage. Faced with one kind of pressure, a species may die out while the ecosystem retains its functioning. This is what happens in a farm field: a forest is transformed into an ecosystem in which young plants are cultivated; yet the field can be more effective in delivering desired ecosystem services such as food to humans. But the farm field is generally a much simpler ecosystem than the forest it replaced. If a new species arrives, it may find the farm field a rich source of food, so that a pest that would not have had much effect in a forest may decimate the farm field. In this scenario, the farm field is much less resilient than the forest. At the same time, the farm field can provide food that sustains its accompanying human community until the next harvest. Under some circumstances, the humans and farm field can become a more resilient ecosystem than the humans and forest would have been separately.

This logic applies to biodiversity. Altering the composition of an ecosystem by removing one or more of its species may or may not affect its ecological functioning, particularly if the species removed is playing roles that other, competing species can pick up. But it is generally difficult to predict in advance.

Ecosystems containing many species, such as a forest, can often be more resilient than simpler ones. As the pressures imposed by humans increase, ecosystems of all kinds can become more vulnerable to disruption. The loss of a single species may not be noticed, but the fragility of the ecosystem to which it belongs increases. The state of the ecosystem can then shift, sometimes abruptly and irreversibly. Consider the Atlantic Ocean off the coast of eastern Canada, where a prolific cod fishery was pursued for centuries. The fishing pressure grew until the cod populations were depleted in the 1990s. Fishing stopped, but the cod have not come back. They were an important predator, and the species they fed on formed a distinctive ecosystem. Once the cod were removed, the ecosystem changed into

BOX 9.1

RESILIENCE AND ECOSYSTEM FUNCTION

The photograph below shows a San Francisco Bay seawall at unusually low tide. This assemblage of shelled creatures is characteristic of what is found on many hard surfaces exposed to tidal waters for part of the day. Some animals, such as barnacles (near the top), flourish by being washed by salt water infrequently, at high tides and in storms. Others, such as mussels (near the bottom), do best when they are underwater most of the time. In between are species, such as the tubeworm, that do well when they are flooded about half of the time.

This ordering of life-forms reflects the climate and hydrology of this particular place in California. It would be tempting to call this collection of life-forms natural, but each species on the seawall comes from a different continent. The barnacle comes from Europe, the tubeworm is from Australia, and the mussel is from the East Coast of North America. Each species probably was brought to California inadvertently in the water aboard a ship or by clinging to a ship's hull, and each then proved vigorous enough to survive in its new setting.

Living on this San Francisco Bay seawall is the Atlantic barnacle, *Amphibalanus improvisus* (upper layer); Australian tubeworm, *Ficopomatus enigmatius* (middle layer); and New England mussel, *Geukensia demissa* (bottom layer).

The native species that would have populated this seawall were displaced, and they may even be extinct in this area of California. Yet the assemblage of non-natives is functioning, with different species reproducing successfully at different levels from the high-tide line. So there is a lesson here. Sometimes the loss of a species may not matter very much if its ecological function is taken over by another one. In this case, the new species are from far away, but this is not always true. A species from a nearby habitat may be able to expand its range into a new area when a species there goes extinct.

The ability of an ecosystem to maintain some important functions even when a constituent species has been lost or when structural relations have been altered is called **resilience** by scientists. Resilience can be seen in social systems, too; for example, the U.S. Constitution has framed the legal institutions of the United States through more than two centuries of often tumultuous change.

a new configuration, one in which the cod could not become abundant again. A large population of fish that had fed humans since at least the seventeenth century may have been permanently depleted.

Biodiversity is observed in places, in specific ecosystems. The conservation of biodiversity has also been a place-oriented activity. In the rest of this chapter, we will describe the biodiversity of the planet using examples drawn from particular ecosystems. The broad patterns we discuss are just that—generalizations that have significant exceptions. In some places, such as the Birds Head Seascape in eastern Indonesia, biodiversity is flourishing. In others, such as the forests of the northeastern United States, once-depleted species have made an impressive comeback over the past century.

The broad picture is a sobering one, however. The state of many ecosystems is declining as human pressures continue to increase. Responses have been increasing in many places of high biological significance, but recuperation has been slow and protection is often incomplete. Some pressures, notably climate change, have effects on all ecosystems, and as yet, no response is available to counter climate change effectively.

STATE: WHERE THE WILD THINGS ARE

The current state of the world's biodiversity is shaped by the fact that the distribution of life-forms is not uniform. The tropics contain both the largest number of different species and the highest biological productivity.

As described in Chapter 5, coral reefs and forests contain a wealth of micro-habitats: small spaces in which the conditions of life are different from those in the areas that surround them. In these complicated settings, different forms of life can find ways to make a living and create distinct communities. Over many generations, competition among species in a microhabitat drives them to evolving into new forms—until sometimes a species becomes a distinct one, no longer able to breed with its relatives in a different habitat close by. In this way, habitats with complex structures foster high diversity. Rain forests and coral reefs occupy small areas globally (Fig. 9.2, page 236), but they account for more than half of all species. These habitats are found mainly in the tropics, where the growth of large structures is made possible by abundant sunlight and, for forests, rainfall.

ENDEMISM: "STATIONARY" SPECIES

Sometimes extinction is only a local phenomenon. Moose and beaver have returned to the forests of North America after being over-harvested so severely that they survived only in areas where human pressures were low or absent. In high-diversity settings, on the other hand, the local elimination of a species can be a global one as well: the species may be unable to survive elsewhere, as the moose did, because it does not exist elsewhere. Instead, the genetic information that is translated into living organisms is lost.

The risk of extinction is related to an idea that biologists call "endemism." An endemic species is one that is restricted to a particular local area. It is the ecological parallel of Gilbert White, the stationary man. This localization seems to be related to the evolutionary pattern of life. Over the past 600 million years, the number of species has increased, as the emergence of some long-lived species such as the mangrove created new microhabitats that provided the staging ground for new species (Fig. 9.3, page 237). Figure 9.4 on page 238 shows the trend in plants over the past 400 million years, which is as long as plants have been present on land. Life not only inhabits a landscape, it becomes part of the landscape and creates habitats for other life-forms.

Where life is dense, such as in a tropical forest, new species may not be able to spread far because their way of making a living is already being done by a competitor in nearby microhabitats. For example, a wood-eating beetle may emerge in one patch of rain forest, only to find that in other patches close by, a wood-eating rodent already occupies its ecological niche. In a temperate-zone forest with fewer species, however, a wood-eating beetle may not have any competitors nearby and may be able to expand over a wide area.

The variety of colors in many American landscapes in autumn suggests that there is a good deal of competition among trees in the temperate zone. Even where the soils are uniform, the species of trees may not be. When a seed germinates and

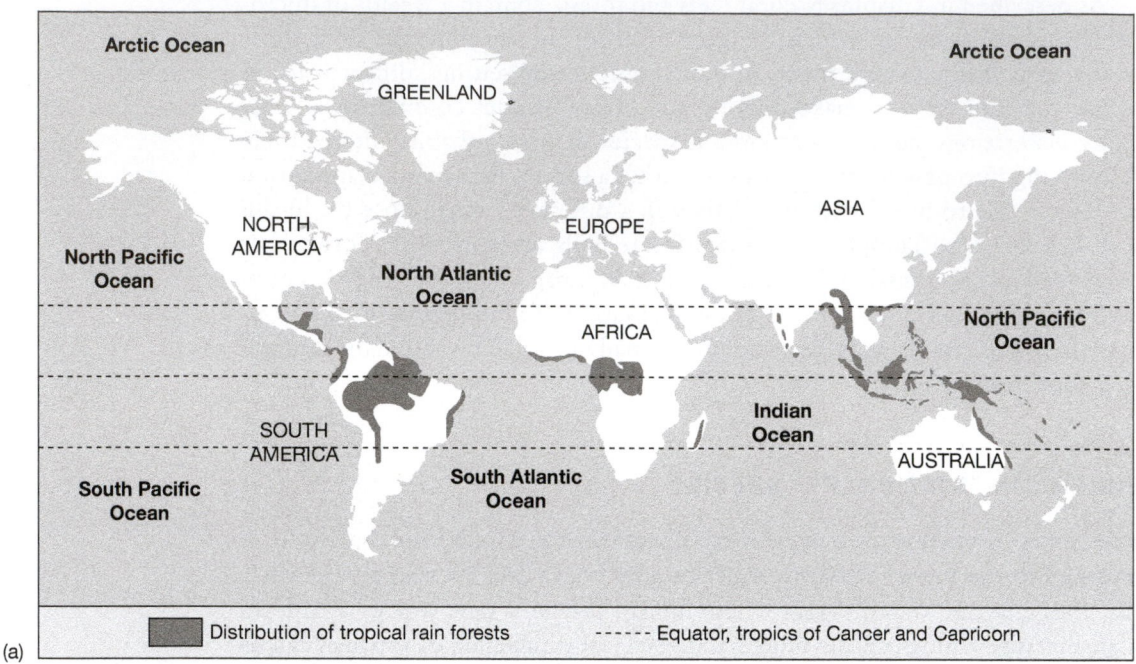

(a)

Distribution of tropical rain forests ------ Equator, tropics of Cancer and Capricorn

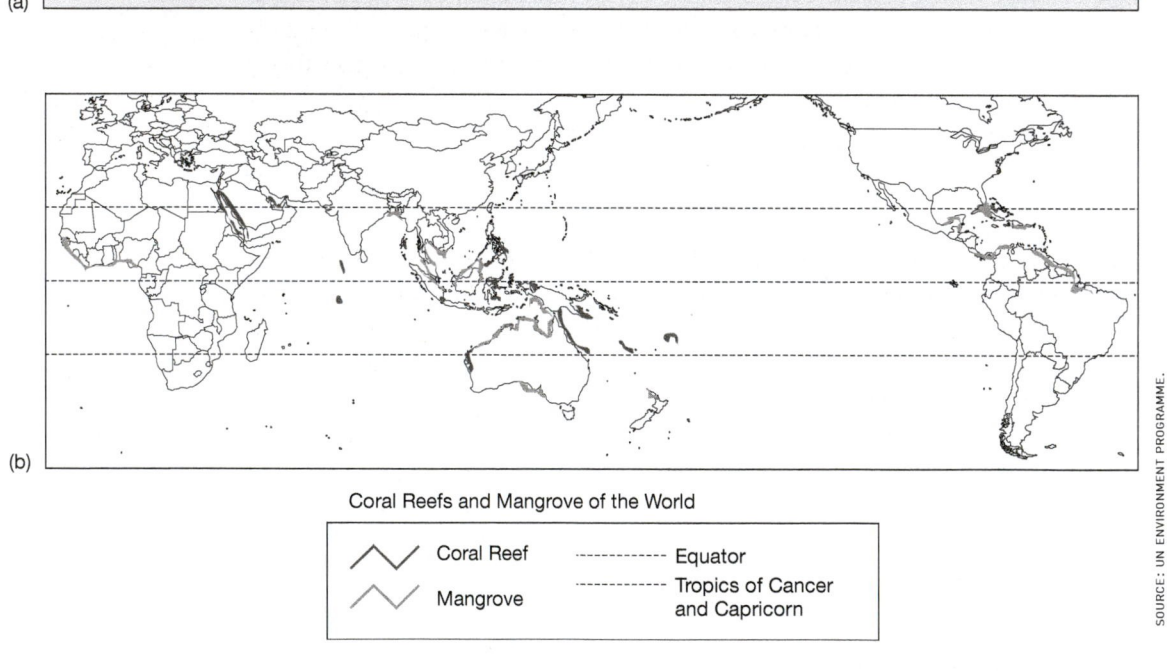

(b)

SOURCE: UN ENVIRONMENT PROGRAMME.

Coral Reefs and Mangrove of the World

Coral Reef ------ Equator

Mangrove ------ Tropics of Cancer and Capricorn

FIGURE 9.2

Tropical forests (a) and coral reefs (b) contain exceptionally high levels of biological diversity.

grows into a tree, its shade will exclude other trees. If that tree is in a shady forest, its seeds may need to travel a long way before they can successfully grow into mature trees that bear their own seeds. Trees compete with one another for sunlight and nutrients. This kind of competition is far more intense in tropical forests and coral reefs, where the number of species is large.

There has been a gradual increase in the diversity of life over hundreds of millions of years. This has been observed in the fossil record and is now found in the ecosystems of the world. The density of species, particularly in the tropics, means that the spatial range of many species is limited. Therefore, it may be exterminated throughout its range more easily and become extinct.

AN INVISIBLE PRESENT

The accelerating loss of species due to human activity has coincided with the dwindling awareness of the natural world by those who are most influential in the world's economy. Consumers in rich nations, and their commercial and

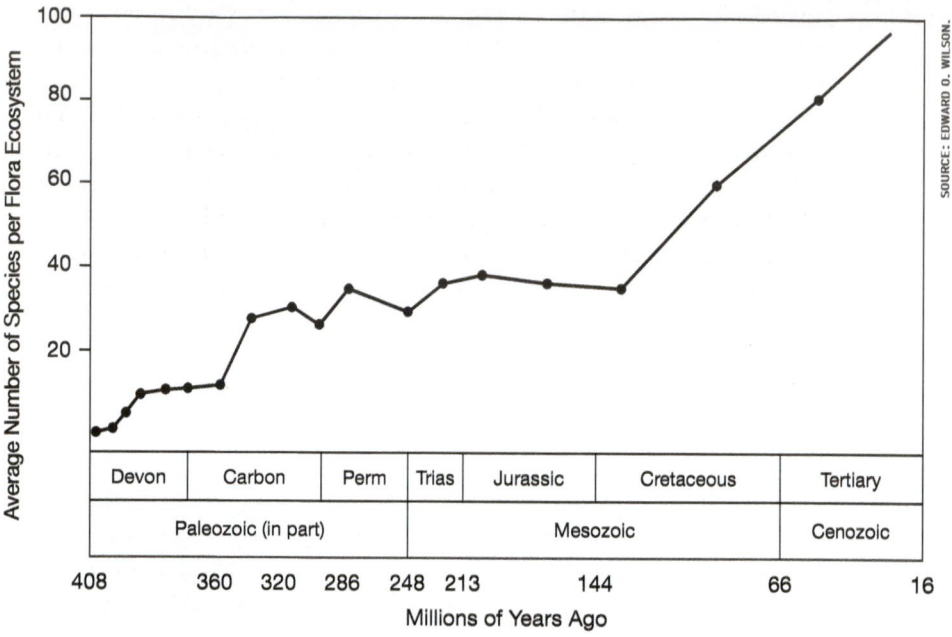

FIGURE 9.4
Biological diversity has risen through time, as illustrated by the increase in the number of plant species in different ecosystems.

governmental institutions, are blinded by the invisible present and the world without edges. We resemble the Mayan elites of Chapter 3—we are driving changes but are unaware of them.

"Shifting baseline syndrome" is one name given to this phenomenon by conservation biologists. It means that we measure change in terms of our own experience. A child growing up at the edge of a suburbanizing area may see rural landscapes transformed into housing and shopping centers. To a child growing up in that same spot a decade later, by contrast, the houses and shopping centers are part of the world that existed before she was born—part of the baseline. Both come to live in human-dominated environments, but the person reared on the frontier of development may be aware of how landscapes can be dramatically transformed because her childhood experience is not the same as that of someone growing up in the same spot after the area was developed.

If you think about the landscapes altered by humans, you might ask the question posed in Chapter 2: What is natural? Is it the patch of woods where the suburban child played before it was turned into a shopping center? How about the forest or the desert that preceded the encroachment of suburbia? Is one of these the "right" baseline? If governments or conservation organizations restore a landscape, should they aim for 1970, or 1870, or 1370? As these questions indicate, determining what is natural is not only a question of science ("What animals and plants populated a landscape in the past?") but also a question of values and social choice ("What baseline makes sense, in light of the costs of protection and restoration and the other opportunities for using this landscape?").

Marine conservationists argue that overfishing has been allowed to continue because the baseline of analysis has been chosen inappropriately. Fisheries were mostly poorly governed open-access systems until the latter part of the twentieth century, and reliable records were not kept until industrial-scale fishing began. As a result, regulators mistook the size of fish populations when record keeping began for the size of these populations before humans began to exploit them heavily. In estimating the capacity of an ecosystem to provide fish or another valuable service to people, we must consider what is known about the history of the ecosystem, rather than assuming that its pristine condition is the one that the older fishermen can recall.

English environmental historians analyzed government records of fishing by the British trawler fleet, which go back to 1884. (Trawling is a method of catching fish by dragging an open net called a "trawl" along the seafloor, which scoops up nearly everything it its path.) They found that the catch per unit of fishing power had decreased by a factor of 17 between 1889 and 2007.[4] This means that it now takes 17 times as much in human and natural resources—fuel, labor, investment in fishing vessels and fish-finding technology, together with government management of fisheries—to end up with a pound of seafood as it did in the late nineteenth century. During this period, British fishing boats depleted fish near the British Isles, then moved on to more distant waters in the Arctic and off the coast of Africa. The baseline shifted, too, and there has been little or no awareness, either among those doing the fishing or those eating fish and chips, that the fish were dwindling. Indeed, regulation of fisheries since the 1980s—after fish stocks had already been drastically reduced—has consistently set catch limits above the levels that biologists recommended as necessary to conserve fish populations. This same kind of response seems to have contributed to the disappearance of cod from the Grand Banks in the North Atlantic.

The shifting baseline is an idea raised in Chapter 2, where we called it the invisible present—the notion that changes that are slow in relation to a human lifespan, such as the retreat of glaciers due to climate change, may nonetheless be ecologically important and be driven by human activities. The analysis of the British data is part of a series of estimates that demonstrate that overfishing has reduced populations of predatory fish (the ones we tend to like to eat) by 90 percent or more since 1900. An interesting result of this decline is that today's British fishing fleet, armed with modern technology and the capability to range far from shore, catches about half as much fish as the sail-powered fleet of 1889, a striking illustration of how a growing technological capability has masked a disproportionate impact of humans on the natural world.

Because we live in and draw nourishment from the world without edges, it can be difficult to see the erosion of biodiversity. Most of the ecosystem services used by people in rich countries do not look like they come from nature at all. Water flows from a tap. Floods that have been moderated by wetlands do less damage, but it is hard to see a disruption that has been prevented. Not seeing this, people turn the wetlands into shopping centers and parking lots, then demand that they be shielded from floods by higher levees and floodwalls. Ironically, as more levees are built, the floods mount even higher—the water has to go somewhere—increasing the damage done where the water finally goes over the top. Right under our noses, the food we eat is obscure: meat is sold as pork chops or hamburger rather than as an entire pig or cow; wheat is baked into bread rather than sold as sheaves of grain; and sugar is blended into soft drinks or ketchup without a sign of the beets or cane from which it came. Indeed, in the world without edges, it is nearly impossible to tell where food items come from, let alone what they looked like when they were alive (see Box 9.2: The Barcode of Life, page 241).

If we do not know where our routine consumption of ecosystem services have an effect, and we usually pay no attention to what species we consume, it is no wonder that people can have such a powerful impact on biodiversity. As the world without edges grows, so does the economic power of humans who know not what they do, and thus the pressure on living things.

PRESSURE: THE SIXTH GREAT EXTINCTION

The diversity of life has risen over geological time, but not always. Several major disturbances have interrupted the gradual growth in diversity over the past several hundred million years. The best known may be the large meteorite that crashed into what is now the Gulf of Mexico 65 million years ago. The worldwide, cataclysmic change triggered by the impact marked the end of the age of dinosaurs and the rise of mammals as the dominant species in most terrestrial environments. As Figure 9.5 on page 242 shows, planetary-scale extinction events, in which large numbers of species were wiped out, have occurred five times in the history of our planet.

Conservation biologists argue that we are now in the midst of a sixth major extinction event. This is an extinction episode caused by humans—an event in which many, or perhaps most, of the millions of species on the planet will be lost. Because about half of all species are located in rain forests and coral reefs, these are the habitats that have drawn the most attention. About 25 percent of the corals in reefs have been lost in just the last 25 years. This damage is already plainly evident and biologically significant in many areas, such as the Florida Keys. These losses are occurring at a pace that may be important to readers of this book: most of the losses will probably take place in your lifetime.

BOX 9.2

THE BARCODE OF LIFE

When you eat a piece of sushi, do you know what you are eating? Using the methods of molecular genetics, it is now possible to tell, which makes detecting deceptive relabeling of fish and other foods easier. The same techniques can also identify field samples, thereby helping to speed up surveys of ecosystems to determine their biodiversity and value for conservation.

The idea, developed by geneticist Paul Hebert of the University of Guelph in Canada, is straightforward: find a marker gene that is unique to each species and then develop rapid and inexpensive ways to identify the DNA in that gene in a specific sample.[1]

Although each species has a unique genome, some genes are widely shared. Hebert looked at the mitochondria inside the cells of plants and animals and found that some genes are present in virtually all living cells. Mitochrondria are organelles, functional units within cells, which play a crucial role: they house the chemical reactions of respiration that enable the cell to combine oxygen with food to generate energy. The same chemical reactions that enable a human to run also power the flowering of a tree or the flick of snake's tongue.

One of the genes that contains the design for this process is called cytochrome oxidase I. Hebert and his colleagues have studied the DNA sequence of this gene in a wide range of animals and plants, and they have found that the variations among those sequences vary from species to species in a way that can serve as a unique signature. This signature has been dubbed a "DNA barcode" by Hebert, after the barcodes that are scanned in supermarkets and stores to identify the packaged products you purchase. Whether the cytochrome oxidase gene variations will in fact prove to be uniquely associated with each species studied remains to be seen. So far, Hebert and a growing team of biologists have identified more than 100,000 species using this method.

In 2009, scientists at the American Museum of Natural History used this barcoding technique to identify the species of fish being sold in sushi at thirty-one restaurants in New York. Half were mislabeled.[2]

1. Paul D. N. Hebert, Alina Cywinska, Shelley L. Ball, and Jeremy R. deWaard, "Biological Identifications through DNA Barcodes," *Proceedings of the Royal Society London B* 270 (2003): 313–21.

2. Robert Lee Hotz, "DNA 'Barcodes' Surface Fishy Imposters on Menus," *Wall Street Journal*, December 4, 2009.

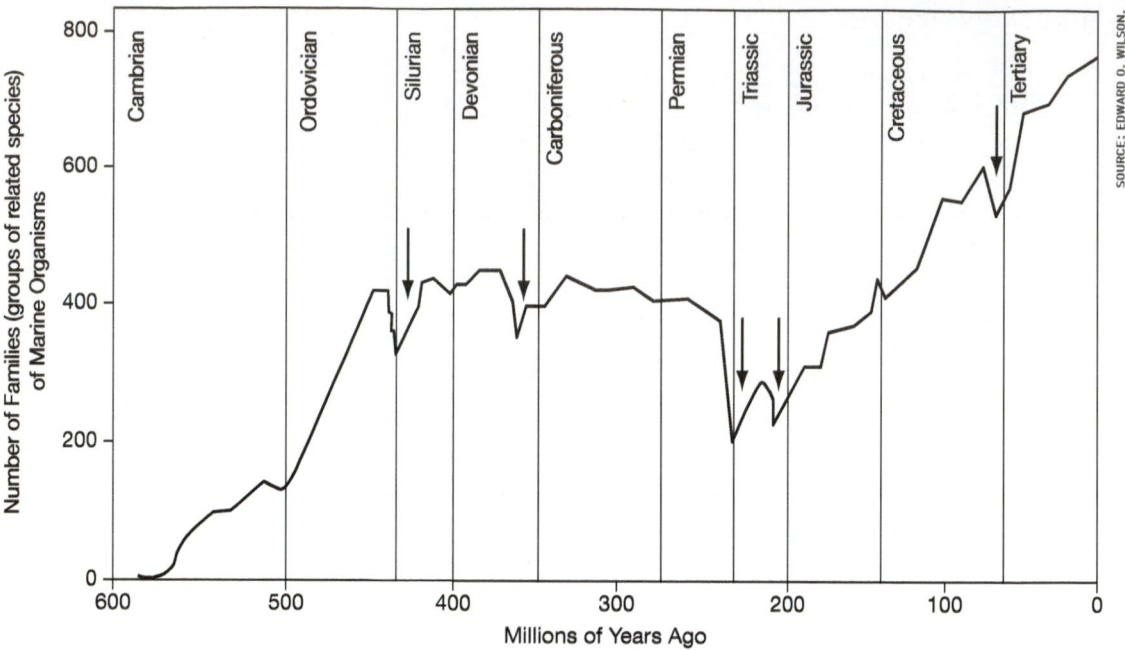

SOURCE: EDWARD O. WILSON.

FIGURE 9.5
The five mass extinctions in the fossil record.

The current rate of extinction is not precisely known, but the Millennium Ecosystem Assessment estimated that known extinctions of mammals were occurring about 1,000 times faster than can be seen in the fossil record.[5] A reasonable projection is that extinction rates might rise by another factor of 10—that is, 10,000 times faster than before humans began to transform ecosystems. This estimate is based on analyses of land that might be cleared for agriculture in high-biodiversity areas in the tropics. More than half of the remaining land that is suitable to be converted to agriculture is located in seven countries: Argentina, Bolivia, Brazil, Colombia, Democratic Republic of the Congo, Angola, and Sudan. The first five are among the twenty-five countries with the highest biodiversity on Earth.[6] Notice that four of the five key countries are in South America, a continent with large rain forests, notably in the Amazon River basin, and with high, steep mountain ranges. The rapid changes in elevation produce many microhabitats separated by altitude.

Our ability to quantify extinction is limited. It is hard to prove that a species has gone extinct. If a species has not been identified, then its absence is likely not to be noticed. Even if it has been identified, the fact that a biologist cannot find individuals in a habitat doesn't mean they are not there, particularly if the species is elusive, such as an insect or a small wildflower that only grows on high cliffs. The ivory-billed woodpecker, a species thought to be extinct since the 1940s, was spotted deep in an Arkansas swamp forest in 2004, but has not been seen again, despite intense searches.

DIRECT PRESSURES

One obvious pressure that humans put on a species is to kill its members faster than they can reproduce themselves. This can happen in poorly managed open-access commons, where the incentives of individual harvesters can encourage behavior that undermines the interest of the group. Bison, once prolific in the Great Plains of North America, were hunted to near-extinction during the nineteenth century following the logic discussed in Chapter 3: because no one owned the great herds, anyone could harvest them, until overharvesting reduced the population from tens of millions of animals down to a few hundred.

Overharvesting can be a product of thoughtlessness as well. Because humans actually use only a tiny fraction of species, our ignorance of our impact on biodiversity is even larger when it comes to species we do not cultivate or harvest. Large quantities of fish that are caught in nets and killed are discarded because they belong to species (called bycatch) that cannot be sold when the vessel returns to port. Dolphins were routinely caught with tuna this way until environmentalists launched an effective campaign to change fishing practices so that "dolphin-safe tuna" could be harvested. Seabirds are also often caught in fishing lines because the birds are attracted to the baited hooks. In some fisheries, the volume of bycatch is equal to the volume of fish brought to market. When shrimp are caught in trawl nets, the bycatch can exceed the shrimp catch by more than tenfold, nearly all of it being thrown away.

A second, even more serious pressure on biodiversity is conversion of habitats by humans, which eliminates the ecosystems that once lived in them. The most serious loss of species on land is likely to be due to the clearing of forest and other forms of habitat modification. Removal of forest or draining of wetland is a step in the conversion of an unmanaged or lightly managed ecosystem into one used intensively by humans for agriculture, industrial facilities such as ports, or urban settlement. In rich nations, roughly three-quarters of the land has been converted to cultivation and urbanization, a legacy of industrialization and industrial-scale agriculture.

Over the past half-century, the rate of land conversion has been highest in tropical and subtropical dry forest, mostly in developing countries.[7] Today, about half the world's work force is in agriculture, and a large number of the people engaged in agriculture are in developing countries, where incomes are low and population growth is high. These are areas where one of the few ways to feed a growing human population is by expanding farmland.

At the same time, the world without edges drives conversion of forest into croplands, often in ways that do harm to local people. The clearing of large tracts of forest in Brazil to support cattle ranching and soybean cultivation, for example, is driven by rising commodity prices and the increasingly tight connection of global markets. Brazilian soybeans are exported to businesses in Europe and China, where they are used to produce soybean oil and provide high-protein feed for pork production. Although this commerce allows Chinese consumers to shift to animal-based diets, it deprives farmers with small landholdings and indigenous communities in Brazil of the land needed for traditional subsistence agriculture.[8]

Figure 9.6 shows farmland on the edge of forest in Madagascar. The forest contains a rich array of species adapted to its complex environment. Many of those species cannot adapt to the simplified ecosystem of a farm field. But without the addition of fertilizers and careful management of the soil, the cleared fields will not support crops for long.

Faced with rapid population growth, the farmers are pressed to clear more and more land, especially as better nutrition and improvements in health care reduce infant and child mortality and the population surges. More and more land is converted, reducing forest habitat and the native species of plants and animals that lived in the forest (Fig. 9.7).

Figure 9.8 on page 246 shows twenty-five biodiversity hotspots identified by ecologists—places where the diversity is exceptionally high. These hotspots are estimated to include more than half of all species. They cover 12 percent of the world's land area, but already hold 20 percent of the world's people. Moreover, the population growth rate in these areas is faster than that of the rest of the world. The growth rate is particularly rapid in tropical wilderness areas, where it is twice

FIGURE 9.6
Aerial view of forest and adjacent farmland in Madagascar.

SOURCE: MILLENNIUM ECOSYSTEM ASSESSMENT.

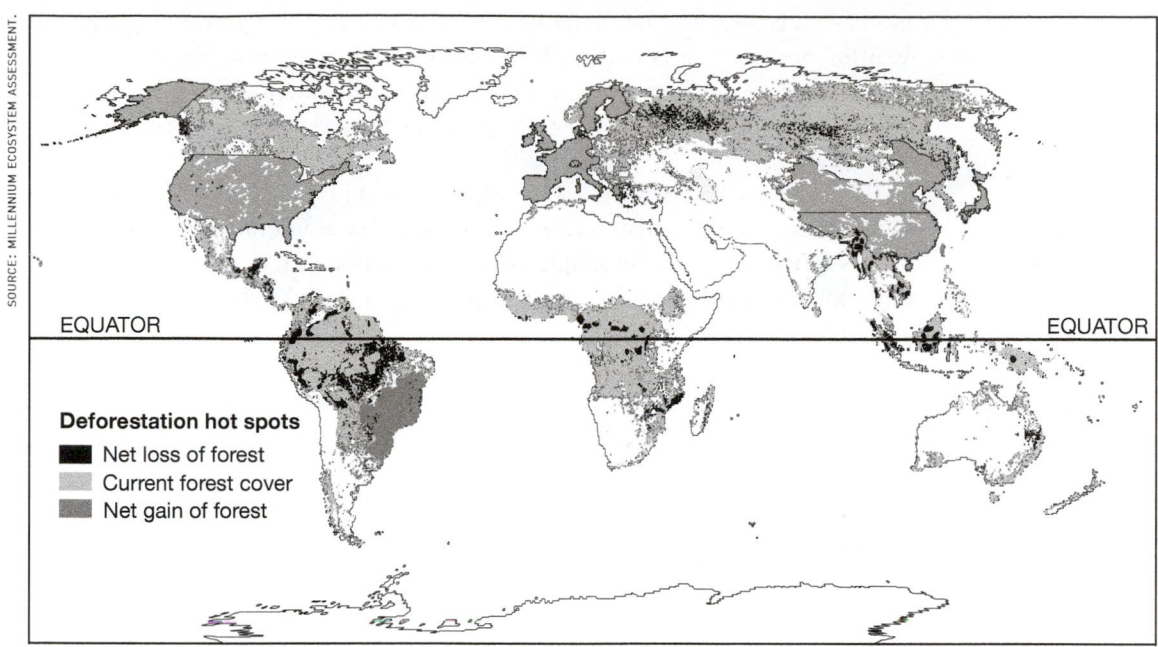

EQUATOR   EQUATOR

Deforestation hot spots

■ Net loss of forest
Current forest cover
Net gain of forest

as fast as in the developing countries generally. The landscape in Madagascar in Figure 9.6 is not an isolated example. These places where high biological diversity has evolved are also where many of the poorest people on the planet are located. Here is the environmental irony: biological riches are often found together with poor people whose population growth rates are high. The English colonists who arrived in America in the seventeenth century were surprised that the Indians seemed so poor when they inhabited such a rich land. Today, after the beneficial end of colonialism, it is no longer acceptable for people in the developed world to say such things, but the irony persists as an international environmental movement struggles to conserve biodiversity.

There is a correlation between the size of an area and the number of species likely to be found within it. And as we saw in Chapter 5, the number of species increases as one moves toward the equator. As more land is cleared, the remaining undisturbed area is cut into smaller and smaller fragments. In the species-rich tropics, rates of extinction are bound to increase, and ecosystems are likely to undergo uncontrollable transformations. We know of ecosystem transformations that have gone badly, such as the loss of the cod fishery off the eastern coast of Canada, or the decline of agriculture in Mesopotamia and other areas subjected to heavy irrigation. These failures suggest that the transformations ahead may produce further disasters.

FIGURE 9.7
Changes in land cover over the second half of the twentieth century, as estimated by Millennium Ecosystem Assessment scientists.

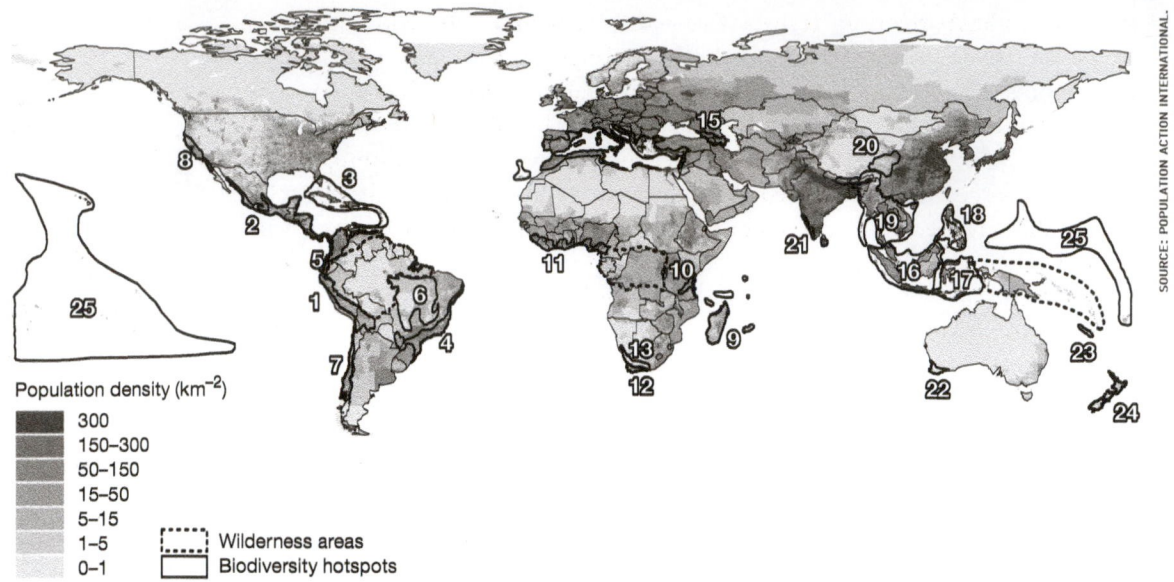

Population density (km^{-2})

- 300
- 150–300
- 50–150
- 15–50
- 5–15
- 1–5
- 0–1

‑ ‑ ‑ ‑ Wilderness areas
Biodiversity hotspots

SOURCE: POPULATION ACTION INTERNATIONAL.

FIGURE 9.8
Biodiversity hotspots include many areas of rapid population growth.

INDIRECT PRESSURES

Beyond the direct pressures exerted when we harvest species and when we convert ecosystems to our own use, humans also press ecosystems indirectly. This happens when we add species to ecosystems that were not present before, when we alter the nonliving components of a habitat by polluting it or by applying pesticides, and when we alter the climate.

For centuries, humans have carried species with them when they travel. Polynesians brought pigs and chickens to the islands of the Pacific Ocean, including Hawaii. In the twentieth century, though, the rapid growth of the world without edges accelerated the transfer of species across oceans, deserts, mountain ranges, and other barriers imposed by geography and climate. Sometimes humans introduce species deliberately, as was the case with kudzu vines in the southeastern United States. Other species escape, such as the domesticated pigs on the Channel Islands mentioned in Chapter 1. Some species have been deliberately introduced without creating ecological problems, as with many of the agriculturally important crops of the world, which originated in places other than where they are now grown. Similarly, salmon native to the North Pacific were released in New Zealand, where they have done well enough to support a thriving recreational fishery.

Others are transferred by accident, such as disease organisms inhabiting the body of an air traveler. One of the great transfers of marine organisms took place in California, when the discovery of gold in 1848 drew wooden ships from all over the world. As the ships arrived, their crews abandoned the vessels to search for gold in the Sierra Nevada, and the boats languished in San Francisco Bay. The wooden hulls of these ships harbored thousands of species of marine worms, seaweeds, and other forms of life that had hung on, literally, for dear life as they crossed the oceans. Like the human crews, many of these organisms jumped ship in San Francisco and transformed the ecosystem of the bay. Over the next 150 years, additional species were brought by humans, leading to the curious shoreline ecosystem described in Box 9.1 on page 233.

Most introduced species die out. They die for lack of suitable food or ways of obtaining it. They are eaten by unfamiliar predators that they cannot evade. They are victimized by parasites or disease. They fail to reproduce because they cannot find mates or nesting sites, or their young die before they can reproduce in turn. Being different is usually a disadvantage in an ecosystem, just as human immigrants arriving in a new place face the pressures of learning a new language, developing new skills, or finding a way to make a living.

Just as some immigrants have been able to prosper in their new environments, however, so have some nonnative species done very well in new ecosystems. Sometimes, the new species has no predator to limit its population, or the new-comers themselves are exceptionally successful predators—as we saw with the golden eagles of the Channel Islands. Local species such as the Channel Island fox can face severe pressures when this happens.

When a new species proliferates rapidly, it causes ecological change—by making other species less abundant, by occupying or changing habitats, and by altering competition among species in a community. These changes can be dramatic. The little zebra mussel, a freshwater species native to Eurasia, was first observed in the Great Lakes of North America in 1988. It has flourished so well that it now covers entire beaches (Fig. 9.9, page 248) and has filled up and blocked pipes used to draw water from the lakes. Like many other shellfish, the zebra mussel lives by filtering food from the water around it. The massive growth of this one species has increased the clarity of the Great Lakes, and sunlight now penetrates deeper into the water—an alteration of the ecosystem that has taken place within a single human generation.

Once introduced, a successful invader species can be impossible to remove. In the Mississippi River lurk Asian carp species that are so fast-growing and prolific, they have rapidly become the most abundant fish in some parts of the Mississippi watershed. Fearful of the impact of the Asian carp on recreational fisheries and ecosystems pressed by the zebra mussel, the Great Lakes states sought to cut off the carp's access to the lakes. In 2010, several Great Lakes states sued Illinois in an attempt to force closure of a barge canal connecting the Great Lakes to the Mississippi River drainage system. But the canal that links a tributary of the Mississippi River to Lake Michigan carries some barge traffic, and the barge operators

FIGURE 9.9
Zebra mussels
carpet the shores
along the Great
Lakes.

persuaded Illinois to resist its neighbors' entreaties. Readers of Chapter 3 will recognize the lawsuit as an attempt to govern the waters of the Great Lakes, a shared commons.

Another significant indirect pressure on ecosystems is chemical rather than biological. It comes through modifications of the chemical constituents in the environment by human action. This occurs deliberately, through the use of pesticides designed to poison species we want to keep away from crops. It occurs as a by-product of industry, in releases of pollutants into the air and waterways, or by the leaching of toxins into groundwater. It occurs as a by-product of agriculture, when fertilizers applied to croplands are washed into streams.

Pesticides and some pollutants have been limited by government regulation, but fertilizer applications have been constrained only by market forces. In the late nineteenth century, an industrial method was devised to convert nitrogen gas from the atmosphere into ammonia. This discovery dramatically increased the availability of fertilizer for agriculture. Today, most of this production starts with natural gas, a fossil fuel. Low-cost fertilizer has meant that farmers routinely apply more than their plants can use. Because unexpected rainfall can wash away fertilizer, for example, farmers apply additional nutrients to serve as a crop insurance policy—but at the expense of some unhappy consequences downstream.

The Mississippi River watershed is sometimes called the breadbasket of America. The vast lands of the Midwest and prairies are used to grow major grains, including corn and wheat, as well as crops such as soybeans. All this is done with the aid of fertilizers. The runoff, rich in compounds of nitrogen, collects from across the watershed and runs into the Gulf of Mexico. The result is a "dead zone" that builds each year as the flood of nutrients feeds marine microorganisms (see Box 6.1: Agricultural Ecosystems in Chapter 6, page 132). As they die, they decompose, a biological process that uses up oxygen in the water at the mouth of the river. The oxygen-starved water, in turn, kills other life-forms. In some years, the area of the dead zone grows to be as large as New Jersey before the end of the growing season in Iowa and Illinois brings an end to the nutrients pouring into the water in Louisiana, allowing the currents to finally sweep the oxygen-poor water away faster than it can be created. The ecosystem then revives, until the cycle is renewed the following spring.

Dead zones have been increasing in number in coastal waters around the world, following the spread of fertilizer application. The seas off the coast of China are experiencing severe oxygen depletion as China's economy has grown, and the use of fertilizers with it. Humans are now putting nitrogen compounds—mostly from fertilizer and the burning of fossil fuels—into the environment at a rate that substantially exceeds the rate at which these compounds are created by natural processes. Such a fundamental shift in the chemical flows in the environment is causing changes in many ecosystems, and these changes are now being investigated in earnest, though much remains to be learned.

One particularly important indirect pressure imposed on ecosystems is global warming. As we saw in Chapter 5, climate is a basic shaping influence on ecosystems. Climate change alters both temperature and the seasonal availability of water. These changes have been hard to recognize amid the normal fluctuations of weather patterns from year to year, although gardeners have noticed that their flowering plants now bloom up to two weeks earlier than they did half a century ago. The change in time of blossoming is not just a curiosity. If pollinators or other species do not (or cannot) shift their seasonal timing as well, the change in flowering times can have wider ramifications.

Climate models of the kind described in Chapter 7 project altered rainfall and more severe storms in many parts of the world, and ecologists are now trying to estimate the impact of these climate shifts on ecosystems that range from farmland to highland forests to coastal waters where fish spawn and grow. And as we saw in Chapter 7, major changes in ice cover in the Arctic and Antarctic have been recorded, with ecological consequences in the Arctic that are apparent to the people who depend on hunting for their livelihoods.

Climate change exerts two additional ecological pressures on the coastal environments where a large majority of humans live. First, as the oceans warm and glaciers melt, the sea level is rising. This means that wetlands will flood more often or be completely submerged, and that coastlines will retreat as the oceans expand. This development has obvious implications for coastal cities. A second pressure comes from the changing chemistry of seawater. The release of large quantities of carbon dioxide (CO_2) into the atmosphere as fossil fuels are burned means that more CO_2 is being dissolved in the ocean, making it more acid. As discussed in Chapter 7, a more acidic ocean affects many species, especially those that have shells or that depend on shelled creatures for food. The acidification of the ocean is a planetary-scale phenomenon that can be addressed only by slowing or reversing the flow of CO_2 into the atmosphere.

The massive oil spill in the Gulf of Mexico in 2010 demonstrated a wider problem facing biodiversity: the accumulation of multiple stresses on ecosystems. The indirect pressures of oil pollution and fertilizer runoff, added to the direct pressures from commercial fishing and the development of coastal areas, all combine to push Gulf species toward extinction. How soon such extinctions will be confirmed by scientists is uncertain, but the stresses are piling up.

Human-dominated ecosystems are almost all subjected to more than one pressure. The resilience of species and their habitats is therefore being tested from several directions at once. When considering how we should address the pressures that are altering the state of ecosystems, we should bear in mind the challenge of multiple stressors.

RESPONSE: SIGNIFICANT PROGRESS, BUT A LONG WAY TO GO

The emergence of environmental awareness and policies has created a large suite of responses. Chapter 12 examines the scope and dynamics of the remarkable family of social inventions that we call environmental policy and activism. Here, we focus on a darker reality: as far as halting the decline of biodiversity is concerned, discouragingly little progress has been made. Humans are continuing to rip the fabric of ecosystems, making it ever more likely that devastating losses of ecosystem services, such as occurred with the disappearance of the Aral Sea, will occur elsewhere (see Box 6.3: A Warning? The Aral Sea in Chapter 6, page 150).

Still, this is not for want of trying—and, in fact, succeeding in many ways. Conservation biologists and other social and natural scientists, environmental advocacy groups and citizens' associations, businesses, governments at all levels, and the UN have all made major strides that would have seemed unimaginable in 1972, when the UN Conference on the Human Environment convened in Stockholm. It is important to look at both what has been achieved and how much remains to be done. Students reading this book are likely to be among those who will shape the outcome of this struggle to tame human enterprise so that a portion of nature can survive, including the human species itself.

U.S. actions to conserve biodiversity within its borders have addressed direct pressures—killing off of species and altering habitats. Some indirect pressures have also been addressed. Policies have now been adopted to control the introduction of invasive species. For decades, motorists entering California have been questioned about whether they are carrying fruit or vegetables that might harbor insect pests harmful to that state's agriculture. Customs inspectors at international boundaries are checking for illegal imports of birds, snakes, and other wildlife. Many ships now flush their ballast tanks in the open ocean, so that the harbor-dwelling species in the water they took on while in port are removed before they arrive at another port.

The first response to direct pressure—keeping members of a species alive—is embodied in the Endangered Species Act (ESA) of 1973, a U.S. federal law that sets out procedures to determine the status of individual species and then provides legal protections that restrict the use of land and other natural resources such as water in rivers. The ESA is credited with bringing back the bald eagle, the national symbol, as well as a number of other species. The ESA has also been roundly criticized by landowners, developers, and many farmers because of the often-drastic limitations imposed on their economic activities by the ESA.

The ESA provides what might be thought of as emergency response: legal measures that attempt to rescue a species on the brink of extinction. Such emergency actions are triggered when the federal government detects a drop in species abundance so severe that it meets the legal definition of an endangered or threatened species. The words "legal definition" are important: protecting a species and the habitat that it needs is a governmental action, invoking the authority of the federal government. Both defining a species as endangered and imposing restrictions on landowners and other interested parties is not only a scientific matter but one that inescapably involves political factors. The well-being of a declining species is an aspect of a commons, and managing any commons requires the exercise of the governing powers of a community.

The ESA is driven by the state of individual species. The law mandates actions to slow decline, but it can do little to improve the health of the ecosystems that endangered species inhabit. Even though the ESA's last-gasp interventions are often sought by activists in an attempt to salvage species beyond the one protected by law, the ESA can focus on only one species at a time. This makes ESA protection usually too late and often too little. It is certainly cumbersome, and those whose economic interests are curtailed find its workings draconian and hard to grasp.

In part, this is how biodiversity conservation must work in the American constitutional system. Conservation means redirecting economic activities. Those activities are mostly ones that put pressure on ecosystems, as with clearing forest for farmland or to build a shopping center. As we saw in Chapter 2, deeply rooted in our concepts of property is the notion of subduing nature and bending it to human use. Deeply woven into our economics is the notion that short-term gains matter more than long-run costs of equal magnitude, a topic we explore in greater depth in Chapter 13. Put these two notions together, and pressure on the natural world becomes deeply legitimate in our culture and pervasive in the choices people make. Calling a halt to these activities, as the ESA does, is an unavoidably wrenching change.

The ESA has worked, in some instances, by prompting last-minute rescues, such as the one that seems to have saved the Channel Island fox. The kind of healthy relationship between humans and nature implied by sustainable development remains remote, however. This is part of the challenge faced by those who would conserve biodiversity. Moreover, the ESA's protection covers only species found in the United States. Elsewhere, protection varies widely, but is mostly weak in the tropical environments that have high biodiversity—at least so far.

In one other arena, policies and institutions have been built with something like sustainability in mind: the regulation of fishing. Fishing is one of the few economic activities in which humans harvest from unmanaged ecosystems. As described in Chapter 3, the open seas have been a commons for a long time. As fishing fleets gradually increased their ability to catch fish, the commons declined. In the 1970s, the coastal nations of the world declared that their national sovereignty extended out to 200 nautical miles—a large expansion over the 12-mile limits observed before. Within that extended economic zone lie most of the commercial fishing grounds.

In the United States, the 1976 Fisheries Conservation and Management Act (FCMA) was passed to regulate fishing and put it under the jurisdiction of regional fishery management councils. These are bodies that resemble the communal management model discussed in Chapter 3. Although appointed by the national government, the regional councils have members drawn from their local areas, and different segments of the fishing industry are represented.

The FCMA did more than create a governance mechanism; it also authorized government subsidies that led to expansion of fishing fleets. This created a situation in which fishers' capacity to catch fish grew, even as the fish populations declined. Faced with these pressures, the regional councils have compiled a mixed record in conserving fish populations. Important fish stocks have declined, and the councils have often been unable to impose restrictions tight enough to reverse the dwindling catch.

A second response to direct pressure is to protect and restore habitat. In this realm, the United States has historically been the leader, declaring the first national park in 1872 at Yellowstone. The national parks have grown into a collection of nearly four hundred sites, some as small as a house lot outside Boston—the home and studio of Frederick Law Olmsted, whom we encountered in Chapter 2—and others as vast as the Gates of the Arctic National Park, which embraces a Maryland-sized chunk of the Brooks mountain range stretching across the northern part of Alaska.

National parks are the most visible component of a large and growing collection of **protected areas**, including state parks, biological reserves, and private lands set aside for conservation purposes. This includes ranch land that owners have agreed to keep as pasture, rather than selling or developing it, in exchange for lower tax rates. And some marine reserves prohibit fishing to protect coral reefs or fish spawning grounds. But legal protections cannot keep out climate change or polluted air.

According to the UN Environment Programme, protected areas have been growing rapidly, although the land area protected has leveled off in the twenty-first century.[9] Marine protected areas are still rising steeply. As can be seen in Table 9.1 on page 254, surprisingly large fractions of land and water areas are now designated as protected areas in all parts of the world. An analysis of these UN figures in 2005 found that 23 percent of all tropical moist forest (rain forest) and 21 percent of coral reefs were at least declared to be protected.[10] This reflects a spreading awareness of the value of environmental protection, together with a rising willingness by international donors such as the U.S. Agency for International Development to pay for environmental protection as a component of economic development.

But this optimistic appraisal has a crucial qualification: the social capability and administrative resources needed to provide effective land or water protection is largely limited to rich nations. In most poor nations, biodiversity remains under pressure on the ground and in the water. Although these nations have areas classified as "protected," protection can be sketchy, particularly in the tropics, where governments are often poor and weak. In many ostensibly protected places, animals and plants are still harvested; land is cleared for agriculture or by logging; roads are built, permitting people to penetrate deep into an area with vehicles; and large blocks of land are cleared or fragmented by roads, putting pressure on species that require large territories, such as the carnivores at the top of food chains.

Such areas provide little of the protection we associate with a national park in the United States. The global database maintained by the UN does not include information on the effectiveness of enforcement and other measures to protect wild species, so there is no estimate of how much of that large acreage actually

TABLE 9.1 PERCENTAGES OF LAND AND MARINE AREAS DECLARED TO BE PROTECTED BY WORLD GOVERNMENTS, 2010.

Region	PERCENTAGE OF AREA PROTECTED	
	Land areas	Marine areas
World (outside Antarctica)	12.7	7.2
Developed regions	11.6	11.5
Developing regions	13.3	4.0
Northern Africa	4.0	4.6
Sub-Saharan Africa	11.8	4.0
Latin America and the Caribbean	20.3	10.8
Caucasus and Central Asia	3.0	0.4
Eastern Asia	15.9	1.6
Southern Asia	6.2	1.2
Southeastern Asia	13.8	2.1
Western Asia	15.4	2.2
Oceania	4.9	2.8

Source: Data from International Union for Conservation of Nature and UN Environment Programme World Conservation Monitoring Centre, The World Database on Protected Areas, January 2011.

conserves species. Because the UN is not a government, it is unable to compel nations to report this data; instead, it is an organization that must rely on voluntary reporting by its member states. What they do not wish to report does not enter the UN's statistics. Thus, the real meaning of this "protection" is not really known.

In 2004, the World Wildlife Fund studied nearly two hundred protected land areas in thirty-four countries and found that only 12 percent had implemented a sensible management plan. In sum, the major threats to habitats continue, to a large degree, and it is at best uncertain whether responses through habitat protection are effective, particularly in the tropics.

This is discouraging, but it also helps to frame the challenge that lies ahead. Conserving species is new; it sometimes runs against the grain of economic development as it has been understood since the eighteenth century. In the many cases in which the economic harms of conservation are imposed on a small minority of economic actors, these actors are often politically powerful and motivated to resist. It is no surprise that hopes and declarations run ahead of implementation.

The growth of conservation has been rapid enough, with sizable funding from rich nations, that there is now controversy over whether saving biodiversity has been done at the expense of poor people. This controversy is one we recognize from our own culture. The way we think about preserving land is by excluding people. This is what a national park or a biological preserve is, a kind of property that is reserved for the particular land use called wilderness.

Yet, unlike North America when the English arrived in 1607, the developing countries are already densely populated. The people who live there think of rain forests and mangrove swamps and coral reefs as landscapes to use, the way the American Indians did in Virginia. The difference in the Americas was that so many of the native peoples died off when they were exposed to Old World diseases. But today, preserving species by excluding people has become troublesome, not only because it is difficult from a practical standpoint, but because dealing with the people living in these high-diversity habitats raises questions of social justice. Indeed, in many places traditional societies have long lived sustainably in local ecosystems, coexisting with the other life-forms. In the coral reefs of the western Pacific, conservation has been shifting to locally managed marine areas administered by traditional communities.

A hopeful example is provided by Costa Rica, a developing country that has pioneered ecotourism, the use of conservation lands for educating and entertaining people from rich countries. Ecotourism provides an economic motive to conserve biodiversity, just as it would be in your best interests to protect the art works in a museum if running it was your livelihood. This is the logic of using privatization to encourage sustainable development.

Another economic motive for preserving biodiversity is that a large fraction of the medicine used in the world today originally came from natural products— and much of it still does. A majority of the drugs most often prescribed come from natural sources—fungi, animals, and plants. Nearly 80 percent of the products used to fight cancer come from natural products, and a similar proportion is found among antibacterial products. The blood-pressure medication captopril, which is made from the venom of an Amazonian snake called the *jararaca*, has earned billions of dollars for one of the pharmaceutical firms that manufactures it. It is reasonable to suppose that among the millions of species that have yet to be studied (including those that have not been identified), many valuable medical discoveries may be made. Another example is paclitaxel, a treatment for ovarian cancer. Paclitaxel was extracted from the Pacific yew, which grows wild in the rain forest of the American Pacific Northwest. Now made synthetically using biotechnology, this highly effective medicine might not have been discovered had nature not made it first.

Some attempts have been made to assign rights to developing countries, so that they will have economic as well as moral reasons to conserve biodiversity. Brazil, the home of the jararaca, did not earn a penny from captopril, and Brazilians are understandably anxious not to let other biological riches from their lands and waters be used without benefiting from them.[11] But an international biodiversity treaty signed in 1992 has not worked as well as its authors hoped it would. This is an arena in which determined innovation in international negotiation and diplomacy is needed.

How does biodiversity matter? First, because there are species, including agricultural crops, that we rely on. As changes in climate or other conditions force changes in cultivation, it is essential to have related crop varieties to draw from, even as biotechnology widens the options to modify and design the genetic makeup of species.

Second, biodiversity matters because we cannot predict the trajectory of ecosystem reorganizations. Some of these, such as loss of important fisheries, have seriously harmed human populations. At the least, it makes sense to proceed with great caution when we can tell that a species is nearing extinction. This is the point of the U.S. Endangered Species Act.

A third reason that biodiversity matters is ethical. As we saw in Chapter 2, powerful evidence visible in the biochemistry of DNA suggests that all living things are related. Even if one does not accept the Transcendentalist vision of nature as sacred, the interconnectedness of species inspires in most people a sense that we also belong to the natural order. And this suggests that we should respect our fellow species and take care before we drive any to extinction.

The trajectory of human development over the past three hundred years, however, strongly suggests that people will need to work harder for some time to come to slow the erosion of biodiversity.

The warnings of conservation biologists have so far raised the profile of biodiversity but have rescued only a small fraction of the ecosystems needed to create a sensible sampling of the library of life. Nonetheless, there has been a dramatic expansion of biological preserves. Making them work is a challenge in which environmentalists in the United States and other countries are making important contributions. To a large extent, this means working in the tropics, often under trying conditions. Getting real-life experience in these settings is important for someone interested in conservation biology, to see if the challenges are ones that are personally worth tackling.

Biodiversity conservation is a grand challenge: the scientists have given warning, but the institutional and technological innovations available are not capable of doing the job, and political will is frail in many places. In this sense, we do not have a solution in sight.

Two things are clear about the grand challenges we have examined in this chapter and the two preceding ones. First, the task that lies ahead is daunting, as people must grapple with climate and energy, urbanization, and the loss of biodiversity. Second, significant responses have been launched to address each grand challenge, with promising innovations undertaken on a limited scale. This suggests that sustainable development is far from being achieved but that it is not a vain hope. Sustainable development is the grandest challenge of all, encompassing the others we have examined. In the rest of this book, we explore the cultural shifts, knowledge, and human resources already available to meet the challenge of sustainability, and we will gain perspective on the conflicts already under way as we move unsteadily toward a world that can be governed in an environmentally responsible fashion.

FURTHER READING

Millennium Ecosystem Assessment. *Ecosystems and Human Well-Being: Synthesis.* Washington: Island Press, 2005.

Wilson, Edward O. "Nature's Last Stand." Chap. 3 in *The Future of Life.* New York: Alfred A. Knopf, 2002.

KEY TERMS

ecotourism

genome

pressure-state-response

protected area

resilience

Chapter Ten

A PERSPECTIVE ON SUSTAINABILITY

"Man is the only animal that blushes—or needs to."
—Mark Twain

THE SITUATION OF THE "MOST SUCCESSFUL" SPECIES

The last four chapters have explored four very different perspectives on the complicated, surprising, and sometimes troubled interaction of humans and ecosystems. One thing we have established is that humans appear to be highly successful in dominating ecosystems. Whether humans can become sound stewards of enough ecosystems is unclear, though it is also clear that we must meet this test if we are to survive, let alone claim to be the "most successful" in any meaningful sense. In this chapter, we draw the perspectives of Part II together to examine sustainability, the question of whether humans are likely to endure and flourish on this planet while maintaining a high quality of life. In Chapter 1 we introduced the idea that sustainable development is "development that meets the needs of the present without compromising the ability of future generations to meet their own needs."[1] In this chapter, we will see why sustainable development is a grand challenge.

To do that, we will review ideas developed in the first nine chapters of this book. We sketch the outlines of the invisible present—how humans have interacted with the natural world over long periods and at large spatial scales. Because of the world without edges, our species has developed the ability over the past two centuries to alter the world at large scales, with impacts such as climate change affecting the entire biosphere. Some of these impacts will be felt for centuries; and some, such as the decline in biodiversity, have already altered the world irreversibly. Understanding the big picture in its large dimensions provides both a summary of the challenges facing our species and a framework for evaluating the responses available to meet these challenges, the task of the five chapters that lie ahead in Part III.

What are the circumstances of the "most successful" species today?

- The rate of population growth is decelerating, but human numbers are still increasing and rapid urbanization will continue.
- Economic growth is transforming societies.
- Human well-being has been improving, despite widening inequality.
- Human domination of nature continues to increase.
- People remain dependent on ecosystem services.

One implication of these trends is that environmental considerations have risen on the agenda of the human race, not because environmental activists are noisy but because the material economy faces real challenges. These challenges show up as

---

Learning Objectives
When you have finished studying this chapter, you should be able to

- put the environmental challenges of the twenty-first century into the historical context of the past thirty, three hundred, and ten thousand years;

- explain the connections between climate change and the way the Earth looks at night from space or an airliner;

- describe how both of these statements are correct: (1) economic inequality across large regions of the world is increasing, and (2) on average, people in every region are becoming economically better off;

- identify the difference between well-being, as measured by the Human Development Index, and average income, as measured by gross domestic product per capita;

- compare the labor you must devote to acquiring the energy you use in your daily life with the labor needed to supply a much poorer person's energy needs in a poor tropical country;

- identify the scale of your competence in an everyday activity such as food consumption or assuring a supply of clean water.

a lack of sanitation for half the human race; as lives shortened by pollution, such as the prevalence of lung disease in China's coal-dependent economy; in abrupt changes in gasoline prices; in the urbanization of the world's population; and in many other circumstances for which the word "environmental" does not seem apt at first. But these are all problems of sustainability, of balancing the needs of the present against those of the future.

The material economy transforms nature. Where there are poorly governed commons, there are environmental problems. Such tragedy-prone commons are everywhere, although not every commons is poorly managed.

Over the next two generations—that is, during the span of the careers and lives of today's students—we are likely to face major difficulties. But there will also be major opportunities. The range of opportunities is implied by the wide range of issues listed above. There is a need for people who understand the pursuit of sustainability in this wider sense, who can bring the knowledge we are sampling in this book to the debates that will shape decisions in every segment of economy and culture.

THE HUMAN TRAJECTORY

How have we humans arrived at this astonishing combination of challenge and opportunity? Humans exhibit a striking diversity—of nationalities, levels of consumption, and ways of life. Yet the genes of all humans lead back to our ancestors in Africa, the technologies that condition our lives owe much to the scientific revolution of Renaissance Europe, we are all dependent on agricultural production and other ecosystem services, and we are all linked to the commercial, technological, and cultural fabric of the world without edges.

The human species came to dominate in stages. Our distant ancestors—whose fossil remains preceded those given the name Homo sapiens—emerged in eastern Africa's Rift Valley, in what is now Kenya. We know about these ancestors from a small number of fossils from sites in different parts of the world. Paleoanthropologists, as those who study these and related fossils are called, are still working out many elements of our story as a species.

By studying the genetic variations in living humans, scientists can estimate the degree to which we are related to one another. By comparison with many other species, we are genetically very similar, from Indonesia to Spain, from the Arctic to the tropics. The fact that our genes are closely related implies that all humans are descended from a population of roughly ten thousand, some of whom migrated out of Africa more than a million years ago.

Our species spread over the world, and by about twenty-five thousand years ago, humans reached the New World, hunting and gathering as they moved eastward from Asia during the last Ice Age. The quantity of ice in polar and mountain glaciers was large enough then that the sea level was more than 100 feet lower than it is today, exposing a land bridge along the track of the Aleutian Islands between eastern Russia and western Alaska. By the time humans reached the Western Hemisphere, the species was already present in all the other continents except for Antarctica. Some remote islands, such as some in the Pacific Ocean, were not settled until after the time of Jesus, two thousand years ago. The South Pole did not have a year-round settlement (a research station) until 1957.

At the end of the last Ice Age, about ten thousand years ago, as the glaciers retreated in the face of warming climates, humans discovered agriculture, a development we discussed in Chapter 6. The domestication of highly productive plants and animals drew our species, which had lived by hunting and gathering for tens of thousands of years, into a settled way of life. This led to extensive environmental changes, as land was cleared for crops and irrigation channels replaced natural watercourses in many parts of the world. The much higher and more reliable production of food from farming also led to large increases in human populations. Permanent human settlements spurred the development of language and the elaborate social structures recorded in human history. The great religions of the world—including Hinduism, Buddhism, and the Judeo-Christian-Islamic traditions of the Bible and Qur'an—all emerged and flourished in agricultural societies.

Historical change accelerated once more during the colonial and industrial expansion of European culture and technology that began with the Renaissance and the discovery of the New World. This process produced the Industrial Revolution in the last decades of the eighteenth century as well as the successful harnessing of fossil fuels, as we saw in Chapter 7. When we met Gilbert White of Selborne in Chapter 2, we learned that the cultural idea of environmentalism also arose at the beginning of the age of fossil fuels.

Environmental sociologist William Catton has described colonialism and industrialization as the two most important breakthroughs in human history—and also as the two main roots of our current environmental challenges.[2] According to Catton, for most of our history, people lived in localized economies and accepted the limitations of doing so. Gilbert White saw a widening set of possibilities in the eighteenth century but decided not to pursue them, calling himself a "stationary man" (Chapter 2). With the discovery of whole new continents, however, it became possible for Europeans to create colonial trading relationships and obtain goods from far away. With the development of energy technologies, our ancestors could also appropriate energy from the distant past by burning fossil fuels. In Catton's view, the transformation is impressive, yet ominous because we have not yet learned to live within sensible limits, the way our ancestors needed to live within their own limits until just a few hundred years ago.

Industrialization brought with it expanding human domination of the natural world, as you read in Chapter 6, together with another round of rapidly increasing human populations, discussed in Chapter 8. Although cities emerged with agriculture more than five thousand years ago, the scale of urban settlements shot up when faster and more powerful means of transportation made it possible to create the city-centered economy of the world without edges.

The ability of humans to affect the natural world escalated as we gained the capacity to organize larger numbers of people and, over the past two hundred years, harness the powers of industrial technologies. We do all this with brains and genetic endowments that seem not to have changed much for more than a hundred thousand years. As the pace of change increases, and as humans do more to intervene in natural processes that we understand incompletely, the chance of serious, irreversible damage also increases. Yet remarkably, the trajectory of the past two hundred years has been one of improving health and growing wealth. Some species have gone extinct, but their absence appears to have caused little inconvenience to human societies, at least so far, as discussed in Chapter 9.

As environmental thinkers such as Catton remind us, however, whether the human species can continue its unprecedented record of success is unclear.

POPULATION IS STABILIZING, BUT URBANIZATION IS ACCELERATING

One measure of the success of a species is its abundance. Human abundance has been increasing dramatically since the middle of the seventeenth century. At the end of the twentieth century, humans numbered more than 6 billion. But it does look as though the end of the demographic transition discussed in Chapter 8 is approaching, and has been doing so for more than a generation. As is shown in Figure 8.4, the rate of growth in human populations has been declining since the 1960s, and the absolute numbers of people added to the population each year have been dropping since 1990.

The projections suggest that the human population may stop growing during the lifetime of today's students at a level between 9 and 11 billion, or roughly 50 percent higher than today's 7 billion. A projection is not a prediction. This is an important point: the population analyses discussed in Chapter 8 do not include specific economic or social factors. Nonetheless, the length of time that the growth rate has been declining suggests that the slowdown is real and still under way.

FIGURE 10.1
Earth at night.

As population grows by another 50 percent, the human species will also become more urbanized. One can see the distribution of cities on Earth by looking at the pattern of lights shining at night (Fig. 10.1). The rich parts of the world are the brightest now, but brightly lit areas are spreading, particularly in Asia, and they are likely to continue to do so.

The stabilizing of the human population will change a lot of other things. The current political debate over the future of Social Security is one example. As the proportion of young people decreases, the proportion of old people increases. Every community struggles to pay for those who are too old and too young to contribute directly to economic output. This is a problem to which no nation has found a solution. In poor societies retirement is a family concern, but in rich ones it is also an issue for governments. You might not think of retirement or Social Security as an environmental problem, but it is a problem of sustainable development. Meeting the needs of the present competes with the desire to afford future generations the ability to meet their own needs.

The broad pattern seems clear: population growth rates have declined as people become prosperous and, in particular, as the economic and social status of women has changed. But this pattern is not entirely consistent. Some nations, such as Bangladesh, have seen declining population growth rates before their prosperity has grown. This can occur when women and girls obtain new access to educational and livelihood opportunities even at relatively low income levels, leading a number of researchers to conclude that improved education for women, rather than overall levels of economic development, may be the real force in slowing population growth. Also, it is important to see that a slowdown in population growth may be less helpful for the environment than one might expect. The continued rapid increase in consumption is increasing what is loosely called the "human footprint," both in rich countries and in countries rapidly becoming richer, such as Brazil or India. Our enlarging footprint can outstrip a slowdown in the number of feet (the number of people added to the population). We will look more closely at consumption in Chapter 14.

As the momentum of the demographic transition has become clearer, climate change has displaced population as the chief concern of environmentalists. Similarly, over the past generation, concerns over loss of wilderness in the United States have broadened into support for conservation of biodiversity worldwide, with a focus on tropical waters and terrestrial ecosystems outside U.S. borders. And we may now be seeing concerns over suburban sprawl widening into an awareness of the environmental and social implications of rapid urbanization, particularly in Asia's booming economies and in the poorest nations. Environmentalists are, in these ways, coming to see the world without edges as the province of their activism.

ECONOMIC GROWTH IS CONTINUING, BUT POVERTY PERSISTS

One of the measures of human success is economic growth. From an environmental perspective, growth includes two very different kinds of change. The first is an expansion of the material scale of the economy—the human footprint. Second, growth also includes increases in the value and sophistication of economic activity. The widespread adoption of computers, for example, has increased the value of many workers' time on the job, increasing profits and sometimes pay. However, these increases may not mean that the impact on the natural world has increased. Raising the energy efficiency of a building is one example: it raises the value of the building in the real estate market but decreases its carbon footprint. Yet overall, the human domination of ecosystems has continued to expand, and economic

GROSS DOMESTIC PRODUCT:

Imperfect but Influential

GDP is defined as the market value of all final goods and services made within the borders of a country in a year. "Final" means that a market transaction belongs in the GDP if a consumer buys it, rather than a business simply buying something as part of its own production process. GDP includes, among other things, the value of food, gasoline, college tuition, and rent. It also includes expenditures made by government on behalf of its citizens, such as for enforcing government regulations or paying for police and fire departments.

GDP is a measure of economic activity, but it is usually understood to mean something much more: a measure of economic welfare or the well-being of the population. This wider meaning leads to highly consequential errors, as we will see below. But it is also nearly unavoidable because of the significance the GDP numbers have acquired since the measurement was first devised in the 1930s as part of an important advance in economic science. The Great Depression of that period afflicted all the countries in the world, with unemployment rates reaching more than 25 percent in the United States. Faced with this collapse, governments were drawn to intervene in the economy in unprecedented ways. This wider role for national government has now become accepted, and the responses to the worldwide recession of 2008 by Presidents Bush and Obama fit the pattern developed in the 1930s.

But if the government is going to intervene in complex markets, it needs to know what is happening economically. This need spurred rapid innovation, led by Simon Kuznets, a University of Pennsylvania economist who developed a system of national economic statistics, of which the GDP is the most prominent descendant. Kuznets was recognized with the Nobel Prize in economics in 1971, and these statistics have become an indispensable part of public policy making in a wide range of fields, including environmental regulation. Perhaps more significant, GDP has become the most widely watched gauge of the economy—and of political performance. GDP growth or decline has, over time, become correlated with the president's approval rating.

Economists and others who study economic activity recognize a wide range of shortcomings in the GDP indicator. GDP does not measure wealth or inequality, and those two important dimensions of economic welfare are not captured by the widely publicized GDP numbers. As an indicator, GDP does not distinguish between economic activities that are undertaken to restore damage, such as the cost of rebuilding after a hurricane, and those that add in a positive way to human life. This produces anomalies: GDP increases when people receive high-cost medical treatments, but not if they are healthy and do not need medicines; GDP increases when people buy burglar alarms but not when crime decreases; GDP increases when air pollution damages crops or causes lung disease, but clean air does not change GDP. From an environmental perspective, this last example points to the fact that environmental harm originating from a poorly governed commons inflicts costs that are not captured in market prices. We will return to this subject in Chapter 13.

If GDP is limited and misleading, however, attempts to replace it have so far fallen short, in part because of the political significance that GDP reports have acquired. Later in this chapter we discuss the Human Development Index, a measure that incorporates GDP as only one dimension of a summary measure of human well-being.

growth is often a reasonable way to estimate the widening impact of our species on the world.

A widely used economic indicator is the gross domestic product (GDP) per capita (see Box 10.1: Gross Domestic Product: Imperfect but Influential, page 265). Consider the trends in GDP per capita over the past three centuries (Fig. 10.2). In this graph, note that the vertical scale is **logarithmic**, so that a straight-line trend represents **exponential growth** rather than linear change. This means that the figure is a description of continued doubling and redoubling. This kind of growth leads to major implications for human domination of natural cycles within a single lifetime, as happened when the beavers of New England were driven to commercial extinction less than a century after the arrival of the English settlers. Often, environmentalists have wondered whether economic growth can continue—a worry that has attracted attention among economists, too, in the wake of the global financial crises that began in 2008.

In the graph, notice that the current GDP per capita for Africa is approximately the same as North America in 1830, the time when the Hudson River School artists were painting and Henry David Thoreau made his sojourn to Walden Pond. It is ironic to think that when New England was roughly as rich as Africa is today, Thoreau was retreating from a society he felt was too materialistic.

The richest populations today include those whose populations are growing most slowly. This has not always been true. In the nineteenth century, for example,

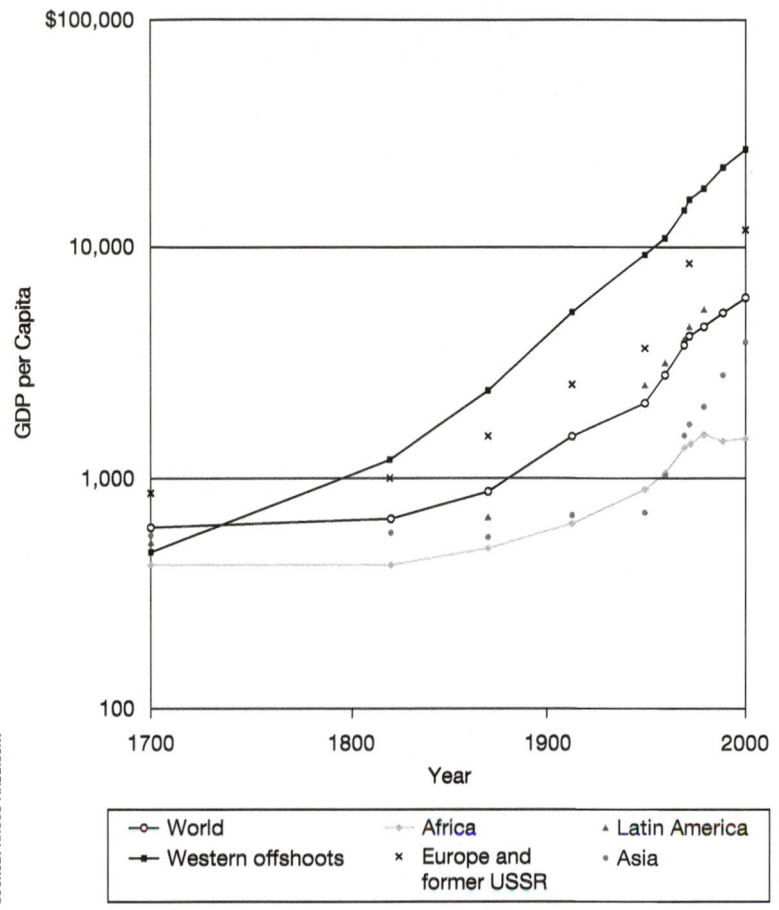

SOURCE: ANGUS MADDISON.

FIGURE 10.2
Gross domestic product (GDP) per capita, 1700–2000. GDP estimates have been adjusted to take into account inflation and differences in purchasing power in different national economies. (Western offshoots = Australia, Canada, New Zealand, and the United States.)

North America was growing rapidly in both population and economic terms, a pattern one sees in India today.

As you can see, the differences among the regions of the world in economic performance have been visible for a long time. The long-term trend, which stretches back to the eighteenth century, is toward widening **economic inequality**. The gap has been opening faster over the past generation. Today, the difference in GDP per capita between the richest and poorest regions is about a factor of 20, whereas this difference was about a factor of 2 in 1800. Figure 10.3 on page 268 shows how the world's 6 billion people in 2000 were distributed with respect to income; Americans are found at the right-hand end—many in the small tail at the extreme right. Notice that the inequality in Figure 10.2 has come about as the rich have been getting richer faster. The poor are getting richer, too, but more slowly.

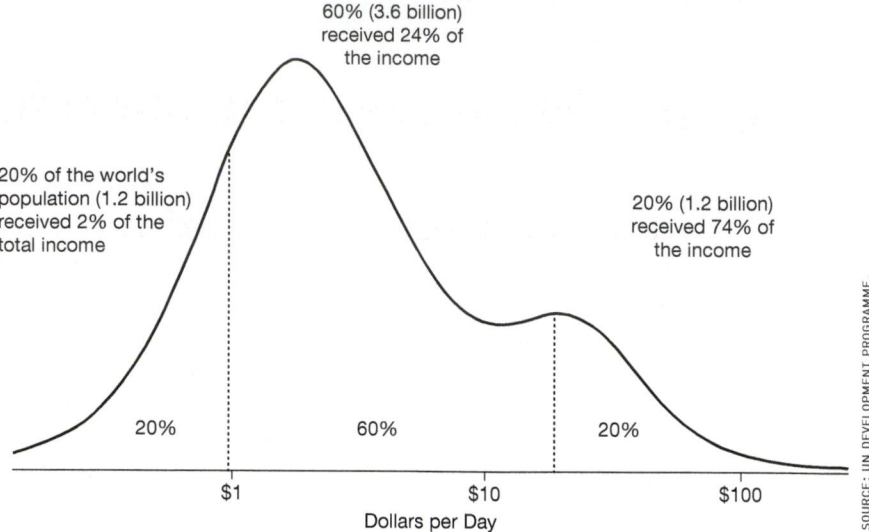

60% (3.6 billion)
received 24% of
the income

20% of the world's
population (1.2 billion)
received 2% of the
total income

20% (1.2 billion)
received 74% of
the income

20%   60%   20%

$1 $10 $100

Dollars per Day

SOURCE: UN DEVELOPMENT PROGRAMME.

FIGURE 10.3
Distribution of
income in the
world in 2000.

Also bear in mind that what these plots show are averages over large populations. The very poorest people in every society, even the richest societies, are much poorer than these figures suggest, and the very richest are many times as rich. The differences in economic circumstances between individual families on the planet is thus far wider than these average numbers indicate.

More than 1 billion people worldwide live on less than a dollar a day. By comparison, the daily cost of attending a U.S. private college such as Dartmouth or Williams is more than 100 times as great. What does this extraordinary figure of a dollar a day mean? People cannot live on this amount of money. Instead, they live largely outside of markets, usually pursuing subsistence agriculture. They literally work for a living, getting food and meeting other material needs by labor rather than by getting paid and buying things.

One aspect of work in a very poor country is gathering wood for fuel—back-breaking labor that takes many hours every day. This work is done mostly by women and children in traditional societies (Fig. 10.4, page 270). Where populations grow but technology does not change, wood gathering puts increasing pressure on forests as trees are cut more rapidly than they can regrow. This also means that the wood gatherers must hike farther from home. Living in the wealthy part of the world without edges, we grasp only weakly the meaning of a subsistence life—both its moments of happiness and the too frequent frustrations and loss. (See Box 10.2: The Volunteer in the Cow Path, page 269, for one view.)

BOX 10.2

THE VOLUNTEER IN THE COW PATH

Living in a developing country for a time is an indispensable experience for an American who wants to be aware of the challenges of sustainable development. In the developing world, one finds three-fourths of the world's people—those for whom meeting the needs of the present are both difficult and urgent. In the tropics, where the developing countries are nearly all located, one finds the preponderance of the world's biodiversity. There, too, are some of the greatest opportunities to shift the trajectory of energy use into directions that slow global warming and promote human well-being—opportunities that will require substantial investment from rich nations. That is, the tropics are an arena in which sustainable development will either succeed or fail, and where the ability of future generations to meet their own needs will be tested, in parallel with the global ambition to improve well-being.

Dorothea Hertzberg is an American who joined the Peace Corps and was sent to Burkina Faso, a nation in western Africa whose Human Development Index score put it 174th out of 177 countries in 2006. One day, her bike broke down, and a man from the local Mossi people stopped to help her. He did not know how to repair the fancy American mountain bike Hertzberg had been issued, so he tied her bike behind his and towed her 7 miles through the blistering midday heat to the village she was trying to reach. She wrote in 2003:

> Two years ago, at the age of 27, I volunteered for Peace Corps service to "give back" to the world. Today, I realize I gained much more in return....When I think back on that moment when I was stranded on that deserted cow path, there was a part of me that was calm, because I knew where I was. I was in a place where you never feel alone or abandoned because someone will always come along to help you; where a starving woman would give her last bowl of food to a stranger; where kids are elated to play with an old tire and a stick. A place where family unity is everything and the guest is paramount.[1]

By living among the people of a developing country, one learns, first, that some of the conveniences of life in a rich country are really important, but others are not. One also gains factual, analytical, and experiential understanding of ways of living that are at once different and profoundly similar to our own.

1. Dorothea Hertzberg, "A Lesson in Giving," *New York Times*, August 23, 2003, A13.

FIGURE 10.4
Villagers gather firewood in Java, Indonesia.

The poorest people today live in ecosystems different than those found in rich countries, however. Moreover, traditional uses of land and water are organized on principles that often conflict with notions of modern government (see Box 10.3: Nature, Wealth, and Power, page 271). For these reasons, the science and agriculture developed by rich countries cannot simply be transferred to the poor. In addition, large cultural gaps—in many instances turned into wounds by the cruel inequities of colonialism—have slowed the attempts of poor countries to enter the modern economic system. Since the end of colonial rule in the 1950s, the institutions of rich nations have sought to bring rapid economic growth to developing countries, most of which are located in the tropics. These institutions include the World Bank, the United Nations, and national aid agencies such as the U.S. Agency for International Development. Economic growth has been ardently desired by the leaders of poor countries and many of their citizens. Yet, as Figure 10.2 shows, these desires have been frustrated, and the gap between poor and rich has widened at an accelerating pace.

Only Asia has seen sustained regional growth, and even there one finds great contrasts between nations. Singapore and South Korea have created developed economies like Japan's, but the Philippines and Burma (Myanmar) have struggled. With the acceleration of growth in India and China over the past generation, however, have come large environmental costs, such as life-threatening levels of air pollution in industrializing cities.

BOX 10.3

NATURE, WEALTH, AND POWER

Although ecosystem services are essential to life and well-being the world over, different societies organize their dependence on nature in different ways. In Chapter 8, we discussed the central role of infrastructure in creating a Second Nature in settlements familiar to Americans. In the western African desert nation of Mali, we see a different approach, yet it is one that makes sense when one thinks of the discussion of commons in Chapter 3.

In the Malian village of Songo, the Dogon people have lived sustainably for generations, although the severe droughts in the Sahel not far away devastated large areas in the 1980s. In the photograph below, notice the parkland agro-forestry system of cultivation in the background: irregularly spaced trees stand amid farm fields bordered by lines of rocks, designed to retard the flow of rainwater from the infrequent storms that visit this arid land. In the foreground, within the village, granaries are clustered together.

In such a landscape, people manage their ecosystem services with care. Grass is essential to the survival of the animals around which the village economy revolves. In the second photograph, in a village not far from Songo, the man on the left is holding a rake made from the branch of an acacia tree. He has been gathering grass to haul off in his cart. But there is a problem: the grass is not his. He is from a distant community. The men on the right, from the local village, are enforcing the

Songo, Mali.

rules of the commons by confiscating the grass he has gathered. This is what happens in well-managed commons: a community lays claim to it, and its members monitor and enforce the rules that keep the use of the commons sustainable.

Community rules are determined by the elders of a village, shown in the third photograph meeting in the sparse shade of a tree in the center of the settlement. But they face a problem. Under Malian law, the land does not belong to traditional communities such as theirs. It is instead public land that is owned by the government. This means that an outsider might come with a government permit to harvest timber or

Enforcing rights to harvest grass for animal forage.

another valuable resource, and the national authorities, acting through the forest service officer in a nearby town, may enforce that permit over the objections of the local community.

Remember that the last of the eight principles of governing commons listed in Chapter 3 is that higher levels of authority must allow local rules to work. Where higher levels intervene, the commons may slide into tragedy. In this harsh environment, the word tragedy is not a metaphor but a description.

Until the late 1990s, international development agencies such as the World Bank favored modernization. This meant strengthening national governments' authority over their own territories so that economic development might take hold. This approach inadvertently fed corruption and incompetence in many governments, however, so there has been pressure in the last decade for decentralization, based on the recognition that top-down management does not work in many situations in very poor countries that lack adequate government administrators.

Yet the way the community organizes itself to govern its commons may not be fair or democratic as an American would see it. The elders in this village are all older men, and the minority ethnic groups of the community are not included in the governing council.

The Songo council of elders.

This puts a different light on top-down rule, perhaps. Consider the civil rights movement in the United States half a century ago. Quite a few white people living in racially segregated communities thought they were managing their commons—drinking fountains, buses, and public schools—just fine. So the imposition of rules in the name of majority rule at the national level can be disruptive to local communities, sometimes for a legitimate purpose.

Legitimacy is problematic in many situations in the developing world, where democratic rule can be dodgy and where local management of natural resources may indeed be far more sustainable when local people are in charge. Whether management in this particular village is better left in local hands is not so clear from the U.S. perspective, however, in part because this apparently isolated village is in fact connected to the world without edges.

In some parkland ecosystems like the ones shown in these images, people gather sap from trees and sell it in markets that put the gum into consumer products we all know, such as the soft drink Mountain Dew. As we saw with the history of beavers in New England, connection of local resources to distant demand can lead to unsustainable harvesting and other practices that put pressure on ecosystem services and biodiversity.

The point for those of us in rich nations to take away is that community governance of commons can work and may be superior to the superficially modern institutions of central governments. This is a possibility that has often been overlooked in the headlong but frustrating pursuit of economic development.

We must not forget that the rise in economic activity has transformed the material conditions of human life. That is, despite widening economic inequality, the life chances of poor people—their **well-being**—have improved substantially in at least some parts of the world, such as East Asia. How do social scientists try to assess this notion of overall well-being? It is widely agreed that monetary income, although important, does not capture human well-being adequately. The UN Development Program, the world's poverty agency, has a simple way to bring together other important dimensions. The **Human Development Index** is a useful summary number that combines income with measures of health and education. The idea is that a population can enjoy significant well-being, even when its money income is not high, as long as people have good health and their children are learning the skills needed for life beyond subsistence agriculture. By this measure, the well-being of people in Albania (GDP per capita just under $4,000 per year) is slightly higher than that of people in Saudi Arabia (GDP per capita of $14,000 per year).

By these measures, human well-being in most regions has increased in recent generations (Fig. 10.5, page 274). Efforts the world over to cure disease, fight malnutrition, and provide basic education have made a real difference to billions of people. These are efforts carried out by pharmacists in small towns in Latin America, by the World Health Organization and the Rockefeller Foundation, and by teachers in villages and cities from western China to the Muslim suburbs of

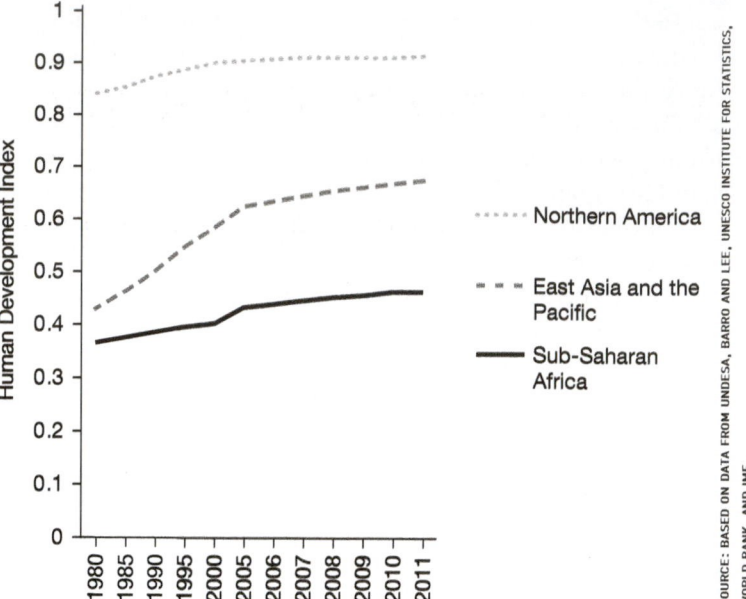

FIGURE 10.5
The Human Development Index in three geographic regions, over time.

SOURCE: BASED ON DATA FROM UNDESA, BARRO AND LEE, UNESCO INSTITUTE FOR STATISTICS, WORLD BANK, AND IMF.

France. The real concerns that persist—especially in sub-Saharan Africa—should not obscure this immense and historically unprecedented progress (see Box 10.4: Health and Wealth over Half a Century, page 275). One of the direct consequences of the dramatic decline in infant mortality in many societies is that families have fewer children because they are confident that the children they have are more likely to survive.

It is important to bear in mind that these increases in well-being have made the most difference in poor countries, where reductions in child mortality and increases in education have greatly raised the life chances of hundreds of millions of people. The Human Development Index is a rough and ready measure, combining readily available data in a simple way. If one looks deeper, however, the picture becomes more complicated and less rosy. Rising economic inequality in the United States, for example, has meant that median income has risen over the past forty years mainly because more women have gone to work. However, their unpaid services, such as child care and household management, still must be done, so the net increase in well-being of families is low and negative in many instances. Surveys of people's satisfaction with their lives in the United States and other rich countries indicate little change over time. That is, a rising average income does not necessarily lead to greater fulfillment or contentment.

BOX 10.4

HEALTH AND WEALTH OVER HALF A CENTURY

The first goal of sustainable development is to meet the needs of the present. How well have we been doing in pursuing that goal? (See video by Rosling cited at the end of this chapter.) The graphs below show two snapshots of the world's nations in 1960 and 2010. They are plotted to bring out four different pieces of information. Each country is shown as a circle whose size is proportional to its population; the circles are shaded by region. In the graph for 1960, the large circle at the upper left is China; you can see how China has grown in population and wealth during this period because it is the largest circle in the middle of the graph for 2010.

Along the vertical axis is plotted child mortality, a statistic that is regarded as a basic indicator of the health of a population. The better the health of a population, the lower the country's circle will be. Along the horizontal axis is plotted the standard economic measure, GDP per capita. You'll see that in both 1960 and 2010, the nations of the world lay along a rough diagonal line. That is, a correlation exists between health and income: in poorer nations more babies perish, and in richer ones more survive.

(a)

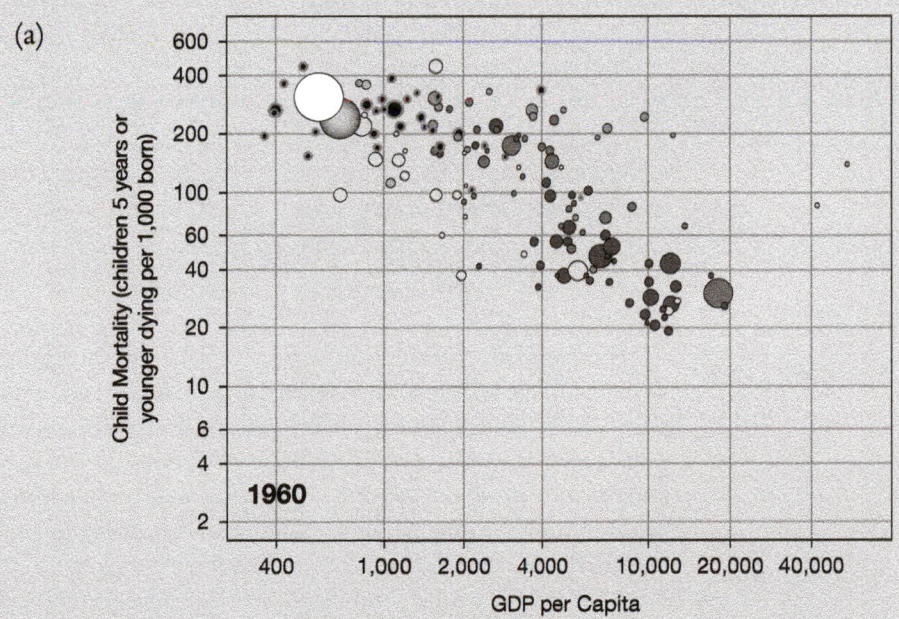

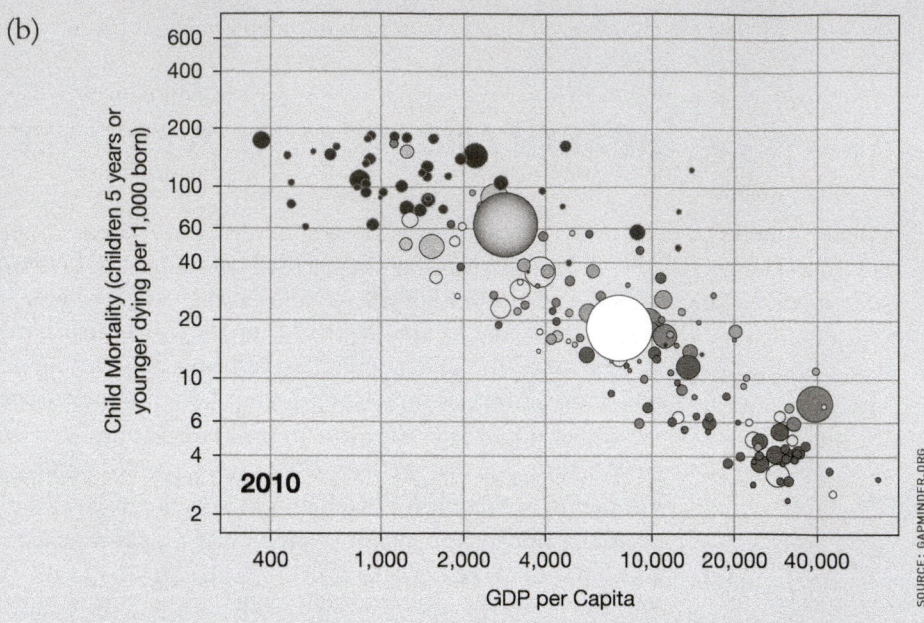

(b)

Child mortality and GDP per capita for the countries of the world in 1960 (a) and 2010 (b).

Now compare the plots for 1960 and 2010. The world has gotten richer (the circles have moved to the right), and the world has also seen infant mortality drop (the circles have moved down). Some, such as China and India, the two largest circles, have moved down substantially. These two nations house more than 2 billion people, or one-third of the human race. So their improvements in health mean that tens of millions of babies who would have died a half-century earlier now survive. But other circles have moved, too. These are large changes in human experience. Notice that the United States, the largest circle in the lower right corner, has fallen behind in infant mortality compared to other rich nations.

The rapid improvements in human well-being among the poorer nations mark a change in the relationship between microparasites and their human hosts. People, particularly young children, live longer more often. This change also seems to reflect an accompanying change in the relationship between macroparasites and people. Improving the lot of people seems to have become an objective of many rulers, not only in their rhetoric but, however unevenly and haltingly, in practice as well. Remember the story of the Maya: the people who suffer as ecosystem services fail and the people who make decisions affecting the suffering are usually not the same individuals. Building reliable linkages, such as democratic accountability, between rulers and those who feel the impact of their rule can make a difference.

Yet in the world without edges, many kinds of actions and consequences have become more widely separated and distant, crossing national boundaries more often than before. At this point in history, the rich live mainly in temperate regions, the poor in tropical places. The poor nations live in ecological and social settings altered, sometimes drastically, by colonialism. A disconnection is evident at the conceptual level: it is difficult for Americans to understand how tropical ecosystems are vulnerable. This makes the relative powerlessness of tropical peoples even more problematic. Not only are those affected not the main decision makers but the main decision makers don't understand the ecosystems their choices are changing. Some relevant differences are summarized in Table 10.1 on page 278. The point is that the differences are great—biologically, economically, and culturally.

As one would anticipate, rising economic activity does more than improve human well-being. It also tends to increase environmental pressures. These pressures are transforming ecosystems, just as the European colonists changed the landscapes of the Americas. But now, when these pressures are played out in tropical settings rich in endemic species, there is widespread but hard-to-see erosion of biodiversity, as discussed in Chapter 9. There are exceptions, however: urban air quality is better in wealthy cities such as London and New York than in less prosperous cities such as Delhi and Johannesburg. Still, indicators such as energy use, carbon dioxide emissions, and solid waste generation seem to increase steadily with income.

What is more immediate, and yet still invisible, is that we continue to depend on the services provided by these modified ecosystems. The poor see their dependence on ecosystem services clearly, especially if they earn their livelihood by subsistence, living directly from nature. Crops, water, protection from floods, and the waste treatment provided by healthy wetlands are all essential to life and well-being. But, being poor, people who live from subsistence agriculture cannot do much to secure ecosystem services if they are imperiled. Instead, they can suffer from forms of enclosure that are every bit as severe as those suffered by English grazers hundreds of years ago, as discussed in Chapter 3. If a mangrove swamp long used by fishermen is cleared to grow shrimp for European food markets, for example, the poor people there, whose lives once depended on catching fish from those wetlands, often cannot do much about it. Their fate is like that of the herders who once depended on the common grazing lands in Europe, who were driven from their customary lands. Many of the survivors became workers in urban factories, a pattern now seen in Asia.

The rich nations can readily harvest ecosystem services and have done so for a long time. We do this through the infrastructure of urban and suburban environments. We take those services for granted. Our food is inexpensive and our

TABLE 10.1 SOME DIFFERENCES BETWEEN THE TROPICAL (POORER ECONOMICALLY, RICHER BIOLOGICALLY) AND TEMPERATE REGIONS OF THE HUMAN ECOSYSTEM.

Environmental issue (and location of book discussion)	Tropical regions	Temperate regions
Climate (Chapter 5)	Equatorial cell: desert, monsoon, rain forest	Mid-latitude cell: clearly marked seasons, Mediterranean
Biological diversity (Chapters 5, 9)	High; many localized (endemic) species	Low to moderate
Impact of land clearing (Chapters 2, 5, 9)	Widespread loss of species (irreversible extinction)	More local loss of species (recovery over decades)
Political history (Chapters 10–12)	Colonial: local culture stressed or extinct; indigenous institutions delegitimated	Imperial: dissemination of culture and institutions; rising influence of environmentalists
Population distribution (Chapter 8)	75% of human population	25% of human population
Population growth (Chapter 8)	Rapid but slowing; more than 90% of population growth is here	Slow, negative in some places
Urbanization (Chapter 8)	Rapid urbanization (migration; land conversion)	Sprawl (land conversion; traffic)
Wealth and income (Chapter 10)	Poor, high inequality	Rich, moderate to low inequality
Consumption per capita (Chapters 10, 13, 14)	Low (chronic hunger) to moderate	High to very high
Energy use per capita (Chapters 7, 14)	Low	High
Economic base (Chapters 2, 10)	Agrarian, often with communal landholding	Industrial and information economies, with private and state-owned property dominant
Dependence on environment (Chapters 6–8, 10)	Direct impact on economic production	Localized impacts on health, aesthetics; large, global, indirect impacts through consumption and investment
Cultural orientation (Chapters 2, 10, 11)	Historically "stationary" and traditional	"Edgeless," highly mobile

nutritional problem is obesity rather than hunger. We also use the bulk of the fossil fuels produced each day, driving climate change. Our consumption propels the world's manufacturing and shapes investment. So much of it is out of sight, though, that the environmental and social implications of industrial production are now largely out of mind, too.

The rich cause environmental problems of which they are only dimly aware. The poor suffer environmental problems that they are powerless to solve. To this paradox, add the challenge of thinking about the future. This is the grand challenge of sustainable development—meeting the needs of the present, including the needs of the more than 1 billion people who live in absolute poverty, but without compromising the ability of future generations to meet their own needs. At present, we are far from meeting this grand challenge. Over the course of your lifetime, it will become increasingly critical to do better.

The grand challenge of sustainable development, as we now see, means that we must conserve the ecosystem functions that deliver services needed by humans. We don't fully know how to do that, except by protecting entire ecosystems, such as watersheds that supply drinking water. As you know from Chapter 9, we are far from being able to conserve large, complex ecosystems, though we have made some progress.

"THE SCALE OF OUR COMPETENCE"

Like the other grand challenges, sustainability is daunting. Clearly, in the world without edges, virtually all people have become citizens of the planet. What is not clear is how to respond.

For many of our students over the years, the reaction to learning about environmental problems has been guilt and anger—guilt at the role each of us plays in the consumerist economy of the United States, and anger that this should be so. Given the logic of the commons, together with the disproportionality between individual consumption and collective environmental impact, the personal changes induced by guilt are not enough to cure the problems that arise from defective institutional arrangements. Moreover, the scale of the world without edges implies that the institutional changes that might be suggested by anger will not be achieved easily or soon. These are frustrating realizations. What can a person do to gain perspective about the problem?

In a provocative 1989 essay entitled "The Futility of Global Thinking," poet and writer Wendell Berry wrote that "our understandable wish to preserve the planet must somehow be reduced to the scale of our competence."[3] Berry derided those who proclaimed an environmental crisis of planetary dimensions. "Nobody," he warned, "can do anything to heal a planet.... In fact, though we now have serious problems nearly everywhere on the planet, we have no problem that can accurately be described as planetary." Berry, also a noted environmental advocate, wrote these words just after the unusually hot summer of 1988 brought dramatic wildfires to Yellowstone National Park and a group of leading climate scientists testified in Congress about the reality of global warming. Berry was not a climate denier, however. He was saying that problems simply cannot be worked on at the global scale. Climate change has not been contained yet, and in that sense Berry has been proved correct so far.

Instead of tackling problems as if they were global, Berry argued, we need to "care for each of the planet's millions of human and natural neighborhoods, each of its millions of small pieces and parcels of land." And in place of economics or politics, we must live at a scale on which it makes sense to act from love: "Make a home. Help to make a community. Be loyal to what you have made.... Love your neighbors—not the neighbors you pick out, but the ones you have.... Love this miraculous world that we did not make, that is a gift to us." These words call for intimacy, for living on the small scale of a country town, set in countryside that the townspeople can care for. "Find work, if you can, that does no damage."

Berry was building on the psychological truth that the fullest range of human communication comes only through face-to-face interactions. This sets limits on our ability to make commitments and to nurture trust, to issue or obey orders, to elicit loyalty or sacrifice. This is why soldiers will die to save their comrades, why people change their life goals to meet the needs of family members, and why morale and motivation are as important as pay and benefits in large corporations. It is why students tend to learn better in small classes than large ones.

Now consider the world without edges. Our activities as consumers affect distant places, most of them places we do not know, with effects we cannot see. As workers, too, our activities typically affect places far away: think of lawyers shaping precedents, or auto workers who never see the cars they make being driven, or teachers who affect their students' career choices far in the future. As investors, too, we reshape landscapes, not only in the houses we buy or the vehicles we drive but also through the companies whose stocks wind up in our retirement funds. Today, when we interact with a firm or a person through the Internet, it is usually unclear where they are or whether it even makes sense to think of them as located in a single place.

All of these examples of economic participation in the world without edges tell us that we already act beyond our competence in the sense of making reasonable decisions from an environmental and social standpoint.

There are, it seems, three alternatives. The first is to live in the world without edges while ignoring the connections we make and the impacts we have. Many of us are rich enough to be reasonably well insulated. We can put our energies into family and career and, if there's time, community. We can live a high-quality life and let the rest of the world take care of itself in response to the markets and political interests out there. This might be called the American Dream. It was how most people who send their children to college lived before September 11, 2001. It is how most still do. But some cannot: the soldiers sent to fight in Afghanistan, like the conservation biologists studying tropical forests and the businesspeople working overseas, inhabit the front lines of America's engagement with the world.

A second alternative is to retreat into a stationary world. One can emulate Thoreau and live off the grid, keeping busy with the honest labor of subsistence and the satisfactions of moral consistency. This alternative may appeal only to limited numbers of people, but it has done so persistently, and these ideas can be seen in the "locovore" movement that has popularized locally grown food and farmer's markets. Living off the grid seems rather close to what Wendell Berry called "work that does no damage." It isn't so clear how this approach would affect the 35 pounds of waste generated for each pound of goods purchased by the majority of American consumers who stay on the grid (as discussed in Chapter 4), nor how those who follow this alternative can affect global warming. As with other commons problems, reducing one's own impact does not solve the problem of everyone else's behavior. This does not mean that it is useless to set a good example, only that a good example is not enough in a commons.

The third alternative is to struggle to enlarge the scale of our competence, as individuals and as a society. One does this through education and institutional innovation, taking the humanity and determination that one can marshal at the face-to-face level and extending it in part to the larger world without edges—not just in our thinking but also in our actions. Historically, the rich nations have expanded their competence through politics and economic reorganization, as we will discuss in more detail in Chapters 12–14. Using the electoral process and the institutions of representative democracy, citizens and their political leaders have ended slavery, child labor, and the contamination of water supplies with raw sewage.

Reinvigorating democracy may seem like a daunting challenge in a time when environmentalists like Berry celebrate individualism and denigrate the potential for effecting change through collective action. But the United States was the world's first large democracy, and it took on the extraordinary concentration of economic power in the Gilded Age more than a century ago. In that light, the largest challenge of our multicultural, multinational world may be lack of trust, an unwillingness to see that we are capable of working together through democratic self-governance to solve the problems we share. That trust, in turn, requires an understanding that we share the planet in common, and its global problems as well. This is a realization that environmental awareness, at its best, has spread among the people of the world.

Already, corporations, nations, and international organizations are expanding their competence in handling problems of commons, as in the extension of public health measures that have lowered the child mortality rate in most of the world. We can see a profit-oriented version in the extension of American corporations into global business establishments, although the humanity of these activities is skewed and highly selective. Yet pharmaceutical companies from the rich nations have begun to meet, with reluctance and difficulty, the challenge of HIV/AIDS in poor nations. Walmart is selling millions of energy-efficient compact fluorescent bulbs. Dupont, General Electric, Duke Power, and a widening range of major corporations have urged Congress to enact legislation to address global warming. California and Europe have set ambitious targets for reducing their greenhouse gas emissions.

All these are remarkable developments. They do not mean that anyone has achieved sustainability, but they point to important steps toward a more sustainable relationship between humans and the natural world. More broadly, human rights, democracy, and environmental protection have spread worldwide—along with Facebook and Starbucks and automobiles—so here, too, there is hope, together with a large challenge.

Understanding the scale of our competence—and the scope of our incompetence—is a continuing task that is integral to living an environmentally responsible life in the world without edges. One cannot become fully competent in the world as we find it. This is why guilt alone is too limited a response. If one is not touched by guilt for the ways in which we participate in the destruction of nature, one is morally in default. Working to enlarge the competence of oneself and the institutions one acts within is one response that makes sense.

We have come to the end of Part II, having assembled a perspective on what we have called, with considerable irony, the "most successful" species in the history of life on Earth. This perspective is itself grounded in history. Early in the twenty-first century, we can say these things about our situation:

- ⬏ *Population is stabilizing*, but economic growth has been the companion of slowing birthrates. Can the total numbers be contained without unsustainable consumption?

- ⬏ *Economic growth is continuing*, but can widening inequality be consistent with social stability?

- ⬏ *Human welfare has been improving in most parts of the world, despite widening inequality*, and there is greater awareness of that inequality among the poor.

- ⬏ *Human domination of nature has increased rapidly*, although this process may be nearing fundamental limits, requiring substantial changes in the trajectory of the material economy.

- ⬏ *Humans remain dependent on ecosystem services*, whose management may require efficient, effective planning and governance.

This is a situation, we have argued, in which there are at least four grand challenges of the environment: climate change, biodiversity conservation, urbanization, and sustainable development.

These are issues for which no clear path to resolution has yet been determined, even in principle. In the remainder of this book, we investigate what people have done in response to environmental problems, and we consider what more might be done.

FURTHER READING

Berry, Wendell. "The Futility of Global Thinking." In *Learning to Listen to the Land*, edited by Bill Willers, 150–56. Washington, DC: Island Press, 1991. Originally published in *Harper's*, September 1989.

Catton, William R., Jr. *Overshoot: The Ecological Basis of Revolutionary Change*. Urbana: University of Illinois Press, 1980.

Rosling, Hans. *200 Countries, 200 Years, 4 Minutes* (video), 2010. Available at www.gapminder.org/videos/200-years-that-changed-the-world-bbc.

Sachs, Jeffrey. "Helping the World's Poorest." *Economist*, August 14, 1999.

UN Development Programme. *Human Development Report* (updated annually). Available at http://hdr.undp.org/.

KEY TERMS

economic inequality	exponential growth	Human Development Index	logarithmic well-being

PART III

Strategies

Chapter Eleven

ENVIRONMENTAL AWARENESS

THE EMERGENCE OF ENVIRONMENTALISM

In this chapter we turn from grand challenges to the search for solutions and the imposing tasks that face a world that values sustainable development but is deeply ambivalent about how to achieve it. In the chapters to come, we will look at environmental problem solving using the mechanisms of government and politics, markets and economics, technology, and changes in consumption. All of these approaches to problem solving are based on the awareness that a problem does exist, and this rise in awareness is the subject of this chapter. It may seem incredible to imagine a society in which environmental impacts were not widely perceived as problems, but the United States was such a society within the memory of people now living.

It isn't that environmental damage was not visible earlier. In the 1950s, the air and water in many cities was dirty; mines and logging left scars on the land; and some species had nearly vanished, such as the egrets of the Florida Everglades and the bison of the Great Plains. Moreover, people appreciated the aesthetic value of landscapes, as evidenced by the creation of large national parks, an idea pioneered in the United States. What was missing was a society-wide awareness that certain production and consumption practices were altering, and often damaging, elements of the natural world that did not have to be destroyed. What was missing was an understanding that pressures of human origin were imposing drastic and often unwanted changes on nature—an understanding, both cultural and political, that we now call environmentalism.

Environmentalism asserts the need for competence—not just to exploit nature, but to manage and preserve portions of the natural world in a sustainable fashion. This is the competence discussed in Chapter 10, a competence often lacking in the world without edges. Environmental awareness, the notion that nature and its fabric of ecosystems should be taken seriously in human affairs, was embraced with astonishing speed the world over. Within a decade of the publication of the popular and influential book *Silent Spring*, the United Nations (UN) convened its first Earth Summit, the UN Conference on the Human Environment in Stockholm, Sweden, in 1972 and created a permanent agency, the UN Environment Programme, headquartered in Nairobi, Kenya.

In this chapter, we examine the American roots of this global change in awareness by looking at three influential people: Rachel Carson, Aldo Leopold, and David Brower. They taught people to see nature and humans in a different way—a way that revived and went beyond the ideas of the Transcendentalists and Hudson River School. The environmentalists' ideas went beyond changing people's perceptions of nature, spurring citizens to organize and governments to respond.

Environmental awareness and its model of social action, which we called classical environmentalism in Chapter 1, is one of the distinctive ways in which the United States has been a cultural innovator and leader in the past century. For more than a generation, classical environmentalism has defined the way the world is rethinking and reworking the relationship between humans and landscapes. This

---

Learning Objectives

When you have finished studying this chapter, you should be able to

⬎ describe how warnings from scientists and citizen outcry sometimes lead to environmental reform;

⬎ identify attempts to apply the land ethic at different levels of social organization— from your own actions or those of your family, to nongovernmental organizations such as a campus sustainability task force, local governments, corporations, national government, international organizations, news media, and social networks;

⬎ understand and evaluate assertions that environmental risks are being taken into account adequately in the choices made by businesses, governments, and individuals;

⬎ explain the claims of a nongovernmental organization (NGO) about its mission, and whether the remedies advocated by the NGO really protect ecosystem services and produce better governance of a commons;

⬎ analyze the activities of an organization, such as the college you attend, in terms of its grasp of the land ethic.

achievement has been both conceptual and social, and it is as remarkable as the changes that emerged from the civil rights and women's movements in the same decades in the second half of the twentieth century. Indeed, environmentalism played a central role in a historical period that recalled the Progressive Era of the late nineteenth and early twentieth centuries, when voters and politicians believed that expert knowledge could be harnessed to achieve a shared public interest. This was the time when the U.S. Food and Drug Administration was created to safeguard the food supply, when the Federal Reserve was created to regulate the money supply, and when women won the right to vote. As we stated in Chapter 1, classical environmentalism is incomplete as a strategy to move toward sustainable development, but the classical approach remains vital and important today, as we will explain in the discussion of environmental politics in Chapter 12.

Classical environmentalism involves an implicit theory of social change, linking the ability of science to identify changes in environmental conditions to the capacity of citizens to press for policy responses to human pressures on ecosystems. The rise of environmental awareness set the stage for NGOs such as the Environmental Defense Fund, founded in the 1960s, to take action on the concerns articulated by Rachel Carson.

Environmental awareness is a necessary component of the institutional changes that lead toward sustainable development. As we set out in Chapters 1 and 3, long-term shifts in human behavior grow from and are maintained by shifts in rules governing human actions—shifts in institutions. Sometimes, these rules are explicit and formal, such as the pollution control regulations we will examine in Chapter 12. Sometimes, the rules are informal, and heeding them becomes habitual for most people. Over the past half century, littering has become an unacceptable form of social behavior among many people, just as smoking has. Usually, there are both formal changes and informal ones. For instance, recycling is more common now than it was in the 1960s, but this is partly because bins for collecting recycled bottles and paper are much more widespread as a result of formal decisions to provide them. Environmental awareness is a foundation for these and other institutional changes, and this is why we begin a search for solutions by looking at how a culture comes to understand that environmental problems are waiting to be tackled.

THE LAND ETHIC

Aldo Leopold, born in 1887, completed his most influential work, *A Sand County Almanac*, shortly before he died of a heart attack while fighting a brush fire on his farm, just northwest of Madison, Wisconsin, in 1948. In this book of short essays, Leopold articulated the idea of a **land ethic**, which continues to shape both environmental concern and the search for sustainable development.

A Yale-educated forest ranger who grew up in Iowa, Leopold became the first wildlife management professor in the United States in 1933, during the dark days of the Great Depression. In 1935, he became one of the founders of The Wilderness Society, which is still a leading environmental group.

Early in his employment with the U.S. Forest Service, Leopold was assigned to a program that sought to eradicate wolves. It was so successful that the deer that the wolves preyed on increased dramatically in numbers, until they ate all the plant foods available and starved. This unforeseen outcome gave Leopold a new understanding of how populations interact in ecosystems. This led, in turn, to thoughts about the human role in landscapes and to a striking revision of the instrumentalist view of nature that guided the Forest Service at the time. Instead of thinking about ecosystems as a source of economically valuable natural resources, Leopold wrote, it was important to understand how humans, too, might be considered part of ecosystems. Nature, he argued, was not just a storehouse of meat, crops, and fiber for humans to use. It is more importantly a community to which we belong, one in which we have responsibilities to exercise.

This perspective meant that ideas about property—ideas that had shaped the American consciousness since colonial times, as we discussed in Chapter 2— needed to be reconsidered. Here is the opening passage of Leopold's essay entitled "The Land Ethic": "When god-like Odysseus returned from the wars in Troy, he hanged all on one rope a dozen slave-girls of his household whom he suspected of misbehavior during his absence."[1] Leopold's point was that—in the eyes of Odysseus' contemporaries—this hanging was not a moral issue. Instead, the girls were viewed as property who were not afforded the human rights that we (like Leopold) now take for granted. If they had betrayed their master while he was gone (for twenty years!), then they were to be disposed of by hanging. One should feel no more compunction about it than one would feel about junking an old refrigerator that could no longer keep food cold.

As Leopold proceeded to argue, we no longer think slavery is a permissible form of human relationship. Our conceptions of right and wrong can change in ways that often broaden our sense of community and reciprocal obligation—a point Garrett Hardin also brought up in his analysis of the commons discussed in Chapter 3. Leopold argued that we were at the threshold of a change in our view of the relationship between humans and ecosystems. "The land ethic," he wrote, "simply enlarges the boundaries of the community to include soils, waters, plants, and animals, or collectively: the land."[2] That is, we need to think of the land—or ecosystems, as we would now say—as an object of ethical attention rather than only as property (Fig. 11.1).

FIGURE 11.1
Martin Johnson Heade, *Newburyport Meadows* (1872–79). A view of the land ethic—humans peacefully inhabiting a landscape.

Two generations after Leopold wrote his essay, we have yet to achieve that transformation in ethical perspective, even if the pace and scale of change so far is notable. His book continues to be influential because respect for ecosystems is now acknowledged as important for individuals and the world. Yet neither the land ethic nor the actions to carry out its imperatives could be said to be in place. Although there is widespread acceptance of the need for effective environmental policies, commitment to them remains shallow. More important, the scale of competence of many of our institutions falls short of being able to realize environmental protection sustainably, or at all. This shortfall, in turn, disheartens people, tempting many to wonder whether any institution can do the job. Yet the actual record is not as discouraging as a lot of politically inspired talk suggests. We have described the substantial progress in many areas while describing the grand challenges in Part II; we will see more reasons for hope and determination in the chapters to come.

"A THING IS RIGHT . . ."

Leopold recognized the world without edges: "Your true modern is separated from the land by many middlemen, and by innumerable physical gadgets. He has no vital relation to it; to him it is the space between cities on which crops grow."[3] What does this lack of "vital relation" to the land mean? In the language of this book, we would say that the world without edges leads people to forget that they still rely on landscapes and ecosystem services, both far and near. The consequence of forgetting this is that humans modify ecosystems, and then we are surprised when those modifications go too far:

> In human history, we have learned (I hope) that the conqueror role is eventually self-defeating. Why? Because it is implicit in such a role that the conqueror knows, ex cathedra, just what makes the community clock tick, and just what and who is valuable, and what and who is worthless, in community life. It always turns out that he knows neither, and this is why his conquests eventually defeat themselves.[4]

This caution is one you see in the concern for biodiversity, a warning to be careful of nature's complexity. Leopold himself made the same point, through a question: "Who but a fool would discard seemingly useless parts? To keep every cog and wheel is the first precaution of intelligent tinkering."[5] In environmental policy, this idea is called the **precautionary principle**. When there is good reason to believe that an action may harm humans or the environment, those who advocate for that action should bear the burden of proof to demonstrate that the risks are acceptable, given the benefits—keeping in mind that risks and benefits often accrue to different groups in society. The 2010 oil spill in the Gulf of Mexico is only one illustration of the degree to which the precautionary principle fails to be applied in judging economic activities, sometimes with dire consequences for both the business enterprises involved as well as for the ecosystems harmed by those activities.

Leopold's call for a land ethic that "simply enlarges the boundaries of the community" to include ecosystems is of course radical. We have spent more than half a century, so far, working on that word "simply." The implications of such an ethical stance remain far from the practices of an industrial economy: "a land ethic changes the role of Homo sapiens from conqueror of the land-community to plain member and citizen of it. It implies respect for his fellow-members, and also respect for the community as such."[6]

The world without edges is a world of human-dominated ecosystems, in which domination amounts to conquest of the land-community. In that world, we are still discovering what it means to be "plain member and citizen." We are trying to preserve land in national parks, and conservation biologists now strive to protect ecological processes such as migration and evolution. Industry and government have made many changes in manufacturing processes and regulations to decrease pollution of water and air. Some are now acting to slow global climate change and to respond to world trade as it spreads exotic species and extends the demands of consumption. Many people now recycle. All of these are ways of exercising citizenship.

But you can see the magnitude of the remaining challenge. In enlarging the boundaries of the human community to include the land, Leopold is calling for ethical consideration of an ecological universe of complex, delayed causation in which the carbon dioxide we emit through our consumption is not visible to us as we consume and doesn't even have a discernible immediate impact in any specific place in the world. We live in a world in which humans routinely act on a large scale and, increasingly, beyond the scale of their competence. Enlarging the scale of competence of human communities is work that is far from completed.

That it isn't easy to respond ethically to human-dominated ecosystems is not surprising, of course. In 1820 Thomas Jefferson wrote a famous letter about slavery, which at the time was still a prevalent form of human domination of other humans. Jefferson described the institutions of slavery this way: "We have the wolf by the ears, and we can neither hold him, nor safely let him go."[7] A nation governed in part by slave owners dare not let slavery go, because its rulers believe themselves dependent on slaves, but neither can the nation move toward a real democracy until it lets slavery go. So it is with the material culture of human domination of ecosystems: we cannot simply let go of industrialization and large-scale agriculture, but it is also apparent that we cannot go on for long with the present trajectory of ever-increasing impacts.

It took the Civil War to end slavery, and a century and a half later, the United States still struggles with the legacies of race. Leopold knew that it would take something more profound still to achieve the land ethic. Here is the principle he advanced to guide that deep change: "A thing is right when it tends to preserve the integrity, stability, and beauty of the biotic community. It is wrong when it tends otherwise."[8] One might think of this as the golden rule of environmentalism: to do unto nature so as to respect the biotic community, a community that includes ourselves. Note the word "tends," however. Taking this carefully and deliberately stated principle and turning it into clear-cut rules is no easier than the dictum of sustainable development. Indeed, intelligent conservation and preservation of the land is sustainable development. Note also Leopold's insistence on beauty: what is right cannot be reduced to material accounting but demands judgment and the hammering out of community values and, in particular locations, a durable sense of place.

A Sand County Almanac was published in 1949, the year after its author died. Through his book, this professor of wildlife management became an influential philosopher-theologian of environmentalism.

But it would take a while for Americans and the larger world even to recognize *A Sand County Almanac*. Leopold's ideas about the ethics that should be taken into account in our encounters with the land appeared as Americans were rebounding from the Great Depression and World War II. It was an era of exuberant consumption, as we built the automobile-dependent suburban landscapes that define American culture in places like Grosse Point, Michigan (outside Detroit), Bellevue, Washington (near Seattle), Fairfax, Virginia (next to Washington, D.C.), and the Los Angeles basin.

For more than a decade, the land ethic lay far outside the mainstream of American life, even as the industrial economy that provided abundant goods for the American lifestyle led to more and more pollution. By the late 1950s, brown smog regularly hung in the air of Southern California. In 1969, the heavily polluted Cuyahoga River near Cleveland, Ohio, had so much oil and grease floating on its surface that the river caught fire. That same year, a large oil spill in the Santa Barbara Channel put images of fouled beaches and oiled birds on television screens across the United States. But by then, change was under way.

SILENT SPRING

Rachel Carson's book *Silent Spring*, published in 1962, launched the contemporary environmental movement. She wrote at a pivotal time. Up until World War I, half or more of the industrial products in the United States were made from renewable resources—plants and animals. This began to change with the growth of the oil and chemical industries in the 1920s and 1930s, but their growth was slowed by the Great Depression in 1929. The economy then accelerated during and after World War II. The war—fought near the end of Leopold's life—was the real beginning of a phenomenal change toward a world that, at least for American consumers, would be dominated by plastics, synthetics, and a dizzying array of new chemicals. During the war, whole new classes of synthetic chemicals—including some developed for chemical warfare—were found to be highly effective weed and insect killers. The manufacturers also enjoyed strong government backing, and within ten years synthetic pesticides had captured 90 percent of the agricultural pest-control market.

Carson was twenty years younger than Leopold. She was trained as a marine biologist and spent the early part of her career as a writer in the U.S. Fish and Wildlife Service, a rare woman in the male-dominated world of federal science. Pressed by family responsibilities, she tried to earn additional income as a freelance writer. It was a tough go until her book, *The Sea Around Us*, appeared in 1951. This book stayed on the best-seller lists for well over a year, won the 1952 National Book Award, and enabled Carson to retire from the government bureaucracy to devote all of her time to writing.

She also began to take an interest in all of those new pesticides. In a three-part series of articles published in the *New Yorker* magazine in 1962, Carson spelled out her concerns. The articles were entitled "Silent Spring," and they were subsequently published as a book. This title pointed to what she felt was an unsettling and unnoticed side effect of pesticides: DDT, the best known insecticide, interfered with the ability of some birds to form eggshells. As a result, eggs broke before the chicks could survive on their own. This is the problem that nearly drove the bald eagle extinct, as we saw in Chapter 1. The result, Carson warned, would be the loss of the songbirds that delighted her and millions of suburbanites every spring. A silent spring would be the unintended consequence of careless applications of pesticides. And the consequences would not be limited to birds. Indeed, as the critique of industrial chemicals widened, the focus came to lie on the risk that chemicals would cause cancers in humans. Today's regulations are aimed almost entirely at protecting our own species rather than wild birds.

It is a sad irony that Carson herself died of cancer in April 1964. But already in that spring when her voice fell silent, the message she had sent had ignited a public debate that continues today. *Silent Spring* resonated with the sensibilities of a growing suburban middle class, bringing environmental ideas into the mainstream of American politics.

Carson chose her audience well. The *New Yorker*, a publication read by affluent consumers from well-to-do, cosmopolitan cities and suburbs, only had a few employees from chemical manufacturers among its readers, and they tended to be executives far removed from the production plants where the pesticides were produced. Not many farmers read the *New Yorker* either; they were firmly against pests and most were in favor of pesticides.

Carson reached a well-educated population, many of them suburbanites who had moved to the country after World War II. Theirs was an American dream sketched along the lines of the pastoral imagery of the Hudson River School a century earlier (see Chapter 2). But unlike the rural figures in the Transcendentalists' paintings, twentieth-century homeowners were deeply enmeshed in the world without edges. They had little knowledge of or sympathy with heavy industries such as chemicals manufacturing or commercial agriculture. They lacked a "vital relation" to industry or commerce, and they were attached to their land for reasons of beauty rather than the material ecosystem services they derived from their surroundings. Their attachment was not much like the appreciation that a farmer or hunter might have of the natural world, an appreciation that Leopold sought to enlarge into an awareness of the biotic community.

Unlike the civil rights battles that erupted soon after *Silent Spring* appeared, environmental politics did not demand that people change their lives in painful ways. The costs of complying with the environmental laws inspired by *Silent Spring* turned out to be modest; perhaps more important, careful research has shown that the benefits of these policies considerably exceed their costs. In an economy that was expanding rapidly, new sewage treatment facilities and pollution control equipment on cars, power plants, and factories all increased taxes and prices, but these were hardly noticed by consumers. So Carson's appeal was easy to embrace. Like the Hollywood screenplays that they watched, in which the drama of righteous democracy was enacted, concerned environmentalists could demand action from government and expect a positive reception from federal officials. The attitude toward government in that era of the Great Depression and World War II was that government's role was to help and to protect its citizens. The anti-tax, anti-bureaucrat anger of the 1980s and 1990s still lay decades in the future.

Smoldering pollution problems could be found in many metropolitan areas and were steadily growing with the expanding economy. Some people were ready to act, including government officials who were eager to solve problems for well-educated, middle-class voters. And Carson ignited a movement and brought a new word into the American lexicon when she wrote, "this is a problem of ecology, of interrelationships, of interdependence."[9]

> We poison the caddis flies in a stream, and the salmon runs dwindle and die. We poison the gnats in a lake, and the poison travels from link to link of the food chain, and soon the birds of the lake margins become its victims. We spray our elms and the following springs are silent of robin song, not because we sprayed the robins directly but because the poison traveled, step by step, through the . . . elm leaf - earthworm – robin cycle.[10]

One can see here a portrayal of Leopold's land-community, in which humans were ignoring their responsibilities as plain members and citizens.

CLASSICAL ENVIRONMENTALISM

What did Rachel Carson achieve in *Silent Spring*? First, as she had done successfully in *The Sea Around Us*, she explained biology in terms a layperson could understand, spelling out how toxins kill or impair cell function. The life of a cell, she observed, unfolds through chemical reactions steered by enzymes:

> When any of these enzymes . . . is destroyed or weakened, the cycle of oxidation within the cell comes to a halt. It makes no difference which enzyme is affected. Oxidation progresses in a cycle like a turning wheel. If we thrust a crowbar between the spokes of a wheel, it makes no difference where we do it, the wheel stops turning. . . . The crowbar . . . can be supplied by any of a number of chemicals commonly used as pesticides..[11]

The impacts on cells could also affect genes: "Some of the defects and malformations in tomorrow's children . . . will almost certainly be caused by these chemicals that now permeate our outer and inner worlds."[12] Moreover, the parallel between chemicals and radiation is exact and unmistakable. The living cell assaulted by radiation suffers a variety of injuries: its ability to divide normally may be destroyed; it may suffer changes in chromosome structure; or the genes, carriers of hereditary material, may undergo those sudden changes known as mutations, which cause them to produce new characteristics in succeeding generations. If especially susceptible the cell may be killed outright, or finally, after a lapse of time measured in years, it may become malignant.[13]

Carson linked the toxic chemicals in pesticides to cancer and radiation, two of the most feared cultural symbols of that time. In 1962, atmospheric testing of nuclear weapons was in full swing, and radioactive elements were found in milk, raising concerns about their potential to wreak harm. And the very success of medical science and public health had revealed novel threats:

> . . . a drastic change has come about in the nature of our most serious public-health problems. Only yesterday, mankind lived in fear of the scourges of smallpox, cholera, and plague—scourges that once swept nations before them. Now . . . sanitation, better living conditions, and new drugs have given us a high degree of control over infectious disease. Today we are concerned with a different kind of hazard that lurks in our environment—a hazard that we ourselves have introduced into our world as our modern way of life has evolved.
>
> The new environmental health problems are multiple—created by radiation in all its forms, born of the never-ending stream of chemicals . . . now pervading the world in which we live. Their presence casts a shadow that is no less ominous because it is formless and obscure, no less frightening because it is simply impossible to predict the effects of lifetime exposure to chemical and physical agents that are not part of the biological experience of man.
>
> "We all live under the shadow of a haunting fear that something may corrupt the environment to the point where man joins the dinosaurs as an obsolete form of life," Dr. David E. Price, of the United States Public Health Service, has said. "And what makes these thoughts all the more disturbing is the knowledge that our fate could perhaps be sealed twenty or more years before the development of symptoms."[14]

This last point brought forward the second major theme of *Silent Spring*. The threats to human life and the quality of life signified by songbirds were invisible and delayed. It might be too late if we waited for the full impact of careless pesticide use to become apparent. These were threats against which we have no evolved defenses because the chemicals do not exist in nature. They were threats with which we had little experience, because the widespread use of pesticides was so recent. We had to rely on the findings of biochemists and geneticists. This meant that science would be an indispensable source of warnings—the premise of classical environmentalism.

Pesticides were disturbingly ubiquitous, even in suburbia. "We have seen that they now contaminate soil, water, and food, and that they have the power to make our streams fishless and our gardens and woodlands silent and birdless."[15] That garden chemicals might carry risks to gardeners and their children was unsettling, particularly to women. In addition, pesticide residues were found in the food supply, as some still are today. The threat was pervasive.

Here, a third theme entered: a government all too ready to lull the public into a false sense of security. "To the question 'But doesn't the government protect us from such things?' the answer is 'Only to a limited extent.'"[16] Carson's careful explanations of the biochemistry of toxicity brought lots of new information to her readers, and it made her point that "Little is done . . . to warn the gardener or homeowner that he is handling extremely dangerous materials."[17] Instead, she charged, "the Food and Drug Administration . . . promotes a completely unjustified impression that safe limits have been established and are being adhered to."[18] Even now, governments' acquiescence in environmental pressures from forest clearing in Indonesia to climate change to foods contaminated with salmonella remains a potent driver of citizen concerns.

Carson concluded:

> There is still a very limited awareness of the nature of the threat. This is an era of specialists, each of whom sees his own problem and is unaware of or intolerant of the larger frame into which it fits. It is also an era dominated by industry, in which the right to make a dollar at whatever cost is seldom challenged. . . . It is the public that is being asked to assume the risks. . . . The public must decide whether it wishes to continue on the present road, and it can do so only when it is in full possession of the facts.[19]

Environmental surprises from nuclear meltdown to the loss of endangered species are still described in the terms that Carson articulated (Fig. 11.2).

Silent Spring expressed the classical model of environmentalism. Science—a wide-ranging, integrative natural science explained in terms that a layperson can understand—warned of subtle dangers. In the technological progress that emerged from World War II lay invisible, delayed, distant, indirect threats to human health and the high quality of life that Americans had moved to the suburbs to find. The cure was an awakened citizenry who would press government to impose effective regulations. This would force commerce and industry to find better solutions.

Silent Spring introduced to its readers the idea of natural controls and organic means of controlling pests. The idea that solutions were available that could deliver

FIGURE 11.2
Fukushima Daiichi nuclear plant, burning and releasing radioactivity, after the plant was crippled by a massive tsunami in March 2011. The accident is one example of the threats to the environment and humans implicit in a large technological economy—threats that accompany the benefits often taken for granted.

a high quality of life to people meant that environmental protection could be achieved with little or no sacrifice.

Of course, the chemical industry disagreed vehemently. The risks were exaggerated, they charged, and the costs of doing without pesticides were underestimated. They also attacked Carson personally, doing so with a degree of ferocity that few scientists had ever experienced. An industry trade group, the National Agricultural Chemicals Association, spent more than $250,000 (equivalent to roughly 6 times that much money today) to attack the book and its author.[20] She was also attacked, notably, for being a woman. For one man who wrote a letter to the *New Yorker*, for example, it was not enough to claim that "Miss Rachel Carson's reference to the selfishness of insecticide manufacturers probably reflects her Communist sympathies, like a lot of our writers these days." Moreover, he sneered, "As for insects, isn't it just like a woman to be scared to death of a few little bugs!" And, just to make clear where he stood, he fumed, "She's probably a peace-nut too."[21] The pattern in which an entrenched industry attacks the credibility of its critics, rather than the substance of the critics' arguments, is one that has since been seen with increasing frequency.

The battle had been joined, however. The activist group that soon became the Environmental Defense Fund was founded by several scientists on Long Island, amid the suburbs of New York City, in 1967. By 1972, DDT had been banned by the federal government's newly formed Environmental Protection Agency, in part due to the Environmental Defense Fund's criticisms. Still, environmentalists did not win all of their battles, and the kinds of fierce resistance that Rachel Carson sparked from many American businesses have by no means died away.

ENVIRONMENTALISTS

Silent Spring was a book about the chemicals that were being sprayed on farms and suburban yards. But classical environmentalism was much more: a way of thinking about social action that described what activists like David Brower were doing already. Brower, who lived until 2000, made his biggest mark in the 1960s as executive director of the Sierra Club. The Sierra Club had been founded by naturalist John Muir in 1892, and it had attracted hikers and climbers to its vigorous outdoor activities. Under Brower, however, it was the indoor sport of politics that made the club *the* quintessential environmental group—the one cited by angry industrialists and government officials and embraced by thousands of people who discovered the environment as a cause in the years after Carson's book came out. Brower went on to found Friends of the Earth, which built a network of environmental activist groups in many different countries, and the League of Conservation Voters, a lobbying organization known and feared in Congress that has played a key role in such issues as fighting proposals to drill for oil in the Arctic National Wildlife Reserve.

What was Brower doing? His Sierra Club campaigns fought to preserve dramatic landscapes against government development. These were scenic places on public lands, including national parks, that a suburban, automobile-driving society began to visit on family vacations in the 1950s and to care about. He lampooned a federal government claim that flooding part of the Grand Canyon would bring tourist benefits by bringing people closer to the rock formations on the canyon wall. The government, he snorted in a newspaper ad, would flood the Sistine Chapel so that the tourists could see Michaelangelo's ceiling better. That sense of humor, combined with a sharply satirical attack on bureaucratic buffoonery, made environmental activism fun. The sense of fighting a good fight resonated with the gathering mood of rebellion and irreverence in the 1960s.

David Brower was not alone. His activism drew upon and contributed to the confrontational mood of the 1960s and its emphasis on community-based activism. The environmental group Greenpeace, founded in Canada in 1971, has been a leading exponent of a confrontational approach imitated by other NGOs. The Natural Resources Defense Council, founded in 1970, emphasized litigation and pioneered the field of environmental law, now a major specialty of legal practice. The Environmental Defense Fund, also a creative proponent of the lawsuit as a means to advance environmental protection, later branched out into sophisticated activism to reform the environmental behavior of businesses. These and other innovations drew ideas and contributed models to a notable period of social activism, in which civil rights, women's rights, and investigative journalism all came to make an enduring mark on American culture.

Environmental NGOs also flowered at the local level, often drawing upon people's sense of place to oppose development projects such as a new Walmart store or to press for cleanup of an industrial site contaminated with toxic chemicals. What may be most significant, though, is that concerns as different as the survival of whales and polluted air all came under a single umbrella: environmentalism.

SCIENCE AND CITIZEN-BASED ACTIVISM

What environmentalists shared was science, loyalty to places, and a belief in the power of citizen action. These were the ideas of classical environmentalism, and they were organized around the ideas of Leopold and Carson.

Unlike social movements anchored in outrage against injustices, environmentalism was initially grounded in an ecological perspective, although issues of justice were raised at the time of the first Earth Day, and they have taken on increased significance in recent years. Human activities take place in a complicated, dynamic, largely invisible natural world, and what we do, on purpose and accidentally, affects the web of life. Advocates for declining species and biodiversity see themselves as speaking on behalf of a natural world that goes unheard, and one can see in their statements a strong component of protest. But mixed in with the indignation is a dose of education designed to explain to their opponents and a wider public why endangered species or important habitats matter to people. Environmentalists combat ignorance and indifference as much as they must overcome prejudice or greed.

The word "ecologist" morphed from its original and still primary meaning of a scientist who studies the relationships among species, into a synonym for environmentalist. Washington State's environmental protection agency is called the Department of Ecology.

The ecological perspective Carson articulated entered the public imagination in a lasting way. Beyond politics is the growing market share of organic produce, the practice of recycling, and ecotourism. We do not live in a society that has outgrown Carson's critique, any more than we have settled the racial divisions that were rearranged (but not settled) by the Civil War. Yet one can say that notable progress has been made in both environmental protection and racial justice.

Ecology, the web of life, and the notion that humans should be "plain citizens" of the land-community were unfamiliar ideas in the 1960s. They entered a culture dominated by an industrializing economy of strong, hierarchical institutions. This was a culture that took pride in being "modern" and scorned the old-fashioned and outmoded. The ecological image was interpreted, as cultural values shifted, as a metaphor for a more decentralized, less hierarchical vision of community order, an order in which history matters and the past has value.

Putting ecology in this cultural light provides a striking contrast to the ideas of Social Darwinism that emerged a century earlier. In that earlier caricature of evolutionary biology in popular culture, it was ruthless competition that seemed to be ratified by nature. "Nature, red in tooth and claw" was a way to justify the pursuit of capitalism without heeding the needs of workers.[22] Aldo Leopold asserted, instead, that nature was better seen as a community than a field of combat, a community whose integrity and beauty should be respected in human ethics. In a pattern that simultaneously brought added energy and increased criticism to the activists, environmentalism seemed to challenge earlier ways of thinking; Carson was a respected scientist, but she was attacked as a communist and as a person who didn't care about "practical" realities.

Both Leopold and Carson stressed the importance of recognizing inadvertent harms and the ignorance of the would-be conqueror of the natural order. Greed was as much overenthusiasm as evil, in this reading, and the cure lay in reform rather than revolution. The path of reform, in turn, lay with the same scientific community that had split the atom, invented DDT, and developed ecology.

What was needed was a scientific understanding of risks. This, too, was an unfamiliar notion. Sometimes, scientists warned us, environmental science could not be expected to provide clear, cut-and-dried answers because damage from toxic chemicals or overfishing did not have symptoms that could be seen immediately, or even all that clearly in the longer term. Rather than showing up as quickly as the mosquitoes that fell out of the skies after a neighborhood was sprayed with a cloud of DDT, the crowbar in the spokes of the wheels of life would silently erode the ability of the mother bird to lay eggs strong enough for her chicks. Trawling the oceans to remove fish would also undermine the food sources of whales, and the burning of fossil fuels would eventually start to melt glaciers on the other side of the planet. None of this was apparent, or could even be understood by direct perception. Instead, technically sophisticated measurements that were collected, assembled, and interpreted by the patient methods of science were indispensable. Yet the assertions of these latter-day oracles, hedged and qualified in the reserved language of probabilities and uncertainty, have now become the basis for large economic decisions and wrenching political choices. By the 1980s, the Environmental Protection Agency had reorganized around the ideas of **risk analysis**, and the agency reinterpreted its regulatory mission as one of managing and lowering environmental risks.

The idea that industrial society carried new kinds of risk was recognized in the 1960s. People knew about industrial accidents and about workers losing their jobs involuntarily, a possibility dramatized during the Great Depression. *Silent Spring* extended this idea to nonhumans: "non-target" species such as bees are affected by pesticides, Carson pointed out, and she also drew attention to ecological damage such as reproductive failure in songbirds.

In the decades since *Silent Spring*, business has increasingly co-opted scientific risk assessment. The tendency now is for conservative critics of environmentalism to stress "sound science." Too often, this phrase turns out to mean that unknown risks should be presumed to be negligible—products are innocent until proved guilty—and that the precautionary principle is an unreasonable interference in business decisions. This is a shrewd strategy. Science rarely provides definitive answers in advance, so if an industry can forestall regulatory enforcement until there is definitive proof that its products are guilty, then the firm may be able to evade enforcement for a very long time, as was demonstrated by the big tobacco firms. The pattern is so common now, and so effective, that it may even deserve its own name—the "scientific certainty" argumentation method, or SCAM.

In addition, as was the case with Rachel Carson, environmentalists are attacked as unscientific extremists who are driven by emotions and who do not care about the economy. There is plenty of emotion among environmentalists, to be sure, but a lot of it comes from a sense of cherished places being violated, threats to human health from invisible toxins, pollution, drastic changes in land use, and the loss of symbolically important species such as redwood trees or eagles. The logic of industrial expansion and world trade has been one that views nature as a collection of resources to be harnessed and of places located along routes of trade and production. Environmentalism provided a language to criticize and resist this utilitarian perspective.

What environmentalism has sometimes failed to do is to take seriously the ideas in Chapter 4 on disproportionality—the fact that many of the most serious forms of environmental harm are not due to economic prosperity, writ large, but originate from a surprisingly small fraction of economic actors who actually create significantly more harm, per dollar, than other companies in the same industry. Environmentalists have constructed many articulate critiques of "industry" and "the economy," but few environmentalists to date appear to have written so eloquently about an industry in which the majority of the harm is actually done by just one or two of the facilities. Perhaps the most painful aspect of disproportionality is the continuation of environmental injustice (see Box 11.1: Environmental Justice, page 304).

Although science occupies a central place in environmentalism, so do nonexpert citizens, as well as characterizations of those environmentalists by their opponents. This is a political claim: the environment is a commons, part of the public space, and hence is subject to the authority of a democratic government and the will of its people. Even when they may be ignorant by the standards of experts,

| BOX 11.1 |

ENVIRONMENTAL JUSTICE

Environmentalism is a universalistic idea: all humans live in landscapes, and no single society can move far toward sustainable development without companion movement in others. This is the lesson of the commons: we are all in the world together, and ecosystems join our fates, whether we want them to or not. The roots of the environmental movement lie in the Progressive political tradition, one that looks to knowledgeable governmental institutions to structure the relationships of the economy so as to serve and reinforce a shared public interest. Understanding the way ecosystems connect people to one another can lead to an appreciation of commons that need to be well governed. Classical environmentalism, like the civil rights and women's movements, makes compelling practical arguments for inclusiveness. The land ethic bids us treat nature as a community to which we humans belong, in the same way that advocates of racial and gender equity seek to build a stronger human community.

Against this background, it is ironic and frustrating that the organized environmental movement, like its governmental agency counterparts, has remained largely white, well-educated, and affluent. More important, the environmental risks borne by minority communities, particularly poor ones, have often been much higher than those in adjoining areas. These unsettling empirical facts have drawn attention to **environmental justice**, the question of how social justice intersects with environmental problems and problem solving.

Years of studies have shown consistently that environmental *concerns*, unlike environmental *activism*, tend to be similar and widespread, whether we are talking about the concerns being expressed by different racial groups in the United States or in different countries, rich and poor, around the world. Over the years, the group that has shown the most consistently pro-environment voting pattern in Congress has been the Congressional Black Caucus, perhaps in part because African Americans have been exposed to many more environmental harms than have most Americans of Asian or European ancestry. A 2010 survey of small-business owners found that African Americans and Latinos were more supportive of clean energy policies than small-business owners generally, although there was majority support for clean energy among these business owners overall.[1] Another 2010 survey funded by an independent think tank, the Bipartisan Policy Center, found that Latino communities in many parts of the United States also show high levels of support for environmental protection.[2]

The importance of science and other kinds of technical knowledge in most environmental issues means that considerable education is needed even to follow these issues. As a result, ethnic groups that are underrepresented among those with technical educations are also underrepresented among environmental activists and in the staffs of environmental organizations. In addition, the recruiters in many organizations—not just environmental ones—often tend to recruit people whom they already know, or people who look like them. This kind of partiality can be so subtle that the person doing the recruiting may not even be aware of bias, but it has made it more difficult to recruit significant numbers of minorities for environmental occupations until recently.

In addition, environmentally objectionable facilities tend to be built in areas that have little political or economic power. In locating garbage dumps and other unpleasant facilities, the language of environmentalism has sometimes been used to keep such locally undesirable land uses (**LULUs**) away from well-off neighborhoods, which have more resources for fighting such facilities than do those that are less well off. On Cape Cod in Massachusetts, wealthy owners of beachfront property staunchly opposed windmills that would mar their views of the Atlantic, even though the wind machines would be several miles offshore and barely visible on the horizon. The attitude of "not in my backyard," boiled down to the acronym **NIMBY**, has been an aspect of environmentalism that has led to the characterization of all environmentalists as elitists.

NIMBY is a sign of a commons problem: a facility may be highly desired by the community as a whole, while also being a highly undesirable neighbor. But the sewage treatment plant has to go somewhere, and the oil refinery is essential for keeping vehicles on the road. So someone has to accept a LULU nearby, and that community winds up bearing a disproportionate negative impact. Even the perception of risk drives down property values, so these impacts can be substantial, even when environmental laws are obeyed. People do not like to bear a disproportionate impact, of course, even though the surrounding metropolitan area or region may benefit. When a proposed site is announced, the neighbors who do not want the facility next door are drawn into NIMBY protests, and the search for a politically feasible location turns into something like a game of musical chairs. Because people who live in well-off neighborhoods are more likely to be well organized and well connected, the undesirable facilities rarely wind up there. Instead, they are likely to end up next to those with little political power. Most of the time, the placement of a LULU is a political matter, and the unhappy neighbors receive no economic compensation for bearing this burden.

Should their cries of "Not in my backyard!" be dismissed as whining? As Frank Popper pointed out, the communities wind up next to unwanted by-products that someone else produced—often for the benefit of still others who are also more affluent.[3] Opposition to LULUs also provides incentives to adopt or develop new industrial processes that do not produce so much waste material. In 1986, Congress established a public database called the Toxics Release Inventory that requires polluters to disclose the volumes of regulated pollutants they discharge. Although the pollutants were legally permissible, disclosure led to sizable decreases in the following years.

LULUs clearly tend to wind up near or in poor communities, but there is intense debate about why. Land values are lower in poor communities, and ethnic minorities are often shunted by prejudice to residential areas adjacent to freeway interchanges and industrial zones in metropolitan areas, which are attractive factors when locating a garbage-collection depot, for example. The dispute is over whether racial bias may reinforce the economic factors. At least in the United States, the evidence is increasingly clear: minority status is an even more powerful predictor of exposure to environmental harms than is poverty. What cannot be disputed is that children suffer more asthma where the air is polluted, and minorities and poor people are disproportionately found in heavily polluted areas, along with higher proportions of people burdened by asthma.

The wider point is that, although environmental protection and sustainable development are needed to protect a common interest and to advance shared objectives, specific groups and people often lose out: unhappy neighbors of LULUs, fishermen and loggers forced into other occupations when their harvesting operations are subjected to regulation, and coal mine and factory owners whose costs go up. In Chapter 12 we analyze the political struggles that accompany the governance of commons. Some of those struggles raise issues of social justice, and it is those that are highlighted here.

For environmental NGOs, there is another question of environmental justice. Demographic forces in the United States are swiftly carrying the nation toward higher minority populations. The environmental movement needs to have larger minority representation for practical political reasons at every level, from the community to the federal government. Lisa Jackson, the first African American to lead the Environmental Protection Agency, strengthened the federal commitment to environmental justice during the Obama administration, redirecting budgets and programs as part of an effort to "widen the conversation about environmentalism."

The Ecological Society of America, the leading scientific professional society of ecologists, sponsors Strategies for Ecology Education, Diversity, and Sustainability (SEEDS), a network of chapters on college campuses that reach out to minority students. This is one part of a wider attempt in high schools and colleges to encourage minority students' interest in science.

There is interest in addressing the question of environmental justice but also a long way to go.

1. Small Business Majority, *Small Businesses and Clean Energy Policy* (report), June 9, 2010, 5–6, http://smallbusinessmajority.org/_pdf/SBM_energy_poll_final.pdf.

2. National Latino Coalition on Climate Change, "Attitudes of Latino Voters on Energy Policy and Climate Change: Results of Initial Multi-State Poll," April 2010, http://latinocoalitionon climatechange.org/news/NLCCC-Latino%20Voters%20on%20Energy%20Policy%20and%20 Climate%20Change%20in%20CO,%20FL%20&%20NV.pdf.

3. Frank J. Popper, "Siting LULUs," *Journal of the American Planning Association* 47, no. 4 (1981): 12–15.

citizens can learn and deserve a voice. Rachel Carson had faith that people would want to learn, and David Brower and his many allies believed that they could get people to act. Out of those views a movement has grown.

Silent Spring posed the questions of environmental hazards as if they were matters of common sense: Why not have pesticide-free agriculture? All it takes is a citizenry that is aware and empowered. To a surprising extent, Carson's claim that this was common sense has come to be accepted. As environmental awareness rose, the ecological perspective Carson championed was echoed in the rediscovered eloquence of Aldo Leopold. Carson reminded us that human technology is embedded in a natural order, and Leopold laid out a way to think through what that means, to link ecology and the other environmental sciences to ethics. Their influential writings complemented each other, helping to inform the social values we now call environmental. The NGOs and policies inspired by Carson and Leopold's ideas, in turn, began to fashion solutions to environmental problems, prodded by activists like David Brower.

Leopold's land ethic was harder to assimilate than Carson's crisp warnings: he did not blame industries but rather the culture of the conqueror. The path to change, for him, lay not in reforming others but in reconsidering ourselves, not only in politics but in ethics as well. Leopold did not shy away from judgment, but his writing did not reach for the polemical energy found in Brower's campaign advertisements.

Whereas Carson cautioned us not to modify inadvertently the biological processes of ecology and evolution, Leopold promoted human values that connect us to the natural world again. Those values, in turn, resonated with the outdoors culture of the Sierra Club, whose members were hikers as well as protesters.

Carson warned that our biological nature would be harmed by thoughtless modifications of nature; Leopold appreciated the continuity between our biological nature and the landscapes we inhabit, raising the consciousness of an industrial culture. Joined to the organizational and political resources of the nongovernmental organizations that have grown by leaps and bounds since 1962, these ideas have created new fields of law, policy, political activism, economics, and psychology; a wide range of practices such as recycling, organic agriculture, and green building design; and themes expressed in music, movies, poetry, philosophy, and religion—an astonishingly diverse set of concerns and activities that we have learned to call "environmental." It's not easy being green, said Kermit the Frog of *Sesame Street*. He was talking about racism, but his words proved resonant with classical environmentalism as well. A lot of us are trying to be greener, or saying we should be, and often it isn't easy.

Leopold is remembered as a voice speaking on behalf of wilderness, whereas Carson is remembered as a critic of technology. Carson defined a way of thinking about public policy—classical environmentalism—that has shaped environmental politics down to the present. Leopold's view is more difficult to put into political demands. Yet the land ethic is a subversive idea, a reminder that humans should acknowledge their place within nature rather than bending nature to human will.

What Leopold and Carson shared is a vision of humans in the landscape, a vision we have come to call environmentalism. The challenge of environmental-ism is to turn the land ethic into competent governance of the large-scale, non-stationary, unprecedented world without edges that we now inhabit and affect. Carson and Leopold reminded us that the world without edges exists, and must exist, within the natural world, after all. It may seem amazing, in retrospect, that humans needed such a reminder. Yet the uncomfortable fact remains that this reminder is as timely today as it was more than half a century ago.

FURTHER READING

Carson, Rachel. *Silent Spring.* Boston: Houghton Mifflin, 1962.

Erikson, Kai T. *A New Species of Trouble: Explorations in Disaster, Trauma, and Community.* New York: Norton, 1994.

Freudenburg, William R., and Susan K. Pastor. "NIMBYs and LULUs: Stalking the Syndromes." *Journal of Social Issues* 48, no. 4 (1992): 39–61.

Leopold, Aldo. "The Land Ethic." In *A Sand County Almanac.* 1949; repr. New York: Oxford University Press, 1966.

Mohai, Paul. "Equity and the Environmental Justice Debate." *Research in Social Problems and Public Policy* 15 (2008): 21–49.

Popper, Frank J. "Siting LULUs." *Journal of the American Planning Association* 47, no. 4 (1981): 12–15.

Szasz, Andrew. *EcoPopulism: Toxic Waste and the Movement for Environmental Justice.* Minneapolis: University of Minnesota Press, 1994.

Taylor, Dorceta. "Diversity and the Environment: Myth-Making and the Status of Minorities in the Field." *Research in Social Problems and Public Policy* 15 (2008): 89–147.

KEY TERMS

environmental justice	LULU	precautionary principle
land ethic	NIMBY	risk analysis

Chapter Twelve

COLLECTIVE ACTION

GOVERNING COMMONS

In Part II we identified four grand challenges of sustainability: climate change, loss of biodiversity, rapid urbanization, and sustainable development. Each of these challenges exposes the limitations of our abilities to manage and govern commons on a large scale and over long periods. Meeting the grand challenges, so that people can move toward a sustainable economy, will require us to expand the scale of our competence, both deepening our capacities and extending them in space and time. In this chapter, we describe how that scale of competence has been expanded by environmentalists, but also how their efforts have fallen short of solving critical environmental problems.

Human life is already managed in significant ways on large spatial and long temporal scales. Major political and economic institutions already address peace and war, education and public safety, threats to health from epidemic disease, and more. The institutions that manage the economy and government also coordinate an extraordinarily complex web of markets for goods and services. The same institutions propel and can worsen serious environmental problems, such as the loss of biodiversity, but their strength and reach makes them indispensable for addressing the grand challenges.

In Chapters 12 and 13 we examine these political and economic institutions with the aim of understanding both their role in creating the grand challenges and their potential for dealing with them. In this chapter, we focus on the way that organizations play a central role in environmental governance. Some of those organizations are business firms; others are governmental agencies. In addition, nongovernmental organizations (NGOs), which advance missions rather than seek profits, play a critical role. The study of politics and markets is the focus of the disciplines of political science, sociology, and economics, and we will explain how these bodies of learning provide essential insights into environmental problems and suggest ways they can be tackled.

Managing human uses of commons is an inherently political process. Recall from Chapter 3 Garrett Hardin's summary phrase: overcoming the tragedy of the commons, he said, requires "mutual coercion mutually agreed to." The word "coercion" signals that governing entails getting people to act in a way that does

Learning Objectives

When you have finished studying this chapter, you should be able to

- identify nonprofit, governmental, and profit-making organizations you interact with each day;

- analyze an environmental policy idea, such as a proposed method to respond to global warming, in terms of the social and political implications of the idea: How does the policy identify a community and a commons in need of better governance? Who would have to change their behavior? Who supports change, and how are they organized? What is the political benefit of favoring, opposing, or staying neutral in the debate over the policy?;

- articulate your own career and life plans in terms of the roles you might be able to play in the economy and in civil society;

- identify examples of concentrated and diffuse interests at play in the governance of your community;

- relate the mission of a nonprofit organization (e.g., a university) to the economic requirements for its survival (e.g., tuition, grants, and gifts from wealthy donors). How does the tension between mission and survival affect what the nonprofit can actually achieve?;

- analyze the complaints of those who criticize environmental regulations and those who oppose environmentalists;

- examine, skeptically and critically, the assertions of progressive politicians and environmentalists that they are improving environmental quality or moving society toward sustainable development. How might those claims advance their own *political* interests? Is having a political interest a reason to distrust a leader?

BOX 12.1   STEALING THE COMMONS FROM THE GOOSE

The power of the community is used to protect private property as well as commons. The protection and even the definition of what is privately owned is a critically important institution, because it draws a boundary between those who have property rights of a certain kind and those who do not. Today, few workers pay attention to the fact that their employers own all e-mail messages that employees send from their company accounts. As a result, communications that they may consider to be private are not protected from surveillance by the firm.

In Chapter 3, we described the enclosure of lands that had been used as commons in Britain until the eighteenth century. Enclosure meant that whole communities, which had relied on commons for their livelihood, were ejected; and many rural people were driven into the cities, where they struggled to find their way in the Industrial Revolution. An anonymous protest poem of the times put it this way:

> They hang the man, and flog the woman,
> That steals the goose from off the common;
> But let the greater villain loose,
> That steals the common from the goose.[1]

Who was the "greater villain" here? It was usually an economically and politically powerful actor—often the hereditary landlord—aided in his claim by the government. The exercise of political influence by the landowners to create the laws of enclosure produced a great tragedy—the destruction of a rural way of life that had persisted for centuries.

The exercise of governmental power to "steal the common from the goose" is also under way today. The development of oil fields in Nigeria, Ecuador, and many other nations has come at the expense of people who had the bad luck to live close to oil deposits and who have suffered rather than benefited from their exploitation. In the coal-rich areas of the Appalachian Mountains, mining firms have sometimes destroyed the landscape and the livelihoods that depended on it. In a process known rather clinically as mountaintop removal, the tops of mountain ridges are blown up and dumped into the valleys, so that the coal in the ridge can be removed. In these and other cases, environmental and social havoc is created, with government acting as a willing partner of those who are appropriating the commons. These actions employ coercion, although such coercion is not always accepted by those being ejected. Nigeria has had a low-level civil war simmering amid its oil fields for decades, and a battle over mountaintop removal has been fought through the courts in Appalachia.

As we discuss the exercise of governmental power to protect environmentally sensitive commons, it is important to bear in mind that the power of government is guided by politics, and that political outcomes are not always just outcomes.

1. Wikipedia, "Enclosure," http://en.wikipedia.org/wiki/Enclosure.

not immediately advance their individual interests. Communities impose sanctions against stealing, for example, even though theft may be in the interest of the thief. Every community protects some of its commons, and questions of power and conflict are joined in doing so.

Institutions of governance are not only abstract rules; governance is brought into being by the actions of people as they carry out the roles discussed in Chapter 3—monitoring behavior, enforcing rules, resolving disputes, and asserting the authority of a community over specific commons. Like other human activities, actual governance is full of imperfections—errors, over-reaching, corruption—as well as diligence, courage, and selflessness. These properties, good and bad, are usually deeply woven into the institutions and cannot easily be changed (see Box 12.1: Stealing the Commons from the Goose, page 312).

In this chapter, we examine the governing of commons in stages. First, we consider the model of classical environmentalism as it has operated in the United States. Classical environmentalism has shaped the creation and course of environmental policy. Although environmental policies have made essential contributions, environmental problems persist, and some, such as climate change, have grown larger. This is due, in part, to the political dynamics of environmental problem solving and the difficulty of overcoming the incentives that guide human behavior in commons (that which belongs to all is cared for by none). The limited but real ability of the environmental movement to counter these incentives is rooted in its ability to organize and sustain a durable presence in environmental politics and policy. We conclude this chapter with a discussion of NGOs and the philanthropic foundations that support them. These components of governing have expanded the scale of competence of American government to handle problems of the commons. The struggle to exercise that capability continues, and as we will see, the competence to engage with the grand challenges is still emerging—often barely emerging. Much remains to be done.

LEGITIMATE COERCION

The challenge of governance is to arrange for mutual agreement, so as to protect the environment and move toward sustainable development. This is where the scale of our competence is problematic. It is comparatively easy to arrange for the messy dormitory common room to be cleaned up, even after a big game weekend. It is another thing to arrange for the cleanup of toxic wastes left behind by a company that employs a small army of lawyers, or worse, has gone out of business.

For more than two hundred years, people have come to accept that coercion is legitimate when it is mutual—that is, when the possibility of coercion is asserted by a democratic government (see Box 12.2: Democracy, page 315). By winning a competitive election, a government can make a claim to mutual agreement in the sense that because the government is chosen by voters it can take action in their name. When that claim is accepted by the people, legislatures and administrative agencies can collect taxes and pass and enforce laws, and courts can resolve disputes, decide on punishments for those who violate the law, and even protect the fundamental rights of people who may be very unpopular with a majority of voters. These are all acts that employ or imply coercion, but it is coercion exercised by a government whose legitimacy is improved when its citizens can choose their leaders. As the slow pace of progress on global warming suggests, the international system, not being democratic—nor, indeed, being a single government at all—does not command the legitimacy to coerce. Instead, nations struggle with a tragedy of the commons, the global commons of the atmosphere.

Within a single nation, however, democracy might be a way to counterbalance the accelerating domination of nature. If coercion can be mutually agreed to, then people can overcome the destructive logic of the commons by adopting restraints on their own behavior and entrusting enforcement, monitoring, and dispute resolution to a fair and effective government. This is what happens with speed limits. Nearly everyone is tempted to speed, but no one wants a ticket and no one wants dangerous roads. Most drivers speed some of the time, but speed limits are enforced and they do lower accident rates. When there is a functioning democracy, the political system can manage commons more effectively than authoritarian governments can.

The hope for a responsible self-government, of course, conflicts with the cynic's suspicion that *any* government is a macroparasite, exploiting people for its own purposes. This conflict runs through a good deal of recent political discourse. Citizens share little agreement about which view of government is better, especially in concrete circumstances. The speeder does not like being pulled over, and firms often complain about intrusive and foolish bureaucrats when an environmental regulation is imposed. Nonetheless, environmental problems are usually approached through the political system, and this is what is assumed in classical environnmentalism.

BOX 12.2

DEMOCRACY

Recall from Chapter 2 that intense exploitation of nature and environmental concern arose simultaneously in the historical process we call modernization. So did democracy—the search for workable self-government of, by, and for the people. The U.S. Constitution was written in 1787, toward the end of the lives of both Gilbert White, the naturalist, and Richard Arkwright, whose factory launched the Industrial Revolution. The Constitution was the first successful charter for self-government in a large nation, one that would span the North American continent less than a century after its basic rules were adopted. The development of democracy has shaped how we think of environmental governance today.

The premodern role of governments was macroparasitic: the monarch levied taxes, conscripted soldiers for war, and provided public order by punishing criminals and repressing political opposition. Since the Enlightenment, the time in which White and Arkwright lived, democracy has taken hold on a widening scale. The concepts and practices of democracy form a rich tradition of social innovation. Key components include citizens who possess political rights, such as the freedoms of speech, assembly, and access to information; equal treatment of citizens within a fairly administered legal system; and the ability of citizens to choose their representatives through open, fair elections. These are ideals, and reality often falls short.

The ideal of democracy itself—that the people can rule—has been a powerful one, and it has led to a broadening of the scope of government to encompass services provided to citizens. The U.S. Constitution was written in part to facilitate trade among the states. It was a document grounded in the idea that economic development was a legitimate purpose of government. Today, roads, sewers, and police forces remain basic responsibilities of government.

From that base, the U.S. government—like those of all other developed nations—expanded its responsibilities to encompass a wide spectrum of commitments. The principal ones form the institutional part of Second Nature: education, public health, social welfare, regulation of the economy, environmental protection, and equality of treatment for women and minorities. All of these commitments arise from the ideas that the citizens should be the direct beneficiaries of the government's activities, and that obtaining the consent of the governed requires treating all members of society in a way that can be regarded as fair. When Gilbert White was born, that notion of government for the people and by the people was as remote from the mainstream of thought and the reality of institutions as Aldo Leopold's land ethic is today.

The idea that democracy might propel historical change has been a bright hope several times: when the United States was founded in the late eighteenth century, in the early years of the twentieth century, in the 1950s and 1960s when colonial empires collapsed, at the end of the cold war in 1989, and during the Arab uprisings of 2011. Each time, we have seen idealism and upheaval, and many new democracies, only some of which proved stable. Like the survivors of a virulent fever, the societies that manage to weather the tumult of democratic change are often resistant to further life-threatening chaos. There are countries struggling to become democratic, such as Tunisia after 2011, and others where democracy seems to have taken hold, such as Taiwan and several nations in eastern Europe.

The American Revolution was fought in the eighteenth century, and western European nations developed parliamentary democracies from the late eighteenth through the twentieth centuries. Japan and India both developed democratic rule in the late 1940s, adding to the surge in the proportion of the human race governed by elected leaders. India is by far the world's largest functioning democracy. The end of colonial rule and the collapse of the Soviet empire in the early 1990s have produced a large array of independent countries, many of which have difficulty satisfying international observers that their elections are free and fair. Still, the trend over time has been an increasing proportion of the human race living under some form of democratic government (see figure). For this purpose, social

Percentage of the world's population living in countries with competitive elections. The existence of competitive elections is an influential indicator of democratic governance.

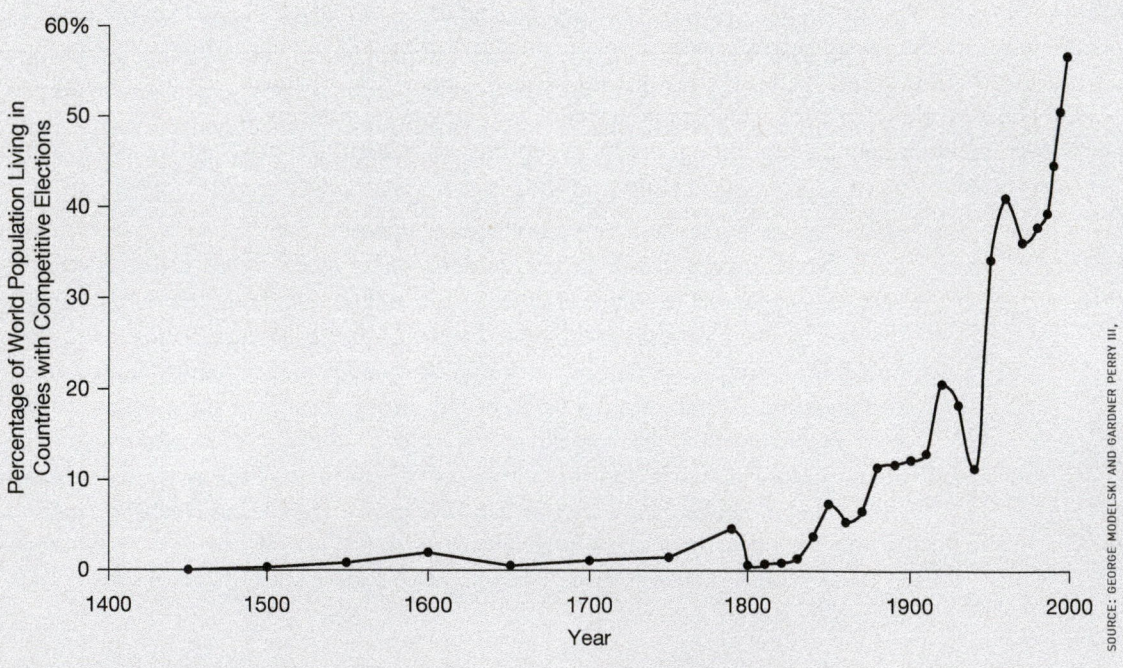

SOURCE: GEORGE MODELSKI AND GARDNER PERRY III,

scientists use open electoral competition as one key indicator that a government is democratic.

To most Americans, it is individual freedom rather than voting that is foremost among the benefits of democracy, and freedom is, of course, fundamental to the consent of the governed that is the basis of democracy. Liberty *is* crucial to the individual person's experience of democratic participation. Free speech and the ability to organize for political action without government meddling or intimidation are clearly important to vigorous political competition. Yet in practice, virtually all societies debate the limits that should be imposed on those freedoms. As a result, the specifics of freedom vary a good deal across democratic political systems. In all of them, the ability to hold governments accountable for their actions has been a significant factor in sustaining individual liberties over time.

In elections, organization is essential. This is why political parties have appeared, often spontaneously, in all democracies. The advantages of the well organized are important: environmental causes have gained support from educated, economically successful people who have the capacity and desire to organize, and who are numerous enough to matter. This is one reason that Rachel Carson's appeal in *Silent Spring* meant so much: it reached people who could make a difference in politics.

The broad appeal of environmental protection and the conservation of nature has meant that, at least since the 1970s, virtually all American politicians, both Republican and Democratic, have claimed to be environmentalists, although the details and implications of those claims have changed a good deal in recent decades. Those working to curtail regulation describe their positions in terms of opposition to environmental extremists ignorant of economic realities. In this way, the politician can claim not to be anti-environment, but only anti-extremist.

The pro-business, anti-regulation stance of the Republican Party has meant that environmentalism has been caught up in the deepening polarization of American politics in recent years. Yet a wide base of political support for environmental protection exists among voters, whether they identify themselves by political party (Republican or Democrat) or by ideology (liberal or conservative).

Over the last half-century, Americans have consistently told Gallup, the best-known opinion survey firm, that paying some price for environmental protection is appropriate (Fig. 12.1).[1] Support for environmental protection wanes during economic recessions, but has been quite robust for decades. The decrease in the strength of support for environmental protection between 2008 and 2010 reflects economic worries during a time of severe recession, as well as the cumulative effect of decades of politicians' charges and advertising claiming that environmentalists do not care about the economy.

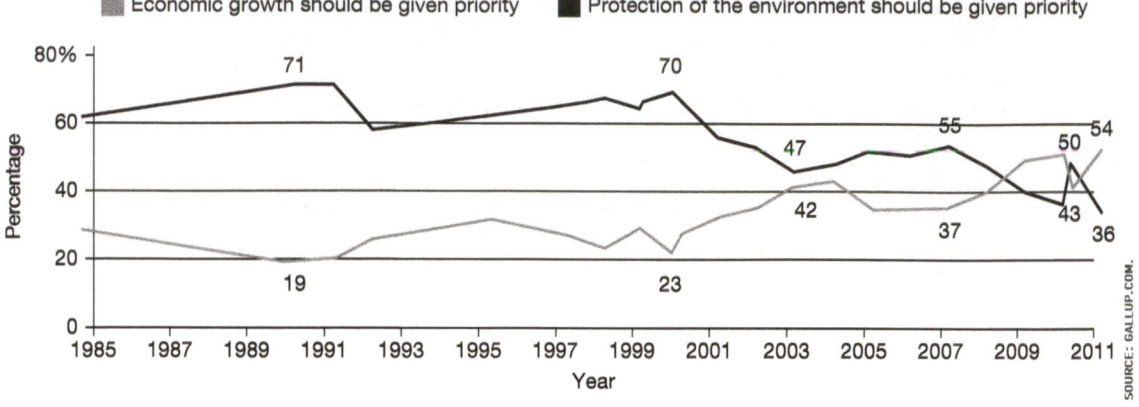

Economic growth should be given priority   Protection of the environment should be given priority

SOURCE: GALLUP.COM.

FIGURE 12.1
Trends in public opinion about the priority of environmental protection and economic growth.

It is no surprise that concrete environmental actions may be unpopular. During the 2010 oil spill in the Gulf of Mexico temporary delays in deep-water drilling met with sharp protest, particularly among those who lived closest to the area where the drilling was being carried out and whose livelihoods depended directly or indirectly on the oil industry. What may be more surprising is that support for environmental protection has been so widespread, and it still is.

An intriguing finding of public opinion studies is that parents recognize that their children are better informed than they are themselves about environmental choices at the personal level—and parents look to their children for guidance on environmentally significant decisions.[2] Recycling, for example, has been adopted by many families after children began scolding their parents about separating waste streams. People in their teens and twenties are both more strongly environmentalist and more confident of their individual capabilities to bring about change than those who are older.[3] We will see in the years to come whether those opinions persist into later adulthood.

The breadth of support for environmental protection indicates that classical environmentalism continues to play a critical role in environmental politics. When scientists warn of environmental threats, voters look to government to undertake mutual coercion to protect the commons. It is from this logic that environmental policy takes shape.

ENVIRONMENTAL POLICY

Environmental policy may be said to have a very long history. The Magna Carta ("Great Charter"), reluctantly signed by England's King John in 1215 under pressure from his rebellious barons, is one of the founding documents of the tradition of constitutional rule—the idea that no person, not even the king, stands above the law. The Magna Carta included language requiring the king and his favored associates to remove their fish traps from the Thames River, so as to reduce the pressure on fish populations. Such rules to govern a commons (see Chapter 3) are a form of environmental policy.

The scope of environmental governance is broad. As we discuss later in this chapter, the laws with the most important environmental consequences do not seem at first glance to be environmental at all—laws that authorize the building of roads or levees, for example, or tax breaks that can make certain forms of energy cheaper or more expensive. When most Americans think of "environmental policy" today, however, they are thinking of a smaller body of statutes having a much shorter history—a raft of laws, beginning with the Wilderness Act of 1964 and continuing through the early 1980s, that reflect the ways in which classical environmentalism modified the existing structure of American politics and law (see the highlights in Table 12.1). These laws covered a broad spectrum—water, air, drinking water, toxics, endangered species, Alaska wilderness, and nuclear waste.

The wide concern triggered by *Silent Spring* signaled to politicians that the environment was an area in which constructive government actions were both

TABLE 12.1 SIGNIFICANT U.S. ENVIRONMENTAL LEGISLATION, 1964–82.

Year ratified	Legislation
1964	Wilderness Act
1969	National Environmental Policy Act
1970	Environmental Protection Agency created; Clean Air Act (significantly amended in 1977 and 1990)
1972	Clean Water Act, Coastal Zone Management Act, Pesticides Control Act (significantly amended in 1996)
1973	Endangered Species Act
1974	Safe Drinking Water Act
1976	Resource Conservation and Recovery Act (management of toxic materials), Toxic Substances Control Act, National Forest Management Act
1980	Superfund established to clean up toxic sites (significantly amended in 1986)
1980	Alaska Lands Conservation Act
1982	Nuclear Waste Policy Act

needed and likely to be rewarded by voters. Some of the champions of environmental causes did appear to reap political gains. The presidential campaign of Senator Edmund Muskie in 1972, for example, was given a significant boost by the key role he played in enacting the Clean Water Act that year. Even politicians with little previous interest in the environment sought to change the ways in which they were seen by the public, depicting themselves as environmental leaders and sometimes bringing about important policy changes. Republican president Richard Nixon, for example, signed the National Environmental Policy Act of 1969 (NEPA) on New Year's Day 1970 with a message to citizens about his support for environmental protection. Later that year, he reorganized several federal agencies to create the Environmental Protection Agency (EPA).

These were actions taken by the most prominent members of the American political establishment for the highest political rewards. Particularly in the first two decades after *Silent Spring*, these choices made a difference. The Clean Water Act was for some time the largest source of discretionary public spending, as federal dollars were channeled to municipalities to upgrade their sewers and sewage treatment plants. The new laws governing toxic wastes put legal liability for cleanup in the hands of property owners, and this led to drastic changes in commercial real estate practices, as potential buyers scrutinized buildings for any environmental hazards that they might contain.

Although the Nixon administration did not intend it, NEPA set off an intense period of litigation aimed at federal agencies. The Act required that, before any agency of the federal government could take actions "significantly affecting the quality of the human environment," the agency needed to study the environmental impact of major projects. These projects ranged as widely as the federal government itself, including everything from building highways to licensing nuclear power plants to the management of public lands. The resulting environmental impact statements were public documents that anyone could read, and they included discussions of ways in which the negative impacts could be mitigated (reduced). The requirement to publish environmental impact statements opened up the decision-making process of government in a new way, spelling out both controversial impacts and things that could be done to reduce them. This disclosure changed the course of many projects, including some that were canceled outright because no plausible case could be made that the benefits outweighed the costs, including environmental costs. This kind of public scrutiny was unwelcome to the projects' proponents, of course, and they resisted. The lawsuits that forced compliance with NEPA formed a central part of environmental law as a new body of jurisprudence.

Most of all, this era created new rules and new administrative bodies to protect citizens and nature from environmental harm. The EPA became a national presence, with ten regional offices working with a growing batch of new, state-level,

BOX 12.3

PLURALISM AND THE STRUGGLE OF GROUPS

Conflict is normal in politics. The standard model of American politics is the theory of **pluralism** initially articulated in the 1950s. In this model, the role of government is to be a referee in the competition among private interests, which are organized into **interest groups**.

Pluralist political systems are biased in favor of interests that can be organized effectively. It can be difficult to organize people around the protection of ecosystems or the health of those who endure high levels of pollution. More generally, in a commons, that which belongs to all can be treated as if it belonged to no one, which is a problem for those who would solve environmental problems in a pluralist political system (see Box 12.4: Concentrated and Diffuse Interests, page 326).

When conflict is persistent, the political alignments that form around it endure; for example, the division between labor and business has been reflected for more than a century in the electoral competition between Democrats and Republicans. These stable divisions are like the trees in a rain forest: they organize the dynamics of interactions for long periods, so that specific patterns of organization can arise and persist.

The iron triangle.

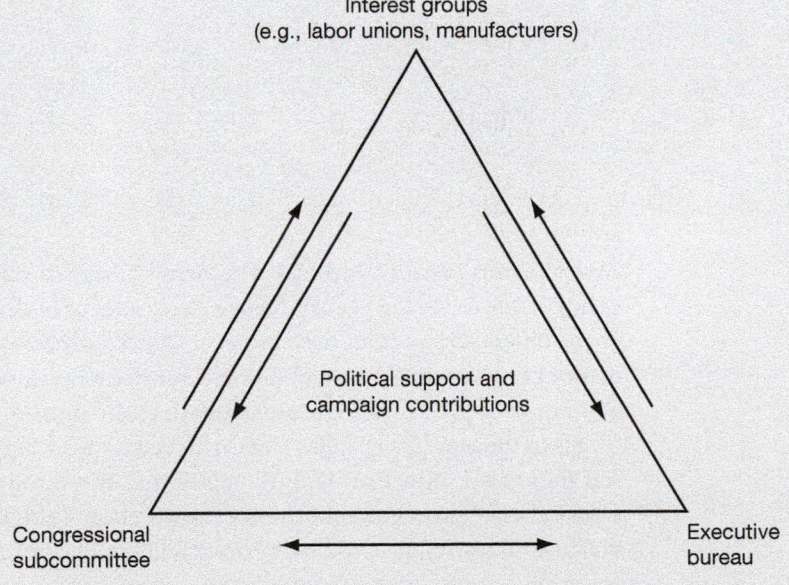

One important class of long-lasting political habitat is sometimes called the iron triangle (see figure)—a mutually reinforcing set of alliances that often works to secure government funding for projects such as interstate highways or major medical centers, or to promote industries such as agriculture, often with the assistance of officials within government agencies who are not acting so much as referees as they are advocates of a particular, well-organized interest. Such "captured" agencies and their congressional allies have powerfully reinforced private economic actors in transforming the natural world, often to the detriment of the environment and of people who lacked the power to stand in the way of "progress." The arrows inside the triangle in the figure indicate flows of political resources—money for election campaigns, statements of support such as letters, news articles, demonstrations, and workers for campaigns at election time. The arrows on the outside of the triangle indicate flows of legitimacy and governmental resources—laws, policies, administrative actions to implement them, tax breaks for favored businesses, and budgets to carry out activities.

The pluralist model describes many issues reasonably well. Corn farmers in Iowa, for example, have a strong interest in obtaining federal support for the growing of corn. Senators and congressional representatives from the Midwest are more likely than those from Connecticut to serve on committees that control funding for the U.S. Department of Agriculture, which has been responsible for administering large subsidies to corn growers among others. These subsidized crops, in turn, are grown with fertilizers that pollute the Gulf of Mexico (see Box 6.1: Agricultural Ecosystems in Chapter 6, page 132). The struggle among groups is a struggle to shape public policy, which is the set of rules that govern the commons of American society. In the pluralist theory, the voters elect representatives seeking to advance a variety of interests. The trade-offs and compromises made by voters and politicians then produce a shared interest. This theory assumes that all interests have an equal opportunity to organize and be heard, an assumption discussed in Box 12.4 on page 326.

environmental-quality agencies. The Army Corps of Engineers, long known as a construction-oriented agency, hired a new cadre of biologists to gather data on the environmental impacts of their projects. Other scientists were hired by the Department of the Interior's Fish and Wildlife Service to evaluate whether certain species of animals or plants were threatened with extinction.

All of these policy changes were achieved by working within the existing political framework (see Box 12.3: Pluralism and the Struggle of Groups, page 321). Classical environmentalism powered a revolution carried out by the conventional means of interest-group politics. As we will see below, though, the very success of environmental politics also set limits on what could be achieved.

HAS ENVIRONMENTAL POLICY WORKED?

The laws created through classical environmentalism have made a difference. Air quality has improved, despite continuing growth in American industrial production and the continuing rapid expansion in automobile use. Today's cars emit huge amounts of carbon dioxide, but there have been major reductions in the pollutants singled out in the 1970 Clean Air Act. The comparison with the developing world is dramatic, as one sees in Mexico City's heavily polluted urban air. In 1970, the air was roughly as polluted in Los Angeles, but it is now far cleaner.[4] However, there are still several days each year when health alerts are issued in Southern California, as is the case in many American cities.

Water pollution control has managed to stabilize water quality in the United States overall, with substantial improvements in many places, such as Boston Harbor. The Clean Water Act provided funding for cities and towns to treat their sewage, in some cases for the first time, before releasing it into streams and bays.

The Endangered Species Act has provided protections that have led to the recovery of some species, such as the bald eagle, the grizzly bear, and the Channel Island foxes you met in Chapter 1. Regional habitat conservation plans have been adopted in high-biodiversity areas of Southern California, as well as other parts of the United States. These provide a regional approach that is meant to allow some land conversion for economic purposes while protecting habitat for species under pressure. But intense conflict still persists over the restrictions on use of private property, and opponents have succeeded in requiring agencies to carry out far more elaborate procedures before taking action to defend new species under the Act. Congressional appropriations for carrying out those procedures have undergone many cuts over the years, so that the list of species being considered for listing has grown ever longer.

Cleanup rules for toxic waste sites have drastically altered commercial real estate practices and markets. Roughly a thousand sites that were sufficiently hazardous to be listed on the Superfund National Priority List have been cleaned up, but there are tens of thousands more. In 1995, in an often-repeated pattern of failing to provide consistent financial support, Congress allowed the major funding source for these cleanups, a tax on the oil and chemical industries, to lapse. Funding was 25 percent lower for the years 2001–4 than it had been from 1992 to 2000, and the number of cleanups completed each year dropped by more than half.

Across the American economy, environmental policies have led to greater complexity. Facilities such as housing subdivisions, shopping centers, offices, and factories must be planned more carefully, with numerous environmental considerations being weighed. These include questions about the need for additional parking spaces when a building changes use from offices to shops, whether asbestos needs to be removed when walls are rebuilt, and how to treat the pollutants produced in a manufacturing plant. All of these and numerous other environmental considerations arise from a commons that was once ungoverned or poorly governed. Urban planning has been transformed, and state and federal laws have required explicit consideration of many implications that had been ignored.

One result has been the growth of a cumbersome maze of permits, planning meetings, and public hearings facing anyone wanting to undertake a significant project in an urban or suburban setting. Is the system imperfect? Certainly. Are all the complications reasonable and rational? Certainly not. Yet many environmental problems are recognized, avoided, or solved through the scrutiny now required. We have extended the capacity of government to use mutual coercion, even though the means of doing so are often awkward and need improvement.

In what may be the most important legacy of classical environmentalism, an environmental movement now exists, unlike in 1962. It is a force to be reckoned with in elections, in corporate decision making, and in public life. This is an immense achievement, largely unanticipated when *Silent Spring* appeared. To an important extent, classical environmentalism has been effective.

ENVIRONMENTAL POLITICS

The major achievements of classical environmentalism, on the other hand, have not yet created a governmental and cultural structure that can move toward sustainable development, or that is yet up to the task of challenging global warming and the fossil fuel consumption that drives climate change. As we will see in the chapters ahead, the struggle to engage with the grand challenges requires economic and technological innovations, as well as shifts in the material strivings of large numbers of people.

Nearly all of the struggles for sustainable development have a political dimension. In some states, the price of a soft drink includes a deposit—a fee that is refunded when the empty container is returned to a retail store. Deposits encourage reuse or recycling of containers. But if no law requires a deposit and one bottling firm charges deposits while its competitors do not, that firm's soda will cost more. This is the pattern of the tragedy of the commons: those who do something that benefits the community—reducing the number of drink containers that are thrown away—are punished for their good deeds. What is needed for better governance of this commons is a rule that requires all bottlers to charge a deposit. Such fees have been adopted in eleven states. They have been successful at reducing litter and raising the proportion of containers that are reused or recycled, at a cost of a little more than a penny per container.[5] But bottling firms, grocers, and others who would have to accept and redeem the containers oppose these regulations and have defeated citizen campaigns for decades by arguing that this form of mutual coercion is too costly. Governing the commons is a political task.

The environmentalist approach to politics is classical environmentalism. The rise of environmental policy demonstrates the power of this approach. By contributing to sophisticated national organizations such as the Wilderness Society and the Sierra Club, citizens from across the country could have a voice in the interest-group politics of Washington, D.C., and many state capitols. The pluralist politics described in Box 12.3: Pluralism and the Struggle of Groups, page 321, turned out to be an arena in which environmentalists could win significant victories, including the major legislation listed in Table 12.1.

In doing so, the environmental movement also has gained powerful enemies who have fought back. These opponents were and are formidable. First, they are deeply entrenched. Most benefited from earlier enclosures of commons that converted public resources to the advantage of specific industries. The coal and electric power companies did not pay for the environmental damage done by the fly ash, mercury, and carbon dioxide they put into the air, or for restoring the ruined landscapes and waterways left behind by mining. Logging companies and farmers using irrigation water paid little of the cost of roads, dams, or canals built by taxpayers, to say nothing of the damage that logging or irrigation do to downstream fish populations.

As these examples show, small but important groups have powerful economic reasons to oppose environmental reforms. These are interests that benefited directly from the enclosure of commons and the channeling of public resources to benefit private parties, such as the fishermen who used the government to stake a claim on a fishing ground. The iron triangle form of political alliance described in Box 12.3 reflects this pattern.

In the iron triangle, government is the organizer and sponsor of the enclosure of the natural world; in this role, government may or may not be the organizer of mutual coercion as well. Governments are widely seen as having responsibilities both to foster economic development and to protect commons. Environmentalists seek to change the balance between these broad missions, as well as to reduce the number of cases in which the commons can be damaged without actually contributing to economic well-being. Rather than lobbying to have Congress increase funding for environmental and resource agencies, such as the EPA or the U.S. Forest Service, environmental groups are far more likely to spend their resources suing the same agencies (Fig. 12.2 on page 327). This approach leads to persistent conflicts.

Here the opponents of environmentalism have two basic advantages. First, the iron triangle already provides an organized network to pursue economic development. These actors have **concentrated interests**, as explained in Box 12.4: Concentrated and Diffuse Interests, page 326. Second, the voices of development have economic resources as well as powerful reasons to defend their interests.

BOX 12.4

CONCENTRATED AND DIFFUSE INTERESTS

Most of the economic actors in the existing economy are already organized to work with government. They belong to a category that political scientists call concentrated interests. Members of these groups already have organizations that enable them to communicate with one another (and reach agreements on their interests) with relative ease, in part because they tend to derive a direct financial benefit from successful lobbying. Firms in polluting industries, the commercial fishing industry, and land developers are all concentrated interests.

Environmental interests, by contrast, have often been **diffuse**—the reverse of concentrated. Access to clean air and clean water is an interest shared by every person on Earth. Yet very few of the people who benefit from that clean air would be able to invest the same amount of money in lobbying for clean air and clean water as the owners of coal-burning power plants who seek to receive exemptions from environmental regulations.

As a rule, governments pay more attention to concentrated interests than to diffuse ones. Representatives of concentrated interests, for example, tend to be much more effective in securing government funding—projects such as interstate highways or new bridges and dams tend to receive far more support than do projects to repair trails or maintain campgrounds in the national parks, or programs that hire forest or park rangers to watch out for people who are hunting or fishing without a license. In addition, diffuse interests such as environmental protection tend to be far more fragmented—split across multiple agencies and jurisdictions.

In the tragedy of the commons, that which belongs to all is protected by none. In a pluralist political systems, interests that are diffuse are harder to organize and to advance than those that are concentrated.

If we inquire into the major sources of environmental harms, many are authorized and subsidized by laws that few people ever think of as environmental—laws that benefit concentrated interests. Major environmental impacts, for example, have been created because of laws for building roads, subsidizing the growing of some crops and not others, and providing far more funding for new housing that sprawls into former farmland than for rebuilding the inner-city housing that is already served by mass transit, to name only a few examples. The problems brought on by these enclosures tend to be diffuse. The role of government in creating these enclosures is often not seen at all because they have been in place for a long time and their effects are hidden in the invisible present.

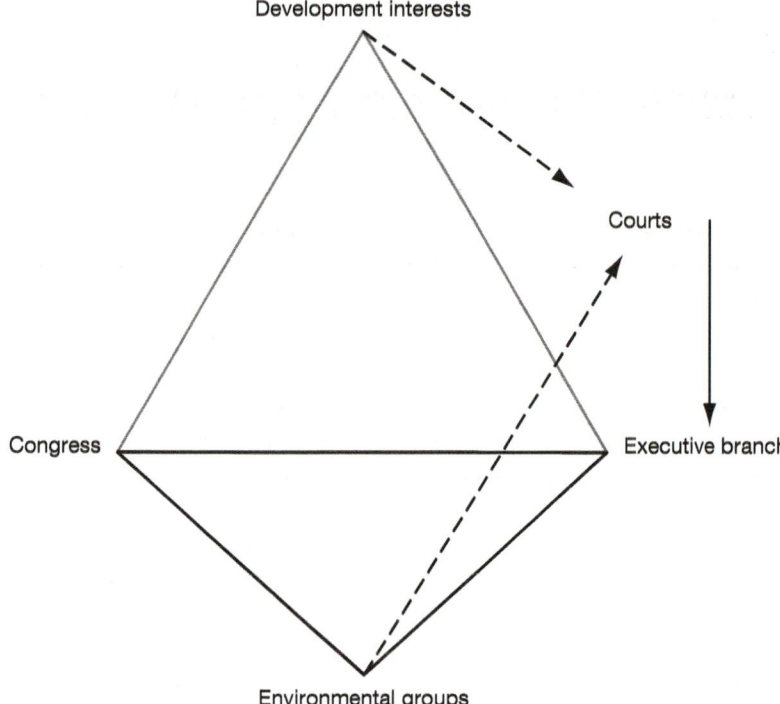

Development interests

Courts

Congress

Executive branch

Environmental groups

FIGURE 12.2
Modifications of iron triangle due to environmental policy.

Environmentalists have talent and passion, and often logic and the facts on their side, too. But the forces of resistance have money to pay for lobbying, advertising, and sophisticated analysis and litigation. Most of all, they have staying power. Environmental groups might be able to seize on a crisis and press for the enactment of a reform. But the implementation of reforms requires complicated regulations, administrative resources for enforcement, and the time and energy to resolve lawsuits stemming from changes in the way things are done. Those who benefited from enclosures of the commons have had both the motive and the means to slow or reverse environmental progress.

ECONOMIC BENEFITS AND ENVIRONMENTAL ELITISTS

Counterattacks from concentrated interests take two forms: emphasizing the social benefits of exploiting the environment, and attacking environmentalists as elitists who care more for nature than people. There is enough plausibility in each of these criticisms of classical environmentalism that these arguments have gained traction in the American political dialogue. Some of the leading voices and much of the funding for these counterattacks come from people and firms who have long benefited from public policies that give them significant economic advantages. They know how things work politically, and they can be potent adversaries of those advocating classical environmentalism.

One of the key themes in the counterattack is to claim that exploiting natural resources leads to broader economic benefits that outweigh the associated degradation of the environment. These are enclosures of common resources that benefit some more than others, but this is not what the critics of environmental protection policies emphasize. Instead, they argue that allowing environmental harms brings jobs and economic returns for society as a whole. Opponents of better gas mileage requirements for automobiles have argued that the requirements would make cars less affordable for American consumers (rather than complaining that they would make things tougher for automobile manufacturers), and that if higher automobile prices lead to fewer sales, it might endanger the jobs of American workers.

Those who oppose environmental policy have a second major tactic: changing the subject by attacking environmentalists. As novelist Thomas Pynchon observed, "if they can get you asking the wrong questions, they don't have to worry about the answers."[6] Instead of trying to defend either the pollution or the benefits they themselves gain from enclosing the commons, the critics portray environmentalists as privileged job-hating snobs who do not care if the economy crashes.

Often, communities fear the pollution, traffic, or social changes implied by a new industrial facility or a big-box retail shopping center, and they cite environmental laws and regulations to delay projects and force concessions. But critics charge that all environmentalists oppose jobs and economic growth, even in cases in which environmentally responsible policy options have clear economic benefits, as with many energy conservation proposals. (Caught up in the polarization, most environmentalists have been slow to recognize that many businesses have actually reacted to environmental policy constructively. These firms are stepping up to their responsibilities by complying with or going beyond regulatory requirements, and developing methods and technologies that lower the cost of environmentally sound approaches to production, marketing, and distribution. We will discuss some of these innovations in Chapter 14.)

The charge of elitism has turned the strength of classical environmentalism against itself. Remember the appeal of *Silent Spring*: Rachel Carson warned well-educated, readily mobilized suburbanites about hazards that could be managed and eliminated by a benign government. In this argument, there is no weighing of the benefits, no estimate of the degree to which the costs and risks may exceed the benefits, and no need to make difficult choices to choose a path toward a more sustainable economy. All this made classical environmentalism vulnerable to the charge that its advocates did not care about the social consequences of their advocacy. That environmentalism is a main element of progressive politics suggests that this alleged lack of concern has only a weak basis, but it has proved difficult to put the charge to rest, particularly as many citizens became suspicious of the motives of government and skeptical of the capabilities of public servants.

It is of course true that all consumers enjoy benefits from the pressures we collectively impose on the environment. This is why sustainable development is a grand challenge. It will not be simple or easy to chart the paths to a sustainable economy. Rather, the paths will require substantial changes in production and consumption, touching on virtually all the economic actors in the world without edges. The fact that much environmental impact is caused by a small fraction of the firms involved (see Chapter 4) means that those firms may be highly motivated to resist the adjustments that may be needed to reach a less polluting, more sustainable future.

Some of those resisting environmental policy are crying wolf: they may indeed suffer, but the impact on the broader society may be matched by larger environmental benefits and progress toward sustainability. It can be difficult to look beyond the rhetoric to see whether the claims of widespread economic dislocation are valid. A major challenge of politics in a large, complex society is doing just that: having a debate about complicated facts and judgments. The pluralist politics of the American government (Box 12.3, page 321) has demonstrated some competence in meeting this challenge—but, as the climate debate has shown, that competence remains limited.

The actual economic impact of environmental policies has been modest. The cost of complying with environmental regulations is roughly 2 percent of GDP.[7] This is a lot of money, but the GDP has grown slightly less than 3 percent each year over many decades. This means that the burden of environmental regulation is about the same amount that the economy grows in approximately one year. The economic *benefits* of environmental protection, ranging from preserving valuable fish species for future generations to improving the health of humans who live downwind from a polluter, are on average far higher. Air pollution regulations return very high benefits because they reduce lung disease and other human health effects, which are easily measured. The EPA estimated in 1999 that the Clean Air Act had cost businesses and automobile owners more than $500 billion to implement over the twenty years since the Act was passed. But the benefits were at least $6 trillion—12 times as much—and could have been as high as $50 trillion. Such impressive ratios have been estimated again more recently.[8] Broader assessments of the economic effects of environmental regulations have shown their impact to be modest or positive. In most cases, even the economic harm inflicted on communities and occupational specialties such as ranching was fleeting.

But public perception has often failed to track these analyses. The story of the timber industry of the Pacific Northwest illustrates a larger pattern. During the so-called golden age of logging—a time that began with the building boom after World War II and ended as environmental regulations were imposed in the wake of the Wilderness Act in 1964—the actual rate of job loss among loggers in the Pacific Northwest was 3 times higher than it was *after* the new environmental regulations took effect. Yet the restrictions on logging imposed in the 1980s to protect the endangered northern spotted owl are still blamed for the travails of the timber industry.

POLITICAL POWER AND LIMITED COMPETENCE

Environmental groups managed to work within the established system of American politics (Box 12.3, page 321), bringing about changes in the relationship between the economy and the public, and affecting nearly every business and every level of government. Although there were limits to their influence, environmentalists did gain power.

Over the past half-century, classical environmentalism has revealed both its strengths and its weaknesses. Its main strength was that classical environmentalism provided a model for action that shaped the laws written during the period of enthusiasm in the 1970s. The creation of environmental NGOs, many of which have flourished for decades, has created a body of expertise that can participate in vigorous debates in Congress, courts, and agencies such as the EPA. Environmentalists have become a presence that cannot be ignored, and environmentalism can rightly claim to have redirected American society and the economy in ways that have measurably increased the environmental responsibility of our interactions with the natural world.

Classical environmentalism also showed three important weaknesses: a heavy reliance on government regulation to impose mutual coercion; a vulnerability to the counterattacks mounted by concentrated interests benefiting from exploitation of commons; and an inability to escape the limitations of a politics based on interests, as the country has tried to tackle the complicated, subtle problems of humans in landscapes.

Classical environmentalism usually led to a complicated regulatory structure that was difficult to sustain over time. Anti-environmentalist counterattacks came to focus on reducing agency budgets, which were easy to describe as representing "wasteful, bloated bureaucracies." This was usually a caricature, but the cumbersome procedures of government often obscured the benefits of environmental policy while highlighting its costs. Although Garrett Hardin focused on coercion that was mutually agreed to, such attacks showed that mutual coercion can be hard to implement. Even if a clear majority of citizens favor strong environmental laws, skilled lobbyists and activists backed by a concentrated interest group may still be able to undermine individual regulations, neutralizing the majority's will.

Second, although environmentalists showed great skill in raising public awareness and concern over environmental issues, they were far less successful in maintaining that interest in the public at large. Instead, concentrated interests have succeeded in reframing environmental regulation as a threat to consumption, economic growth, and the American way of life.

The net effect has been stalemate: although the majority of American citizens have continued to report to pollsters that they care about the environment (see Fig. 12.1), **regulated industries** have dug in, effectively shifting the political balance. Public support for environmental protection often translated into actual environmental protection measures during the 1970s. But since then, as the electorate turned slightly more conservative and elections became much more expensive, environmental politics has been a seesaw struggle.

The industries determined to ward off environmental regulation increased their lobbying expenditures and their campaign contributions to political allies, gaining considerable influence in Congress. Environmental law soon had practitioners representing developers and regulated industries. They sued the EPA and other administrative agencies at every level of government, challenging regulations on legal grounds and attacking the science or engineering underlying regulatory decisions. Starting in 1981, Republican administrations became outspoken supporters for regulatory rollbacks, giving regulated industries some notable successes, such as a decades-long deferral of the cleanup of coal-burning power plants.

Leading environmental advocacy groups, such as the Environmental Defense Fund, the League of Conservation Voters, and the Sierra Club, have been outspent in Washington, D.C., but these groups have proved to be effective in advancing some policies, such as wilderness protection, and in defending many more, including air pollution regulation and the protection of endangered species. For the most part, however, since the early 1980s, environmentalists in the United States have focused on trying to defend earlier policy gains, rather than making new ones.

This history reveals a third limitation of classical environmentalism—its reliance on finding stark dangers, such as the pesticides that killed songbirds and caused cancer. As some of the most dramatic threats of environmental harm have been brought under the umbrella of public policy, it has been harder to rouse public anger, even though the solutions devised have sometimes fallen far short of protecting public health or ecosystems. Although climate change is under way, some of the public and many politicians persist in thinking that a winter of unusual snowstorms and colder than average temperatures means that global warming is a hoax.

The deeper problem here is in a politics based on the reconciliation of interests. Environmental problems are virtually all complicated: for example, pollution is emitted at one place but its damage can be felt far away, as is the case with fertilizer that washes off farm fields only to lead to choking growth of algae in coastal wetlands downstream; or an ecosystem is disrupted by clearing forest or overfishing, but humans only notice a decline in valuable species years later. Usually, the people who feel the impacts are not the same ones who cause those impacts. Often, those who cause the impacts do not realize they are doing so, and they already have made large investments or other hard-to-change commitments when the complaints come.

Governing commons that are overexploited in such ways requires more than an articulation of interests, as is done by the pluralist form of democracy (Box 12.3, page 321). It requires means of deliberation and problem solving that can accommodate both social and technical complexity. No society, democratic or not, has achieved a satisfactory way to do this. A centralized, authoritarian government such as in the former Soviet Union had so little effective communication of complexities that it resulted in large-scale disasters such as the death of the Aral Sea (Box 6.3: A Warning? The Aral Sea in Chapter 6, page 150).

In the United States, a lot of information swirls about in environmental controversies, but decision-making processes often cannot sort out reliable information from disinformation or rumor. Lobbyists on all sides are tempted to spread half-truths and worse, making it that much harder for leaders to respond intelligently. Moreover, the combination of antigovernment suspicion and partisan polarization means that deliberations within official processes, such as the setting of environmental regulations, is often subjected to corrosive legal and political challenges. This is not to say, of course, that the complicated discourse of bureaucrats, businesses, and environmental activists leads to results that citizens should always respect. It is to say, rather, that it is difficult to figure out which of those outcomes will actually move behavior toward sustainable development, even when there is ample transparency and opportunity for citizens to vote.

There is a darker side of environmental politics, too. The benefits of environmental protection have not been evenly distributed (see Box 11.1: Environmental Justice in Chapter 11, page 304). When the British landlords took over the shared common grazing lands during the Enclosure Movement, they were taking land not from other rich or powerful people, but from those who were poorer and less able to defend themselves politically. In more recent years, the worst environmental insults tend to be found in poor neighborhoods, not rich ones. Classical environmentalism has had only modest effects in countering this form of social injustice.

This issue is far from simple. The environmental and social justice movements interact in important ways, often reinforcing each other. In particular, the benefits of clean air and clean water accrue to all segments of the U.S. population, saving tens of thousands of lives each year and yielding major improvements in public health, especially in dense urban areas with substantial low-income communities.

Still, since Rachel Carson's appeal to the readers of the *New Yorker*, the most successful environmental activists have been like the most successful activists on other issues; they have been well educated, well off, and well-organized. Those who are well off create far more garbage than the less well off, but the well off are far more effective at preventing garbage transfer stations and power plants from being built close to their own homes, and only a tiny fraction of those who are well off are active members of environmental organizations. The indirect result can be environmental injustice, as the less well organized and less powerful poor have their neighborhoods invaded by facilities that provide most of their benefits to richer neighborhoods. Here, too, the world without edges can prove to be as close as the nearest sewage treatment plant, which is more likely to be located in a poor neighborhood than an affluent one.

ORGANIZING FOR SOCIAL CHANGE

How do environmentalists make a difference? They bring about social change through organizations. We will spend the rest of this chapter explaining this abstract statement. It is an important statement to understand, because making a difference this way is an essential component of the search for sustainable development.

Each of the grand challenges of sustainability presents a challenge of social change, of altering both formal rules, such as environmental laws, and informal practices, such as consumption habits. Social change is almost always resisted: established ways of doing things almost always benefit at least some people, and they seek to keep on doing things the old way. Pressure for social change often emerges in the form of political opposition. Social change can be started by a small group and it can be led by an individual, but for change to become effective, it must be organized, and usually sooner rather than later. If environmental policy is to make a difference, it has to result in social change, and its implementation propels additional change in turn.

In 1970, no electric utility company provided assistance to its residential customers so they could use energy more efficiently and lower the amount of pollution produced by generating plants. Now all utilities do, and many encourage installation of solar panels that can send surplus power back to the grid. This shift is a result of changes in regulatory law that made it profitable to encourage energy efficiency. This was a change that took place on many fronts: in the utility companies, among their customers, in regulatory agencies and Congress, and in the media, which described the political debates and helped to educate consumers about how they could reduce their electric bills by improving the insulation of their homes and water heaters. Environmental organizations such as the Natural

Resources Defense Council pressed for change on all these fronts. As this example demonstrates, environmentalists do more than politics: they work to create a different way of doing things.

Particularly in the United States, where individualism is prized, the tendency can be to overlook the fact that, in large societies, people act through organizations—government agencies, business firms, religious institutions, and political parties, among others. We pay attention to the stars of a movie but usually walk out of the theater as the hundreds of people involved in putting together a motion picture have their names scroll across the screen. The language of politics is individual, but the reality is organizational. Nelson Mandela symbolized and led the struggle for a democratic South Africa, but it is the African National Congress that governs. Individuals can make a difference—and yet, for the most part, they do so through organizations.

Organizations established for the purpose of social change are usually nongovernmental, for the simple reason that they are formed to alter things that are supported and defended by the existing government. Some profitable businesses can bring about changes by persuading people to buy their products or services, the way the cell phone has become widely adopted, bringing considerable social change in its wake. But many of the NGOs that tackle environmental challenges are trying to do so in a way that does not produce earnings, so they are also often called **nonprofit organizations**. The terms "nonprofit" and "nongovernmental" define this genus of social organism negatively, by what it is not. A positive definition would emphasize organizing people around value commitments; these commitments are usually described in the mission of the organization. The mission may include profit-making activities, but these are not primary as they are in a business enterprise.

The need for interests to be organized alters the picture presented by classical environmentalism in an important way. A citizenry may be aroused by events, but in an information-intensive world, this kind of arousal rarely lasts very long. Activist organizations stay focused, because this is what they do—they specialize. Moreover, NGOs seize upon anger in their constituencies, whether those constituents are environmentalists or mining firms, to secure resources. In the terminology used in Box 12.4: Concentrated and Diffuse Interests, page 326 NGOs can be seen as attempts to develop concentrated interests around some of the most diffuse of all human concerns, including planetary-scale environmental systems. Even during times of heightened public concern, however, only a tiny minority of the people who care about environmental preservation—and an even tinier minority of those who benefit from such preservation—can be expected to join in letter-writing campaigns, call their congressional representatives, or even send in annual membership contributions. The great majority are **free riders**: people who may benefit from the work of NGOs but do not contribute resources to it. This problem is found in almost any kind of organization that is designed to bring its benefits to a larger group beyond its own members. You will recognize a logical parallel to that of the commons: missions that benefit all are sometimes supported by none.

As small and as outnumbered as such organizations may be, however, their fund-raising, organizing for electoral campaigns, letter writing, and other forms of activism provide an important institutional mechanism for translating intense but temporary concern into lasting institutional change. In that sense, NGOs *are* the environmental movement, the durable face of environmentalism in the public arena and in government. Of course, when we complain about industry lobbyists—many of whom also technically work for NGOs, even though those NGOs do their work on behalf of profit-oriented firms—we are portraying the durable presence brought by organization in a less favorable light.

The central role of organizations provides an insight into the scale of our competence. Organizations increase the scale of human action. No single person can build a jet aircraft or elect a political leader or manage a national park. These are inherently organizational tasks. Some are done well, some not. Many of these organizational capabilities generate environmentally destructive consequences. Yet organizations are also human creations, and the record of the past half-century is that many organizations have learned to reduce the environmental damage of the economy.

"Find work," author Wendell Berry wrote, "that does no damage."[9] It has turned out to be possible to work toward steering organizations toward less damage. Organizational work is often difficult and unrewarding (as are other forms of labor). But a lot of worthwhile work remains to be done as humanity faces its grand challenges. Finding effective and creative ways to change organizations is a central task.

CIVIL SOCIETY AND SOCIAL CAPITAL

Nonprofits and NGOs have been around for a long time. The first European university, at Bologna in what is now Italy, was founded in the eleventh century, and the institutional Christian church traces its origins to Jesus and his apostles. There are even older surviving nonprofits beyond the West. Over the past generation, other nonprofits have risen in visibility and importance, from international organizations such as Amnesty International, to universities whose faculties contribute public policy ideas, to hundreds of community land trusts, to special-purpose temporary organizations such as a city's Olympic organizing committee. Some nonprofits also organize across national boundaries, such as the China Sustainable Energy Project, which links the Chinese government's ambitious efforts to increase energy efficiency with expertise from developed economies. In many cases, in short, NGOs can have very large goals.

Nonprofits form the organizational backbone of **civil society**—the parts of the social order that are not part of the government or the for-profit part of the economy. It is out of civil society that environmental activism emerges. In the political arena, civil society movements have played a key role in launching democracy in Poland, South Africa, and Tunisia. Civil society is the contemporary expansion of the concept of communities, as social interactions—not just market transactions or formal governmental interactions—have reached into the world without edges, transcending places. Civil society includes spatial communities, such as land trusts, whose mission is to protect valuable landscapes; and it also encompasses large groups in many places, such as the Republican Party, or nonspatial groups, such as the Jon Stewart Intelligence Agency, a web-based fan club.

In a settled place where everyone knows everyone else, people know where they are in a social sense—whom to trust, whom not to trust, what to be careful about saying, and so forth. This is a component of the sense of place discussed in Chapter 2. We now inhabit a world in which we deal with many strangers, and the rules are often uncertain. Much spam e-mail tries to lure people into providing information that can be used to defraud or harm them, disguising its intent by claiming to be from a proper institution such as a bank. This misuse of the rules of social interaction breeds mistrust. As in other commons, bad behavior is not only harmful to victims but it undermines the community as a whole. This erosion of trust is an example of what social scientists call the loss of social capital. Instead of assuming that a message comes from someone who has a legitimate question or offer, we have now learned not to open e-mail attachments from strangers.

The meaning of the abstract idea of **social capital** is brought out by something that happened a while back at a liberal arts college in a small town, a place that comes about as close to Wendell Berry's model of a face-to-face community as is likely to be found in the contemporary United States. A new assistant professor, fresh from her doctoral work at a large urban university, drove to campus on a foggy morning. The day turned bright en route, and she neglected to turn off the headlights she had switched on to navigate through the murk. Upstairs in her office, she was beginning class preparations when the campus police called. They told her the lights on her car were on. They also apologized for being slow to call: she had arrived so recently that they had yet not had time to transfer her license plate number from the form she had filled out into their computer database. Amused at the intimacy of the "bureaucracy," the young professor started down to the parking lot. On the stairs she encountered the department chairperson. "Oh, that was your car!" he said. "I saw the lights on and tried to turn them off, but you had locked the door."

There were formal, big-city ways—a computer database, a police force on the alert—providing one layer of social capital to protect absent-minded professors. A parallel second layer was the informal social capital of the unlocked door and the colleague looking out for people in the community, even if he doesn't know precisely who they are. Both kinds of relationships are valuable forms of social capital. Too often, as with spam e-mail, we lack either kind of social capital as we struggle to find our way in the world without edges, both as individuals and as communities caught up in the turbulent changes of globalization.

Even in a modern, urbanized world, civil society organizations can do more than conserve social capital—they can actively create it. They create social capital by building networks of associations in which people can learn to trust one another by interacting repeatedly. In nonprofit civil society organizations, people participate because they want to, not because it's their job. This provides a basis for building trust that is different from the formal relationships that arise in the workplace or in the governmental arena.

This does not mean that life in civil society is happier. Conflict and duplicity can be found in all kinds of human relationships. A society with a vigorous civil society is more resilient, however, than one in which the civil society is weak—as it was in Iraq under Saddam Hussein's secret police, for example. Civil society provides an alternative way to articulate problems and to address them. Moreover, nonprofits are often deliberately inclusive in their missions, reaching out to "groups too poor to purchase from the market and too politically weak to matter to the state."[10] One of the problems of the Internet is that its civil society is frail and is still being built. (The attempt by the editors of Wikipedia.org to make it a reliable source of information offers an example of an attempt to build trust.)

CREATING CIVIL SOCIETY

Nonprofits can both complement and provide substitutes for market and government. Like government, nonprofits seek to advance the public good and address problems, such as pollution, that arise from the operation of markets. Unlike government, nonprofits lack powers of coercion or regulation and must rely on persuasion (including advocacy aimed at changing policies of the state). Unlike the market—and unlike most pro-industry lobbying organizations—nonprofits lack financial profit making as an incentive driving the organization of activity. The purposes that motivate donors and volunteers and staff of nonprofits differ from those found in government or business: charity, voluntarism, and sometimes religious fulfillment, all focused on a mission that contributors to the nonprofit can define as "making a difference." The sense of commitment and challenge that emerges from this situation is different, though not completely different, from what one sees in business or government.

Nonprofit organizations are the institutional manifestation of civil society, and they commonly have several of these characteristics:

- ⬊ Their missions are nongovernmental but community oriented (sometimes by *creating* a community, as a fan club does).

- ⬊ They organize mutual aid among members, as in an organic-food cooperative.

- ⬊ They do charitable works through social service and social welfare missions, as one sees in the Rotary Club, an association of businesspeople.

- ⬊ Their work often overlaps with religious institutions, as in the case of the Catholic Relief Services, an international humanitarian organization.

- ⬊ They provide a base for articulating critiques of existing institutional arrangements of political and economic power. The Center for Responsive Politics, a nonpartisan research group in Washington, both studies and criticizes the effects of campaign contributions and lobbying expenditures, making the information widely available through its website, www.opensecrets.org.

- ⬊ They supply a base for mobilizing people, particularly those who may not have access through formal institutional channels. Labor unions have done this on a large scale.

Nonprofits have grown rapidly. In the United States, the most important population of nonprofits is that set of organizations recognized as public charities under Section 501(c)(3) of the Internal Revenue Code. These nonprofits can receive donations that are tax deductible, so that donors can claim their gifts on their tax returns and not pay tax on them. There were more than 800,000 such public charities in 2000, a population that was 77 percent larger than in 1989, when an earlier survey was done.[11] The federal government estimated that nonprofits contributed 4.2 percent of the gross domestic product.[12] For comparison, this is considerably more than is spent to comply with environmental regulations.

In other countries, too, the growth of nonprofit organizations has been rapid. A study found that "a veritable 'global associational revolution' seems to be under way throughout the world, a significant upsurge of organized private voluntary activity in virtually every corner of the globe."[13]

In the United States, more than eight thousand nonprofits identify their missions as environmental, and in 2000, they spent $8.2 billion to advance those missions. About half of that funding came from donations from individuals and other organizations. These environmental nonprofits constitute the environmental movement in an organizational and political sense. This is the case despite the **free rider problem**: the total membership of environmental groups is a small fraction of the populace who tell opinion pollsters that they are environmentalists; self-described environmentalists number roughly half of the U.S. voting population (see Fig. 12.1). The nonprofits are the public face of environmental advocacy, and they

- educate the public and raise awareness, through science, policy analysis, school curricula, and publicity;
- try to increase coverage of issues in influential media, so as to compel reactions from government leaders and high-profile commentators—a function called **agenda setting** by political scientists;
- advocate for legislation, press for administrative actions, and litigate—all steps aimed at altering institutional rules and practices;
- produce leaders, some of whom go into government or other leading institutions, such as Nobel Peace Prize winner Wangaari Maathai, who was deputy environment minister in the Kenyan government when she was awarded the prize in 2004 for work she did in an environmental NGO she had founded earlier in her career;
- generate resources to keep advocacy going, by building membership, fundraising, obtaining grants, and forming corporate ties; and
- monitoring the implementation of some policies and evaluating their impacts and success.

NGOs provide a way to sustain the determination and persistence that are essential to meeting environmental challenges. Scholars have begun to study NGOs and to identify factors necessary to marshaling social will and bringing about social change. Because NGOs are not profit-making enterprises, you might wonder how people are motivated to join, to stay, and to contribute to them. Motivation, like its companion trust, is a central determinant of what competences can be fielded in political struggles at large scales. We do not expect neurosurgeons or airline pilots to serve without pay, but conservation biologists of comparable expertise and devotion do their work with very little financial compensation (see Box 12.5: Agents, Incentives, and Making a Difference, page 340).

PHILANTHROPY, CHARITY, AND INVESTMENT

Nonprofits receive an important part of their resources from **philanthropic foundations**. Under American law, private individuals can create and endow independent foundations, a class of NGO that operates in accord with special

BOX 12.5

AGENTS, INCENTIVES, AND MAKING A DIFFERENCE

NGOs are agents of their constituencies. The National Rifle Association claims to speak for gun owners, but what about the policemen who are both hunters and advocates of gun control? Universities act on behalf of their students, but students are rarely consulted before tuition is increased. People who act on behalf of organizations are called **agents**; they may or may not act in the interests of the organization's members, who are called **principals**. Often, it may be unclear what the principals' interests are. Even if a board of directors cannot agree on a course of action, the president still needs to meet a payroll and keep things going. Members of Congress often vote on issues that their constituents back home know little about. This is a recurrent dilemma: agents need to act, even when their principals are divided, indifferent, or ignorant. The situation with principals and agents is somewhat reminiscent of the commons: the problem lies in aligning the agents' interests and incentives with those of the principals.

What draws people into organizations? There is a simple but helpful classification of the reasons why people participate in organizational life:

- *Coercion or duty.* One or both of these motivate prisoners or nuns, as well as students who see no acceptable alternative to going to college. Either of these motivations can lead to half-hearted participation, as we see among students on every campus who are not sure why they are there but enroll nonetheless.
- *Material compensation.* These are the motivating factors we are all familiar with—pay, benefits, stock options, and other forms of compensation that can generally be valued easily in financial terms.
- *Making a difference.* Some organizations, including most nonprofits, offer individuals the opportunity to achieve, or to join in achieving, a goal that they could not achieve on their own. This is what draws some to become activists, even though no one makes them do it and they are unpaid. This is what impels members to join, volunteer, or donate to environmental groups. Public service in the government or military often motivates people in this fashion as well.

Environmental politics is often a struggle between people motivated by making a difference and those energized by material rewards. These are very different kinds of motivating forces, of course.

More generally, where organizational purposes collide with ideals, one should expect tensions. Some businesses struggle for legitimacy; for example, it is difficult to pay enough to motivate a smart young person to work for a tobacco company. Conversely, cleaning up mine waste is a daunting challenge, and even idealistic environmentalists need resources to pursue it. The president of Tiffany & Company (who is also a trout fisherman) decided to contribute to the cleanup of the American Fork River in Utah, which had been damaged by mining.[1] This surprising support catalyzed an unusual alliance of NGOs, government, and the private sector to undertake a cleanup on both public and private land. Social values affect which of these incentives work—and thus, which organizations survive and do well. The Tiffany executive decided to make a difference, and he and other fishermen benefited along with the watershed that nurtured the trout. There is an ecosystem of organizations, and this shapes the politics of the environment.

Every organization is a human community, and organizations have internal ecosystems, too. A university is a community, not just an organization with a president who speaks for its students and faculty. There is always the possibility of internal strife and arguments over what the organization should do to realize the aspirations of its members. For a business, this is normally a matter of profitability and labor relations—of material rewards and how they are distributed. But in a setting such as a political party or an NGO, in which people are motivated by making a difference, the arguments often focus on principles and practices. These arguments have a very different rhetoric than profit and pay. But the disputes are no less important and often no less bitter.

Most discussions of environmental politics and public policy emphasize facts and institutions like Congress or laws. Often, the conflicts can be explained by the differing missions of the organizations pressing their views. An environmental group may be making arguments in defense of a poorly managed or unrecognized commons, while property owners are seeking to protect their right to develop their land as they see fit. These differences are important to understand. Often, however, they do not explain the human dynamics that lie just under that surface of facts and official explanations. To dig deeper, look at how the organizations work (internally and externally) and at the tensions between agents and those they claim to represent. Political science and sociology are two useful approaches to the study of social ecosystems.

1. Felicity Barringer, "Unusual Alliance Is Formed to Clean Up Mine Runoff, *New York Times*, August18,2004,www.nytimes.com/2004/08/18nationnal/18mine.html?scp=1&sq=tiffany%20american %20fork%20river&st=cse&emc=etal.

provisions of the tax law. Foundations make grants to NGOs. Both the grant and the foundation are tax exempt, so long as the foundation gives away 5 percent of its endowment each year for charitable purposes. Foundations with large endowments have substantial staffs to administer grant making. Foundations are governed by independent boards of trustees, who make grants to NGOs with wide latitude. Their freedom to fund a wide range of activities—from ballet companies to environmental advocacy to community hospitals—gives foundation trustees a flexibility that is virtually unique in public life.

There are about sixty thousand grant-making foundations in the United States. These foundations gave nearly $43 billion for all causes in 2009.[14] This is a lot of money, although it is a much smaller amount than government or corporate expenditures. The EPA alone spent just under $10 billion in 2010, which was far more than the grants given for environmental causes by philanthropies. Charity has a history as long as human communities. Charity embodies the competence for generosity of an individual, often a ruler or aristocrat, or of a group of like-minded individuals, such as the thousands who donate to the Red Cross. With the rise of industrial capitalism, very large private fortunes were made rather than inherited. Andrew Carnegie, the founder of U.S. Steel, opened the way to a distinctive form of giving in an essay entitled "The Gospel of Wealth," published in 1889. He argued that it was the duty of those who earned large fortunes to see that they were properly used. To him, proper use meant directing private wealth toward providing opportunities for people, of whatever means, to improve themselves. Carnegie endowed several thousand public libraries in towns and cities across the United States, and Carnegie library buildings are still in use today.

John D. Rockefeller, the founder of Standard Oil, the predecessor of Exxon-Mobil, the world's largest oil company, developed the idea of "scientific philanthropy" further by creating the Rockefeller Foundation, an organization aimed specifically at investing in social capabilities. This is giving to make a difference in the long run, going beyond charity directed at meeting present needs only. The Rockefeller Foundation created the academic field of public health in the early twentieth century by building programs at Harvard and Johns Hopkins Universities.

Together with the Ford Foundation, created from another industrial fortune, the Rockefeller Foundation laid the groundwork for the Green Revolution in the mid-twentieth century, by funding an international program of plant breeding that created high-yielding varieties of rice, wheat, and other grains. Although the Green Revolution has created some serious environmental impacts of its own, the large increases in agricultural productivity have enabled large poor countries like India to become self-sufficient in food, transforming the problem of hunger in the last third of the twentieth century. There are still hungry people, which is obviously a serious problem, but starvation is now less common than it was earlier in the life of our species. More broadly, the Green Revolution developed a form of advanced science—new seeds and methods of growing—that could be imported into traditional agrarian societies disrupted by colonialism and by the demographic changes caused by falling infant mortality. In that sense, the Green Revolution was responding to the international public health revolution touched off by Rockefeller a generation earlier.

In their best forms, Green Revolution innovations took into account the reality that poor people in the tropics inhabit ecosystems very different from those in temperate countries. In other cases, the innovations have been less well-suited to the needs of the world's poorer nations. Many of the new varieties of crops require large quantities of chemical fertilizers and pesticides to produce their high yields, raising concerns about costs as well as about potentially harmful environmental side-effects. Poor farmers, unable to afford the fertilizers and pesticides, have sometimes reaped even lower yields with the new, scientifically developed grains than with the older strains, which in many cases had been developed by local farmers over centuries, providing better adaptation to local conditions, including resistance to pests and diseases.

Despite the drawbacks, the impetus behind these major changes in health and agriculture was humanitarian. The changes also pushed traditional societies— many of which had been sustainable in an environmental sense—into the world without edges. The story of natural resource management policies in the African nation of Mali in Box 10.3: Nature, Wealth, and Power in Chapter 10, page 271, describes one aspect of this transformation of traditional societies.

Philanthropy is investment, but its returns are not primarily financial. Instead, the aim is to create capabilities—social capital—that might not be created or sustained by either the market or government. Philanthropies give money for a range of activities: scientific research, analysis and advocacy to make and change policies, empowering specific constituencies, and demonstration projects that show the feasibility of doing things in new ways and build a demand for them. The children's health insurance policy adopted in 2009 by the federal government was initially developed and demonstrated by a philanthropic foundation.

The story of the environment program at the William and Flora Hewlett Foundation in California shows how one foundation has put together a strategy to tackle climate change. During the first decade of this century, environment program director Hal Harvey built the foundation's grant-making program on a vision of striking simplicity: "Two things matter: Coal and cars," he told the *Wall Street Journal* in 2007. "Two countries matter: China and the U.S."[15] One solution is aimed at these four targets—the policy design called cap-and-trade, which will be described in detail in Chapter 13. The Hewlett program funded an international network of scientists, lawyers, pollsters, and campaigners to bring cap-and-trade into American policy and into international climate treaties beyond the Kyoto Protocol. The foundation's grants come with performance targets, and their budgets include funds for evaluations. The aim is to find out what works and to adjust the investment profile accordingly. What is important to see here is the simplicity and clarity of the strategy. Is this the right way to move climate policy? We will know before too many more years whether Hewlett's investments pay off. There is risk, obviously, in this strategy. But the Hewlett approach—bold, wide-ranging, simple, and measurable—provides a good example of how philanthropic investments are being made.

Philanthropy, in short, is a form of investment, and its entrepreneurs are NGOs. In fact, foundations provide only a small share of the funding of nonprofits. Membership dues and fees for services, such as those charged for admission by a museum, supply a larger fraction of the revenues. What foundation funding aims to do is to fund innovation, and when that works, its grantees can bring about significant change.

As noted earlier, people in nonprofits are motivated by missions, rather than by profit or the exercise of governmental authority, but they still need to sustain their organizations by meeting payrolls and securing other resources; organizing staff, members, and often large groups of volunteers; and conserving assets such as the art belonging to a museum or a nature preserve managed by a land trust. The relationship of nonprofit to philanthropic donor is shaped in part by the tension between mission and organizational maintenance. In many instances, an environmental group can team up with business firms to achieve a specific objective, but does the NGO sell out its independence when it does so? What if its business partner then wants the NGO to reshape a campaign in a way that helps the business? What if a senior executive of that business has become a significant donor and joined the NGO's board? Some foundations provide some of the NGOs they fund with grants specifically designed to identify organizational weaknesses and to develop solutions.

As with other attempts to extend the scale of competence of human institutions in the world without edges, nonprofits and philanthropy still have much to learn. There is nothing parallel to the global financial and legal structures that give business enterprises their enormous capabilities; there is nothing parallel to the diplomatic or military resources of powerful nations. But as you can see in the example of the Green Revolution and in the ambition of the Hewlett Foundation's energy and climate program, steps are being taken to deal with the grand challenges. Those who do this work believe in what they do; perhaps some of the readers of this book will be among them.

Half a century after *Silent Spring* galvanized citizens around the world to action, environmental problems persist. Some have been managed successfully through classical environmentalism, but grand challenges endure. We can now see how both success and frustration take shape. Environmental problems arise in commons. In a dynamic economy, many of these commons problems are novel ones, such as the intricate confluence of forces that nearly drove the Channel Island fox to extinction in the history we related in Chapter 1.

Classical environmentalism turns out to be necessary but not sufficient as a means of attacking the grand challenges of sustainability. Science does in fact supply important warnings of threats that humans cannot readily perceive: toxins that are invisible and odorless, with harms that are long delayed; pressures on natural systems hidden in the invisible present, such as overfishing and global warming, harm to distant ecosystems and people that is obscured by the world without edges. These environmental problems signal the presence of poorly governed commons, such as water and air being polluted by those who bear only a tiny fraction of the risk they impose on ecosystems and people. Commons problems require institutional or technological solutions. It may be possible to prevent the pollution at its source, but even in those cases, someone's behavior needs to change. That change needs to come through shifts in the rules and incentives faced by the people abusing the commons.

Changes in rules and incentives usually require governments to implement them. Politics is thus necessary. When environmentalists are part of concentrated interests, such as suburban voters worried about air pollution, significant changes can be achieved, as with the expansion of mass transit systems. But when those resisting change belong to concentrated interests, as is the case with the energy firms opposing climate change legislation, the politics of interest groups may not lead to an effective solution to environmental problems.

Recognizing and responding to environmental problems is a social process in which many actors—including the people affected, the scientists observing natural systems, and NGOs—play key roles. Their role is not only to observe but to contribute to the conflict-laden processes of government, because the proper management of commons is unavoidably a question of governance. For this reason, politics and the messy, noble, and sometimes amusing struggles of self-governing societies are central to the solution of environmental problems. As we will see in the next chapter, the decentralized coordination of human behavior that is made possible by markets both drives the environmental impacts of the world without edges and suggests important tools for managing those impacts.

The land ethic is not embodied in the institutions of our society or economy. For that reason, deadlock and delay frequently result. The prevalence of conflict has made environmentalism less popular, as people sense that environmental protection is more than just disciplining distant corporate polluters. Yet the essential question remains: Where a valuable commons is at stake, how can mutual coercion be mutually agreed to?

The environment is defended and contested now, often with institutional supports in law—and perhaps equally importantly, with a base in culture and values. Environmentalists do not win all the time and, as the defenders of wilderness say, their losses are frequently irreversible. Yet even having a fight about these matters is a new phenomenon historically—one that has not disappeared even in conservative times. The struggle over environmental policies is a component of society's attempt to extend the scale of environmental competence. Its reach is still imperfect and limited, but there have been real advances.

FURTHER READING

Freudenburg, William R., Lisa J. Wilson, and Daniel O'Leary. "Forty Years of Spotted Owls? A Longitudinal Analysis of Logging-Industry Job Losses." *Sociological Perspectives* 41, no. 1 (1998): 1–26.

Olson, Mancur. *The Logic of Collective Action: Public Goods and the Theory of Groups.* Cambridge, MA: Harvard University Press, 1965.

Rosenbaum, Walter A. *Environmental Politics and Policy*, 8th ed. Washington, DC: CQ Press, 2010.

KEY TERMS

agenda setting	diffuse interests	nonprofit	principal
agent	free rider	organization	regulated industry
civil society	free rider problem	philanthropic	social capital
concentrated interests	interest group	foundation	
		pluralism	

Chapter Thirteen

MARKETS

COORDINATING HUMAN CHOICES

A key element of the world without edges is the market, an institution that plays a central role in organizing many of the activities of human societies. Trade is as old as human society itself. But over the past five hundred years, improvements in transportation, finance, and technology have increasingly knitted the world into a single economy. This is, not coincidentally, the time when our species emerged as a potent environmental force. It is the world without edges that enlarges the scale of human influence far beyond the face-to-face interactions that we identify with moral competence. It is the world without edges that weakens the human perception of our "vital relation" to the land, as Aldo Leopold put it. If we are to understand how to extend the scale of environmental competence, we need to understand how markets work.

"A LOW-GRADE CHRONIC INFECTION"

Historian William McNeill portrayed the end of feudalism in Europe as a time when the market was "implanted like a low-grade chronic infection in the tissues of imperial bureaucratic empires."[1] By this he meant that trade across political boundaries provided links and information about distant places, so that even autocratic rulers could not control the expectations of their own populations. The desire for trade with Asia and the Near East led eventually to the voyages of discovery and to the exploration and settlement of the Americas by Europeans.

The concept of the market held out an extraordinary promise—voluntary transactions in which both buyer and seller would be better off as a result; otherwise, the transaction would not occur. This combination of "win-win" outcomes

---

Learning Objectives

When you have finished studying this chapter, you should be able to

↘ demonstrate your understanding of markets by explaining how the components of a good—such as a hamburger's beef, bun, and ketchup—can be assembled through *voluntary* transactions alone;

↘ identify the role of government as a complement to markets in the hamburger story, as well as address the safety of the food, the use of pesticides and fertilizer in producing grain and tomatoes, the claims made by sellers of products, the disposal of the product packaging, and the value of the currency or credit cards used to purchase the components and the burger itself;

↘ explain how a governmentally defined cap on total emissions uses price to coordinate the behavior of firms that emit pollutants;

↘ analyze the problem of ecosystem services that do not have market values, by considering the real, but unpriced value of flood control from a forest that is left standing rather than logged;

↘ discuss the way in which markets connect human activities across widely separated

spaces, creating the web of relationships we call the world without edges.

↘ identify environmentally significant aspects of the food you consume that you do *not* have any information about—that is, to spell out some environmental aspects of your food consumption that you know must matter (such as the impact of fertilizers on watercourses) but about which you, like nearly all consumers, remain uninformed;

↘ give an account of your decisions (including the factors that you routinely ignore) in a supermarket or restaurant when you choose the food you eat, including economic, environmental, and social factors. Note that nearly all consumers can and do ignore factors that they would otherwise think are significant; what is it that enables them to do so?

↘ recognize concrete examples of the abstract theories of markets and interest-group politics, and explain how abstract ideas such as demand or interest affect behavior in consumer choices or when deciding what to say in a letter to a congressional representative.

and personal freedom was alluring to nobility and peasants alike in feudal societies. You can see how tasting this kind of commercial freedom might also kindle an interest in political freedom, helping to set the stage for the rise of self-rule as a powerful vision of the political order.

Markets transformed imperial societies in the way that the evolution of mito-chondria gave rise to a new class of living things. Mitochondria are the cellular power plants where oxygen is converted into the chemical energy needed by the cell. Evolutionary biologists think that mitochondria started out as independent cells, whose capability to harness and produce usable energy led to their incorpo-ration as part of the standard equipment of eukaryotes. Eukaryotic cells have, in turn, become the building blocks of the dominant species of life, the multicellular plants and animals. Societies that have relied on markets have turned out to be qualitatively more effective at harnessing the material world—that is, transforming the environment—than those that have tried to avoid markets. Trading societies are like eukaryotes, better able to direct the resources of the material world to the purposes of their constituent members. Societies that have constrained the role of markets significantly have not fared well. This fact has had political consequences, including the end of the cold war in the early 1990s. The Soviet Union was an imperial bureaucratic empire that could not compete effectively for the support of its own citizens once they became aware of the life-styles and comforts to be had in the West.

What is also significant is the intellectual reach and persuasiveness of eco-nomic theory. Economics provides a conceptual model of human behavior that is influential across the social sciences. Economics is important because it is a major dialect in the language of public policy. If one cannot understand and use the terminology of economics, one has trouble seeing how economic thinking influ-ences, validates, and can often change public policy.

What does learning economics have to do with environmental concerns? First, most environmental problems are the result of human decisions. Second, many human decisions are made in response to economic incentives. And third, eco-nomics provides us with a set of theories to understand decisions, particularly in a society with a materialistic culture.

PRICES AND CHOICES

In every human society people trade with one another. Today, in nearly all places, this is achieved through markets: people buy and sell things, making choices about what to buy by agreeing on the price that the buyer will give the seller in exchange for the thing sold. This act enables people to turn the complexities of nature and society into a simple choice, one implemented by swiping a credit card or laying down bills and coins. Billions of these simple choices each day, in turn, organize the transformation of nature by industry, commerce, and consumption. The facilitation of choices by prices is, accordingly, the most powerful mechanism in people's routine reshaping of the natural world.

The analysis of prices is a core concern of economics. Microeconomics, the theory that spells out how prices are determined, is powerful but limited as a model of how choices are made. With a small number of assumptions (some not so evidently plausible), it turns out to be possible to provide an orderly account of a surprisingly wide range of human behavior. In a world becoming ever more complex, the relative simplicity of economic theory is alluring to the overloaded mind. In part for this reason, belief in economic theory can drift into ideology: beliefs about how things *ought* to be—or *would* be in an idealized, theoretically pristine world—can persist even in the face of contrary evidence or moral discomfort.

The conflation of theory and ideology is particularly tempting for Americans because our political culture emphasizes individual liberties and the idea of freedom. As we saw in Chapter 12, this emphasis puts into the background the institutional structures of governance and social organization. In a similar way, the idea of a "free" market—in which buyers and sellers strike bargains without interference—obscures the fact that the very existence of markets depends on a system of property rights, contractual obligations, and financial institutions (such as the ones that assure the validity of your credit card to the sales clerk). These constitute a dense texture of social institutions without which no markets could exist. The environmental implications of the human economy both arise from those institutions and are hidden by them, as the simplifications of the market produce an illusion of independence, when it is actually interdependence that characterizes the world without edges.

Figure 13.1 depicts the economic interaction between business firms and households, including people's activities as workers and consumers. Firms combine inputs to make goods, which are purchased by households. Economists often label inputs as "capital, labor, and land," though these terms are defined broadly to include *all* factors that influence economic output. **Capital** includes the stock of tools, buildings, and equipment that are employed in the production process; labor is supplied by workers; and **land** stands in for the flow of resources and services that come from the environment—trees turned into paper for copiers, iron-rich rocks refined into steel girders, and so forth. The history of the Industrial Revolution is filled with struggles between the owners of capital (capitalists) and the workers who supply labor. The history of industrialization also shows scant understanding of a land ethic beyond the

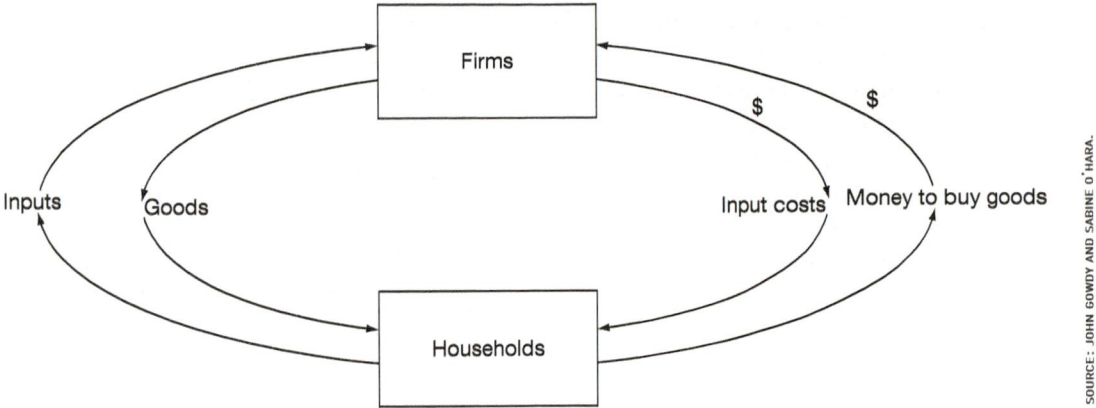

SOURCE: JOHN GOWDY AND SABINE O'HARA.

FIGURE 13.1
The circular flow of economic activity.

economically efficient exploitation of resources advanced by Progressives in the early twentieth century. Even now, the value of many natural resources is ignored in the accounting conventions that guide economic decision making; for example, the economic cost of climate change is not taken into account in the price of fossil fuels. From the trajectory of economic development, however, has come the array of means for meeting the needs of the present that we discussed in Chapter 10—an array that shows some signs of moving toward a more sustainable profile of development.

Economics explains how it can be that individual households and firms, acting with no explicit coordination other than the marketplace, can bring about an efficient allocation of resources—that is, an outcome that allocates goods and services to their highest-valued uses as measured in monetary terms (see Box 13.1: Reasoning Graphically about Economic Choices, page 352, and Box 13.2: The Magic of the Market, page 354). The guiding principles are simple. Households make choices to maximize their "utility"—that is, to meet their varied needs and desires in the way that provide the greatest level of satisfaction. Firms are assumed to maximize their profits, a financial measure of their ability to perform economically. Maximizing is the pursuit of more utility or profit, often but not necessarily measured in dollars. This is the economic model: people know their own interests, and they trade in pursuit of those interests.

In that pursuit, buyers and sellers, meeting in a market, come to agreements that benefit them both. The process is described in Box 13.2, page 354. The point to notice is that this is almost like magic: *self*-interest turns out, under certain idealized conditions, to generate *collective* benefits. Indeed, in the kind of idealized market described here, net benefits to society are maximized at the competitive equilibrium. (For an interesting but problematic example, see

BOX 13.1

REASONING GRAPHICALLY ABOUT ECONOMIC CHOICES

Economic analysis is done to a significant degree using analytic geometry, which combines ideas from geometry with ideas from algebra and calculus. This is an unfamiliar way to think for many students, so it is worth spending time looking at how it works. If you were interested in how water flows down a valley, you might make a map of the valley so that you could study how the water would flow if it were all at the surface. (In most places, the streambed allows water to percolate underground, too, so the water's flow is not entirely at the surface.) Similarly, economists try to map the contours of markets so that they can see how the goods, services, and money in Figure 13.1 (page 351) flow among firms and households.

Such a map is shown in the first figure below. Instead of east–west and north–south, the axes of this map plot two variables that describe a market. The horizontal axis measures the amount that is brought to the market by firms producing something to sell—pounds of coffee, say, or hours of legal effort offered by a law firm. The vertical axis measures the economic cost of providing those goods or services.

Firms decide how much to produce by maximizing profit.

The map in this figure shows how firms approach their most basic business decision about how much to make and bring to market. In a competitive market, a firm cannot control price, so the revenue the firm will earn from its sales is given by the straight line labeled "price × quantity." The higher the volume of sales, the more money is brought in.

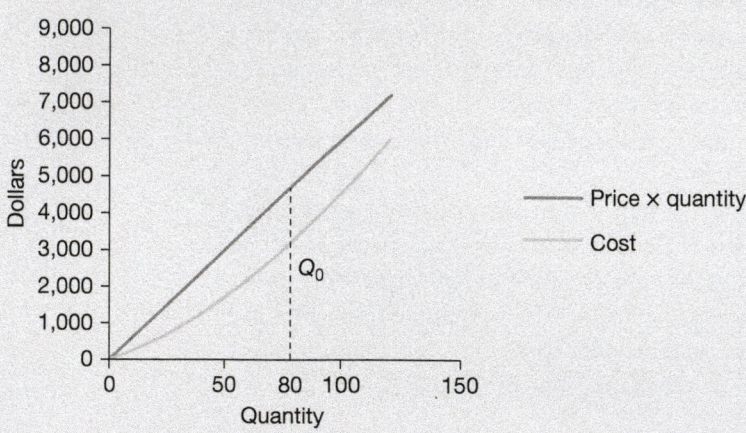

However, a firm is guided not just by how much revenue it brings in but by the profit it earns. The profit is the difference between revenue and production costs—that is, the vertical distance between the straight line plotting the revenue and the curved line showing the cost of production (point Q_0 in the figure). (Think of our map analogy. The revenue line is like the bottom of a valley, whereas the production cost curve is sort of like a ridge. The steeper the ridge, the faster water will flow down to the bottom of the valley. In this way, the first figure identifies the steepness of the valley for the firm—or, as an economist would say, the size of the incentive to bring different amounts of its product to market.) It is reasonable to think that firms are trying to increase their profits—this is what shapes the decisions they make about how much to produce.

What about the buyers? Buyers are the households shown in the second figure. They are making purchases that will bring them benefits that can be valued in monetary terms. It is sensible to think of buyers as seeking the best deal—the largest net benefits, given the price they have to pay. This net benefit is called "consumer's surplus" by economists, and it is the vertical distance between a point on the curve labeled "benefits" and the straight line representing the expenditure by buyers to sellers—that is, the price of the good multiplied by the quanitity purchased. The consumer's surplus is maximized at point Q_0.

These two graphs point toward a solution to the questions about how much will be bought and how much will be sold. As you can see, these are not two questions but only one: the amount bought and sold are the same. The answer provided by economic theory is that the amount bought and sold is the amount at which *both* the consumer's surplus and the firms' profits are largest, as shown in the third figure.

This is an amazing result: both buyers and sellers maximize their goals at the same point. Box 13.2, page 354, describes how this can happen, by examining a different way of mapping the economic dynamics.

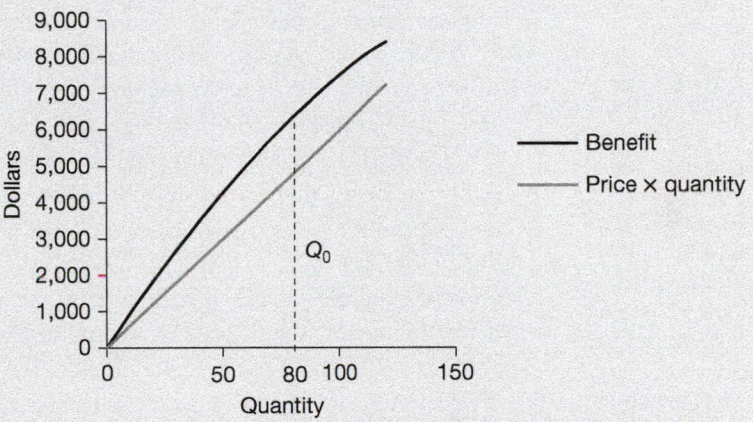

Households decide how much to buy by maximizing their net benefits.

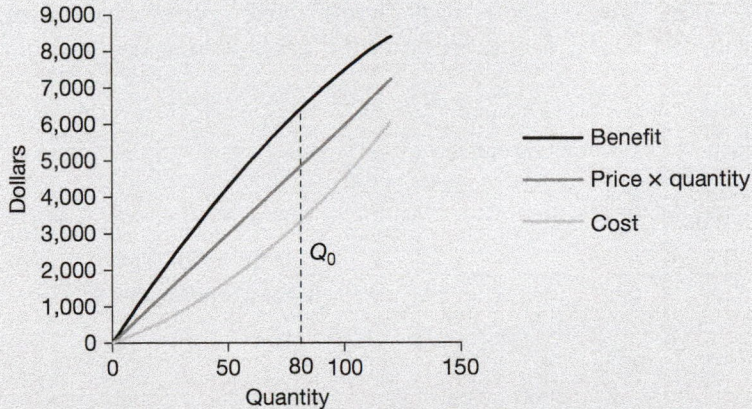

Market equilibrium is the point at which the profits of firms and the net benefits of households are maximized.

BOX 13.2

THE MAGIC OF THE MARKET

How does it happen that the amount sold and bought in a competitive market occurs at the very point where both buyers and sellers maximize their economic gains? The answer is that the price provides a signal concerning both the value of goods to consumers and the cost of producing those goods to businesses. The result is that the price mechanism balances benefits and costs from the perspective of society as a whole.

If you think about this statement, it is both plausible and mysterious. It is plausible because it rings true to everyone who has ever approached buying or selling carefully—as people do when they consider paying a college tuition or taking a job (perhaps to pay for educational expenses). As sellers, we agonize over whether the price we would be paid—or the wages we get if we are selling our labor—will be worthwhile given the cost, in time and energy, of devoting part of each day to the job. As buyers, we wonder if we could gain more net benefits by attending a more expensive school, where we might gain a more prestigious degree or a better education—and what we would need to sacrifice to pay for it. In making choices as buyers and sellers, we do aim at securing maximum net benefit.

But it remains unclear how it can happen that both buyer and seller can achieve maximum net benefit at the same time. Or, more accurately, under what conditions this surprising outcome can be achieved. For when it can be achieved, buyers and sellers do not need anything but the opportunity to make individually rational decisions given the going price in the market. In a complex global economy, it is essential to know when a market of this kind should be encouraged.

Market equilibrium— price = marginal cost = marginal benefit. Market equilibrium is reached at the quantity and price at which a buyer's marginal benefit and a seller's marginal cost intersect.

The map of the economic landscape in the figure on the left shows how price works to achieve market equilibrium. The horizontal axis plots the quantity produced—just as in the figures in Box 13.1 (page 352). The vertical axis shows something new: the value *per unit* that is sold. Notice that, for the buyer, the expenditure required to purchase a single unit is simply the price.

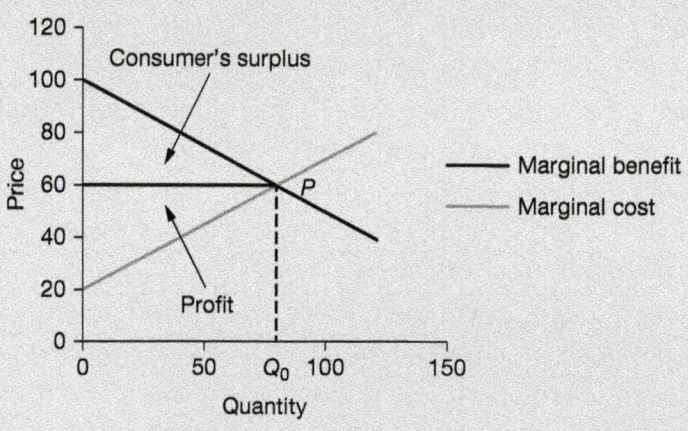

Now look back at the first figure in the previous box. The curve that represents the costs the firm must pay to produce the goods sold slopes upward, gently at first and then more steeply, because when the quantity made is small, it is easy to make more, but costs increase as quantity does. For instance, if a farmer has vacant land, he can easily expand the lettuce he plants. But as the farmer plants a larger area, the water needed to irrigate the crop may become scarce, raising the cost to bring the lettuce to market. This is shown as the upward-sloping cost curve.

The figure above plots this slope as the line labeled "marginal cost." Conceptually, this represents the incremental cost, or the increase in cost that occurs when the level of production is increased by one unit. As in the first figure in the previous box, the farmer wants to increase his profits—so he keeps increasing production as long as the price is higher than the marginal cost. (Of course, the farmer makes decisions during the growing season based on his guess about what prices will be when the crop is harvested and brought to market.)

Now consider the householder who wants salad. This buyer values lettuce but she must also take into account the cost of salad dressing and other things she needs. The higher the price of lettuce, the less net benefit she will derive from buying a head of lettuce. As you can see in the second figure in the previous box, the consumer's benefit curve also bends, but it is steeper when the quantity bought is low, and it flattens out as the quantity increases. This means that the slope of the benefit curve decreases as the quantity rises, which is shown by the line labeled "marginal benefit" in the figure above. In conceptual terms, the marginal benefit represents the incremental benefit the consumer derives when her consumption increases by one unit. *Marginal* benefit decreases because, for most goods, the satisfaction derived from each unit of consumption falls as the quantity consumed gets higher and higher. On a hot day, for example, one glass of lemonade is likely to provide a high level of satisfaction. But a second, third, and fourth glass would provide less satisfaction as the drinker's thirst was slaked.

The map of marginal cost and marginal benefit now shows the destination toward which both seller and buyer are headed: a situation in which the quantity that consumers desire to buy is equal to the quantity that businesses want to sell, given the prevailing market price. Where the two lines cross, each actor has done as well as they can: the price equals both marginal cost and marginal benefit. Both buyer and seller have maximized the net benefit they seek.

The figures in the previous box and the one above describe an ideal situation in which sellers and buyers have cost and benefit curves that are shaped in a certain way that leads them to interact through the market to arrive at an outcome that is individually optimal given the equilibrium price. More important, no one else is involved: the production process does not entail pollution or exploiting workers. So the buyer and seller can achieve shared benefits without affecting another party. (This means that there are no commons problems.) When those conditions apply, market transactions are efficient from a social perspective as well as for the buyer and seller.

An **efficient allocation** of resources is one in which everyone—households and firms—are gaining the maximum possible benefits, given the benefits that are enjoyed by other parties. It is in this specific sense that resources (e.g., money, skills and effort of workers, machinery, buildings) can be efficiently arranged. Economists call this efficient allocation **Pareto optimality**: a situation in which no one can be made better off without making someone worse off.

Even though neither buyer nor seller has a good idea of the other person's benefit or cost curve, the market mechanism allows them to interact in ways that lead to the discovery of a mutually satisfactory price. Both can improve their situation by going ahead with the transaction. How do we know this magical process of mutual discovery works? Because billions of willing trades take place each day. Not every willing trade actually results in higher satisfaction, of course. What economic theory provides is a description of how willing trades can occur, not a guarantee that every trade is regarded as a gain after it happens. But as every consumer and producer knows, the gains do occur often.

It is important to be clear on one aspect of economic efficiency. Pareto optimality does not include considerations of fairness or equity. If the producer is poor and the household is rich, as happens often with coffee, for example, it may be better for society that the coffee farmer earn a larger profit than an unfettered market might allow. And consumers may be willing to do that. With "fair trade" coffee, sold at Starbucks and other retailers, a premium price is charged for coffee brewed from beans distributed through a system that provides higher prices to low-income farmers. Fair trade fits the Pareto criterion, but so might a cheaper cup from a supply chain in which the farmers do not earn as much.

The figure above provides a kind of aerial view of an economic transaction. The economist says that if we approximate the reasoning process with marginal values, we capture the essence of what is going on, even though it is not a very realistic description of the experience of buyers and sellers. This is a fine example of theoretical science at work, using highly simplified and nonobvious assumptions. These assumptions include the focus on marginal value as the key factor in making choices; the ignoring of equity and indirect environmental implications; and the heroic simplification of the mind games of bargaining into the tidy intersection of lines on a graph. With these extraordinary assertions, amazingly, one can explain a very wide range of economic behavior.

Box 13.3: Carbon Offsets, page 358.) Think of other cases in which individual actions lead to good collective outcomes. Democracy and competitive sports are two examples to consider. How do these activities combine individual motivation with institutionalized rules to produce a positive result for the participants and their communities?

A market is the complement and antithesis of a commons. In an ideal market, self-interest benefits its participants rather than producing tragedy. This can occur when the costs and benefits of economic decisions fall strictly on buyers and sellers with no third-party effects, or externalities (discussed in more detail below). The promise of the market is that it can organize productive human behavior without an apparent need for government. This is an illusion, however, as comparison with sports makes clear. What differentiates a basketball game from a brawl is the fact that basketball has rules that are enforced by referees and specify standard heights for the hoop, the size of the ball, and what constitutes a foul. What is required is an agreement among the players that what they are doing is playing the game of basketball. Similarly, participants in markets agree to play by the rules, and these rules are enforced by government in a variety of ways. This is why stock fraud or adulterated milk lead to prison terms, and why some levels of pollution are sanctioned by fines and legally binding orders.

The mutually beneficial outcomes of the market arise through the coordinating mechanism of price. Note the dollar signs in Figure 13.1. If households and firms know the **prices** of goods and services, they can make choices about buying or selling. This is an enormous simplification. Environmentalists and other critics point to how much information is left out. The price of electricity, for instance, does not usually include the costs imposed on the natural world, or on people, from burning coal or building and operating a windmill. Generally, whenever the process of providing a good or service to buyers involves a poorly governed commons, the price does not reflect the environmental costs of production and distribution. As we saw in the discussion of ecosystem services in Chapter 6, misuse of commons is the usual situation rather than the exception.

Yet if many prices do not include a full accounting of the environmental impact of economic activity, one should still be impressed that this single parameter is enough to coordinate human activities as well as it does. Attempts to supplement price information, as in the marketing of organic food, require the consumer to care a lot and to pay attention consistently. Information is hard to obtain, or as economists say, getting information is an important transaction cost. Prices provide a powerful shortcut—one that oversimplifies. But markets work to a degree that cannot be ignored.

As markets became significant forms of social organization, they altered the terms of human experience. Our interaction with the material world is no longer that of the subsistence farmer or hunter-gatherer. We work to earn money, and

BOX 13.3

CARBON OFFSETS

The purchase of a carbon offset is an unusual, and instructively troublesome, example of a market transaction and the way it works to shape the allocation of resources in a more efficient way. A carbon offset is an agreement between a buyer and a seller. In exchange for a payment, the seller agrees to carry out activities that will remove a quantity of greenhouse gases from the atmosphere or prevent their emission. The amount removed is matched to the amount emitted in a specific activity, such as taking a trip by airplane.

A company called Native Energy sells offsets measured in tons of carbon emitted. The company invests in renewable energy resources such as windmills. If an environmentalist couple is planning a wedding but worries about all the greenhouse gas emitted by their guests on the way to the celebration, they can send a payment to Native Energy. In exchange for the payment, the company will make investments that are estimated, over the life of the projects, to displace an equivalent amount of greenhouse gas emissions.

What a carbon offset firm does is to coordinate the wants of a large number of energy users. The bride and groom find it impossibly difficult to arrange for trees to be planted to offset the travel associated with their wedding. The offset firm lowers the **transaction cost**, making it possible to counter one's burning of fossil fuel by allowing the wedding planner to easily arrange the tree planting. But notice that the total number of flights and car trips has not decreased: the guests still get to the wedding.

If the offset transactions work as advertised, there will be less greenhouse gases in the atmosphere than without the offset. For those buying the offsets, their satisfaction may be increased, particularly so long as the offset is inexpensive.

Notice, however, that the buyers face no *economic* incentive to lower the consumption of fossil fuels. They are choosing to buy the offsets, and in doing so, they are not giving up the opportunity to gain from burning fossil fuels. In fact, they may feel they have a moral license to keep on emitting greenhouse gases profusely because they are buying offsets. This possibility has led critics to compare offsets to the sale of indulgences by the medieval church, a practice that provided forgiveness of sins in exchange for a suitable donation to the Church. What about the firms selling offsets? If their business model works, they actually become dependent on people who emit greenhouse gases.

An underlying problem is that buying offsets is voluntary. That is, there is no mutual coercion and no effective restraint on the overuse of the commons. The atmosphere remains an open-access commons until binding rules are imposed on everyone. The offsets do nothing directly about this. Yet the offsets may be significant socially and politically as a demonstration to reluctant politicians that influential people feel strongly enough about climate change that they are willing to pay to offset their own emissions. This might mean that these same people would also support the difficult political choice of agreeing to mutual coercion. An example of how that mutual coercion can be done using prices to steer behavior is discussed in Box 13.5: Cap-and-Trade, page 368.

we spend most of what we earn. Income and spending have become the way we judge and know about our own footprint in nature. With some exceptions, the owner of a wood stove simply considers how much a cord of firewood costs, not how many hours of sawing and splitting and stacking she needs to do to heat her house next winter. She judges her desire for that firewood against her income, rather than whether she wants to remove a tree from a particular stand of forest. In this and dozens of similar cases each week, we act in a way that far exceeds our environmental knowledge. This is how most of the gross domestic product is allocated.

By providing a model of how people make choices in a market, and then testing its predictions, microeconomic theory has provided considerable insights into what people actually do. Notice, however, that some productive organizations, such as universities, do not appear to be maximizing profits, at least not in a simple way. Moreover, a lot of our choice-making behavior is not sensibly described as looking for the best deal in a market. Consider how an athlete allocates her time in training, or how friends decide how to hang out. It is not clear in either case how the idea of "more is better" would be translated in practice. So one suspects that the theory of microeconomics is not a complete explanation of human choices, even if it is a powerful one.

MARKETS AND NATURE

What is left out by economic theory? In one sense, nearly everything. Consider Figure 13.2 (page 360). Market exchange seems to occupy a small part of the world being modified by humans. Yet on closer inspection, we can see that the economic analysis illuminates ideas we have met before. Think of this figure as a stack of placemats of different sizes. All human activity takes place in the biophysical world, and so, of course, does all economic activity, which is a subset of human activity.

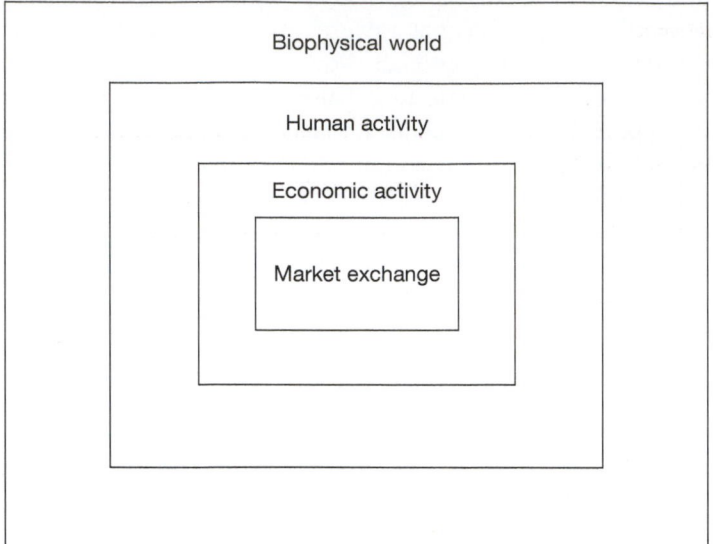

Biophysical world

Human activity

Economic activity

Market exchange

SOURCE: JOHN GOWDY AND SABINE O'HARA.

FIGURE 13.2
A schematic view of economy-environment interactions.

The market lies within the biophysical world, not apart from it. In that sense, by setting off land as one of several separate economic factors to be combined for productive purposes, the image drawn by economic theory is misleading. The environment—land—is a separate factor of production in the sense that the environment is a store of resources and a source of ecosystem services contributing to the economy. But the environment is biophysical nature, the context for all human activities, including the market. The market can run for a time as if the biophysical world were only a set of separate resources and services, but this is likely to be unsustainable if market transactions ignore the connections between the economy and nature. Poorly managed commons are ignored connections: too many animals are put on the pasture, guided by the economic return to the herders, but the grass cannot sustain them all. Markets anchored to inadequate institutions fail to recognize important connections.

In the outermost layer of Figure 13.2, the biophysical world, is a portion that lies mostly beyond human activity. Some of this is what is called "wilderness." The biosphere, for example, includes subsystems such as the polar regions. But as ecosystem scientists and environmental historians emphasize, even the most remote and seemingly untouched parts of the biophysical world are now influenced by human activity. Virtually all of the biosphere is now part of the human-dominated world, the world without edges. The Arctic is polluted by industrial chemicals released thousands of miles away, and the poles are now undergoing climate change at a pace and scale that far outstrips the effects seen in temperate or tropical settings. These are examples of the human domination of nature described in Chapter 6. Modification of ecosystems in the pursuit of human benefit is organized in large measure by economic activity.

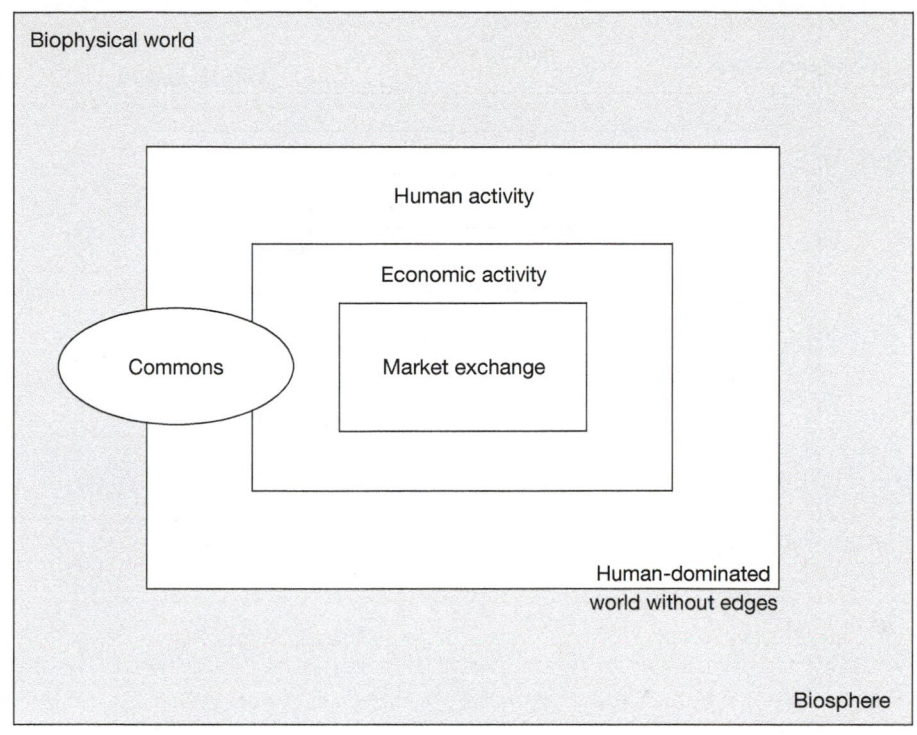

FIGURE 13.3
The biosphere, commons, and the world without edges in the economy.

Now consider Figure 13.3 and recall the discussion of commons—arenas that no single individual owns, but that are visited and used for economically valuable activities such as grazing sheep or dumping untreated industrial wastes. Although such uses and abuses of commons are economically significant, they usually take place beyond the range of market exchange. For that reason, the governance of commons described in Chapter 3 focused on rights and duties, on institutions that are political rather than economic in character. In the world without edges, however, many market transactions cross scales (Fig. 13.4, page 362), linking human activities far beyond the boundaries of traditional communities of the kind that could once govern commons well (see Box 10.3: Nature, Wealth, and Power, page 271).

Many economic activities, such as agriculture in the hills of New England, have left their traces in today's landscapes, a phenomenon we called the invisible present in Chapter 2 when we looked at a regrown forest that still bears the signs of farming. The physical effects of market exchange flow across all the boundaries in Figure 13.4, but economic theory focuses on what happens within the innermost boundary. A great deal of insight about human choice-making can be gained from

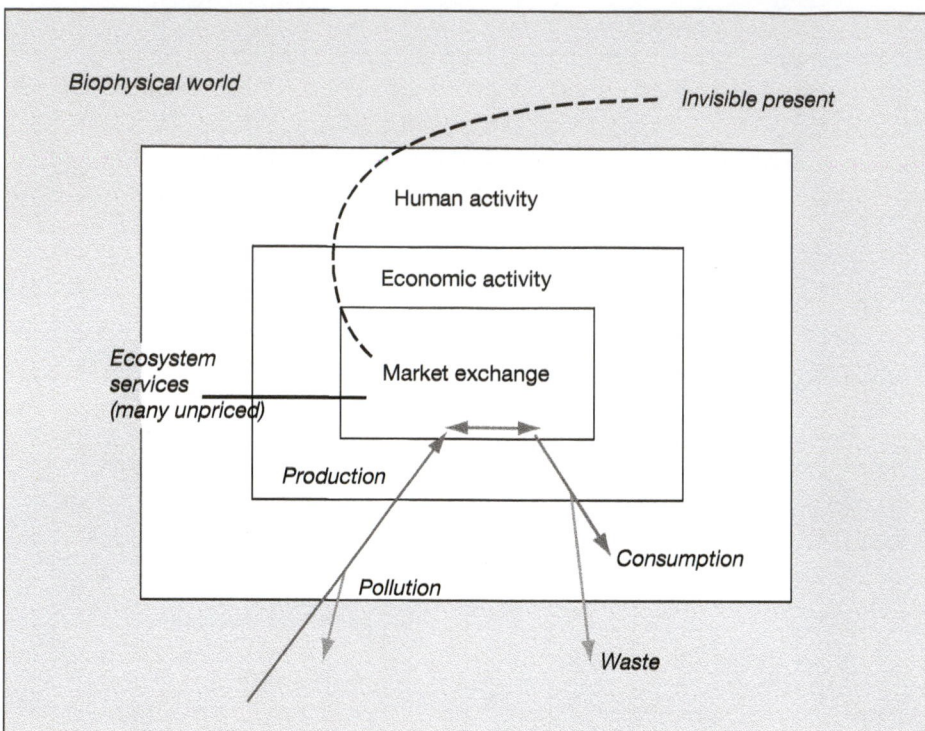

FIGURE 13.4
Economic activities cross boundaries, often in ways that are not taken into account in market exchange.

looking inside the market exchange box. But the way that some things are ignored, including pollution, waste, and ecosystem services, skews the perspective if one mistakes the market for all of reality.

GETTING PRICES (CLOSER TO) RIGHT

As we have already observed, the prices of many ecosystem services fail to take into account the damage done to the natural world when humans appropriate the flow of a river, or cut down trees on a steep slope and raise the risk of floods and erosion. Economists say in these circumstances that market price does not incorporate the **social cost** of the modifications to an ecosystem when its services are diverted for human benefit. The social cost is a cost borne by humans when they are harmed by changes in the biophysical world that are of human origin. The damage is what economic theory calls an **externality**, a cost of producing or consuming something that is not included in the accounting. Externalities can be understood as **third-party effects**—intended or unintended impacts of economic decisions that fall on people who are not party to the transaction.

Externalities can be positive as well as negative. This is why businesses often help to fund cultural amenities such as museums and concert halls. These amenities make the community more attractive to workers and customers, so they can be worthwhile to local businesses. Yet charging visitors an entrance fee that includes the full cost of operating a museum would keep many people away, so even a hard-headed business can find it valuable to subsidize the museum so it can charge fees that are below cost.

The existence of negative environmental externalities, such as pollution, is a major topic in environmental economics. When human activities are carried on without considering all of the relevant values being traded off, the use of ecosystems is not efficient. Information that is not reflected in prices can matter, both to nature and to people relying on ecosystem services. Ignoring this environmental information leads to inefficient choices, a phenomenon economists call **market failure**. (Another difficulty with valuing nature's services comes up when we want to ensure that those services will still be available in the distant future; this is discussed in Box 13.4: Valuing the Future, page 364.)

The costs of externalities can sometimes be internalized by nonmarket means. One way is through government regulation, which limits the choices available to producers, raising their costs and the prices they must charge. Another is by capping the supply by public policy, and then allowing the limited supply to be traded (see Box 13.5: Cap-and-Trade, page 368). Figure 13.5 shows the relationships among prices, costs, and output for a product that is made by a process that generates pollution. The demand curve is the sloping line descending toward the right. As the price is reduced, demand rises, so that more needs to be produced.

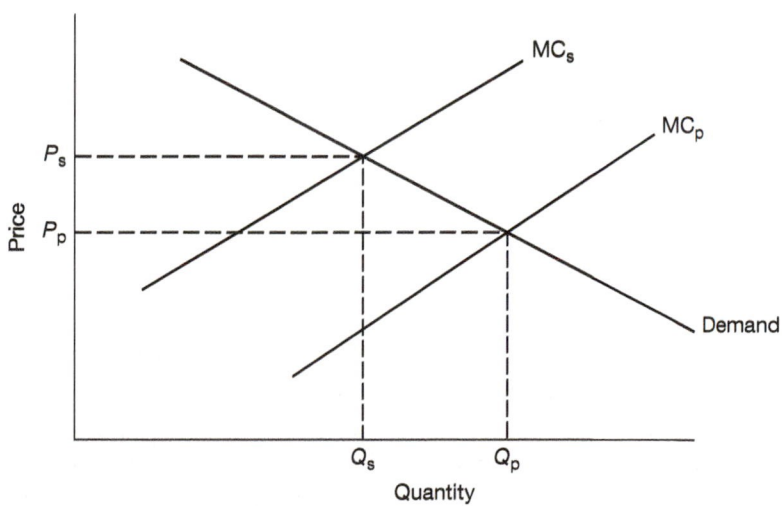

FIGURE 13.5
Social cost and externalities.

BOX 13.4

VALUING THE FUTURE

One important and widespread market failure is found in that distant country called the future. This problem is called **discounting** in economic theory. Consider a forest, which is growing at 2 percent a year in the value of the timber that can be harvested from it. The owner can also put the money obtained from logging the forest in a bank account that yields 3 percent a year. If the price of timber and the cost of logging are constant over time, then the money in the bank grows faster than the value of the trees, and the owner should log the forest. Correct?

An example of discounting.

The owner can decide between cutting the forest now and deferring the cut until the money is needed, say, for a child's college education. Suppose the family owns 100 acres of forested land; the standing timber might reasonably be

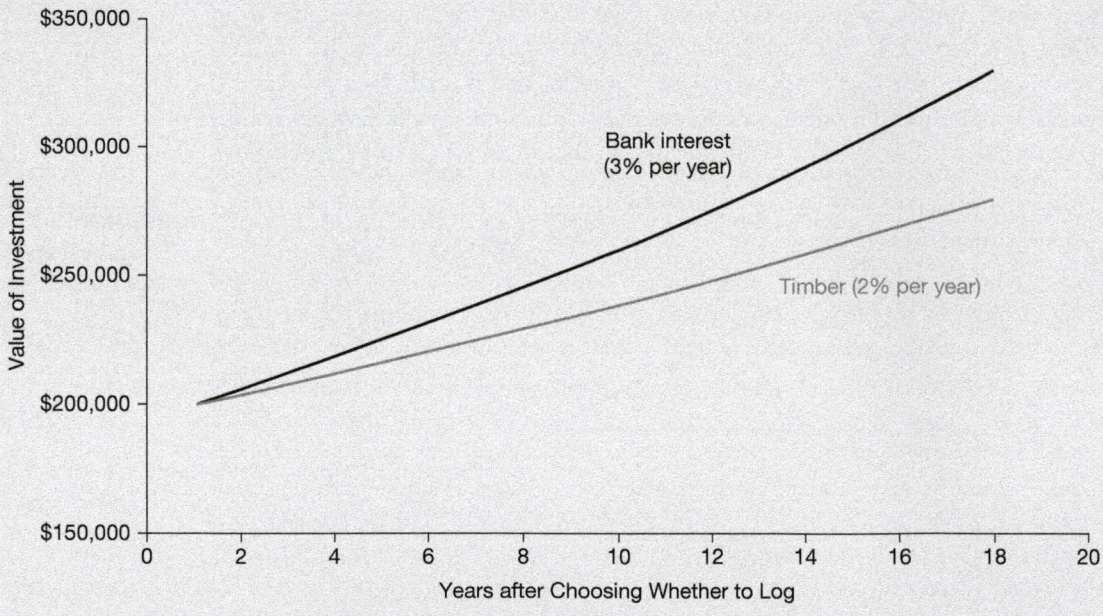

worth $2,000 per acre, or $200,000 total. The family decides what to do with the land when a new baby is born. With the assumptions above, by the time the child reaches college age, the difference between the value of the uncut trees and the money earning interest in the bank is more than $50,000—a substantial scholarship fund (see figure). Of course, this is only an example, though the numbers are representative of returns on forests and on prudent investments. In any event, what matters when making the choice is what the landowner *thinks* the difference in return will be, given the options of cutting and investing or waiting to cut.

The puzzle here is one of human behavior, not just economic theory. The problem lies in the fact that we cannot easily make transactions with people in the future. How much are those trees really going to be worth when one's children are ready to go to college? We don't know, so we create an estimate based on how much more wood will be in the trees. But this doesn't tell us what people will be willing to pay then. Because we are ignorant of the future price, we settle for the comparisons that economic theorizing tells us to use. This isn't very satisfying, but it is often the best available estimate. Bankers, for example, rely on discounting when they make loans. And in this way, the assumptions of economic theory are translated into real choices.

What should we do, as a society, to signal to the forest owner that there is value in undisturbed ecosystems? How might people who are not owners recognize this value in the way they treat the owner? One way is to provide tax advantages to owners who do not cut their forests. This is a way of spreading the cost of the benefits of a growing forest over the society that benefits from it, rather than expecting the landowner to bear the entire cost. Agricultural and forest land preservation laws in many states now provide such tax advantages. The rules may not enable a landowner to gain financially from logging at a later date to pay college expenses, however.

The economic analysis shows where the interest of the forest owner lies, in what is a temporal tragedy of the commons. The owner can act in his own interest, but to the detriment of the environment and community. So the question can be reframed this way: What kind of community norms should be adopted to achieve sound management of commons over time? There is no theory with the intellectual prestige and political influence of economics to define the appropriate form of restraint. This is one major reason the environment is in trouble. But environmentalists should not blame economists when they are describing something real in human behavior. Instead, environmentalists should enlist economists in the search for usable solutions.

If the pollution damage—the social cost—is ignored, the cost of production is related to output in the marginal cost curve MC_p. As we described in Box 13.2: The Magic of the Market, page 354, the marginal benefit curve and MC_p intersect at the point at which price equals the marginal cost. This is the level of price and output that economic theory says would result from exchange in the market.

Now, consider what happens if society—acting through the government—includes the social cost (actually, an estimate of it) by imposing regulations or limiting supplies. Now, the marginal cost curve rises to MC_s. That is, at each level of production, the cost is raised by government intervention in the market.

If production costs are raised by these means, while demand stays the same, the amount used declines, from Q_p to Q_s. What is happening is that buyers, facing the higher price, adjust their purchases. The adjustment may be different for each buyer, but the overall effect is to lower the amount sold.

How does the government know by how much to raise prices? The economically efficient solution would be to raise the costs just enough to incorporate the external costs that have been ignored. This means charging polluters for the monetary value of the harms they impose on third parties. But remember, this information is difficult to get. Even if governments could get accurate information on social costs, the costs would change as ecosystems change. The value of water left in a stream, for example, goes up during a drought, when the fish and plants would face lower flows even if humans were not withdrawing water. Two general strategies are used to deal with this problem.

The first is to adopt an environmental regulation, even if it leads to an economically inefficient outcome. The regulatory rule is set so as to achieve a specific level of protection. The U.S. Environmental Protection Agency does elaborate analyses, for example, to estimate how many cases of asthma and other respiratory diseases will result from a small change in the levels of diesel emissions from trucks. Based on those analyses, agency officials determine a socially acceptable level of respiratory disease. Generally, this level is not zero. Based on that health target, regulations are written to assure that the pollution-control equipment on trucks will keep the emissions low enough to meet the targets on crowded roads. Vehicle manufacturers and truck owners must then absorb the elevated costs. These social costs are now included in the cost of producing the transport services supplied by trucks.

Note that it is usually impractical to reduce emissions to zero. As the regulations are tightened, the cost of achieving lower and lower truck emissions soars. So regulators usually adopt rules based on conservative estimates, to try to make sure that the levels of damage to health and environment end up being lower than they estimated. Not all of the social cost is internalized, however, and some children still will get asthma. Of course, no one knows who they are in advance, and the kids and their parents generally cannot prove that it was truck emissions that caused their illness in specific instances. (As we have discussed in Chapter 11, asthma rates are disproportionately higher in poor urban neighborhoods with the dirtiest air.)

The second strategy is to create a market in which polluters pay a tax or price for each unit of pollution they emit. This is what was done with the cap-and-trade policy described in Box 13.5: Cap-and-Trade, page 368. What is sold in the market in that example is the right to pollute, a tradable emissions right. The emissions right is a kind of property, created by the government when it defines a limit on the total amount of pollution allowed (the cap). The cap creates a scarcity: before, polluters could emit wastes without paying, but now permits are required to emit the wastes. By allowing the rights that fit under the cap to be traded, the rights to pollute are apportioned in a way that minimizes the total cost of pollution control, given the environmental quality target established by regulators.

Note that the resulting outcome is not necessarily socially desirable or economically efficient, even if it does minimize pollution control costs given the level of environmental quality that is attained. The total cap may be too high, allowing too much damage to the environment. Or it might be too low, so that the prices of emissions rights rise so much that their cost stifles business activity without an equivalent benefit to society. The government, by setting the cap, is setting the point at which the value of less pollution is counterbalanced by the restraint on economic activity. This is a social judgment that attempts to balance two competing rights: the right of those affected by pollution to be protected from harm, and the right to pursue one's economic interests, which usually entails pollution. In using a cap-and-trade approach to global warming, the objective would probably be to lower the cap over time, so that the economy could adjust to increasingly stringent limitations on greenhouse gas emissions without major dislocations along the way.

Damage to ecosystems cannot be eliminated, but by considering externalities, damage can be managed. By using decentralized policy instruments such as cap-and-trade, the management of social costs can be harmonized to a degree with the goal of cost effectiveness. Sometimes policy measures that use prices and markets are not feasible or are undesirable in other ways, because economically efficient allocation is not the only or highest goal of environmental policy.

In Chapter 12 and this chapter, we have discussed the two major means by which contemporary societies enable people to act far beyond their capabilities as individual persons: through governance and markets.

The market enables individuals, households, and business enterprises to pursue their economic interests through market exchanges where prices are set by demand and supply. The market provides an approach to economically efficient allocation of resources, one powerful enough to drive the astonishing growth of the material economy over the past two centuries. Markets are self-organizing systems: households and firms respond to incentives and organize their own behavior

BOX 13.5

CAP-AND-TRADE

One way to approach the problem of social cost uses the dynamic of the market to guide polluters toward a social goal. Here is how it works under the 1990 Clean Air Act for sulfur dioxide (SO_2), an air pollutant that plays a key role in acid rain and related air-quality problems. A large portion of the SO_2 is emitted by big industrial facilities such as coal-fired electric power plants. Under the Clean Air Act, a national cap, or limit, on SO_2 emissions was set, and every facility that emits SO_2 had to have a quantity of emissions rights equal to their emissions. Rights were distributed to all of the emitters at no cost and in proportion to historical emissions, and a schedule for lowering the national cap was announced.[1] Power plants that were cleaner than average did not need all of the emissions rights they had been issued, and they could sell them to facilities that needed additional rights as the cap was lowered over time. Because the emissions rights can be traded in a national market, this policy design is called cap-and-trade.

A firm that owns a dirty plant faces a choice: either buy enough rights or clean up its plant so that it does not need to buy as many emissions rights. This is a business decision, one that can be made using conventional methods of business management once it becomes clear how much the emissions rights will cost, now and into the future. An environmental choice has been internalized as a cost of doing business, but in a way that enables producers to make decentralized decisions in response to the cost of polluting.

The cost of an emissions right should reflect its value to businesses—that is, the marginal cost of cleaning up a dirty plant. When the Clean Air Act amendments were passed in 1990, analysts anticipated that the permits would cost several hundred dollars per ton of SO_2, but the actual costs turned out to be far lower.[2] It turned out to be relatively inexpensive to comply. Prices have risen as the national cap has been tightened, but they are still less than half the original estimates.

In this case, the social cost is still being judged by government. As the program has gone forward, the national cap has been brought down, and the damages are being increasingly included in the market price of electricity and other goods that require the emission of SO_2 in their production. It is not necessary to determine the optimal level of pollution, so long as cost-benefit studies show that the benefits to society are substantially higher than the cost being imposed through the emissions permits. This has been the case for the Clean Air Act program.

The enforcement costs of cap-and-trade are lower than would be needed to police a regulatory program. Inspectors verify the actual emissions of the holders of the emissions rights. This means checking that the monitoring instruments are working properly, something akin to inspecting the pumps at a gas station to make sure they are delivering the correct volume of fuel. This is simpler than checking complicated pollution-control equipment installed to comply with regulations that often require varying levels of cleanup at different hours and seasons.

At the same time, cap-and-trade has limitations. The SO_2 emissions rights market is national, so the right to release a ton of SO_2 in New Mexico can be used in a power plant in West Virginia. Yet SO_2 is blown by prevailing winds and tends to be a more serious problem in the eastern United States than in much of the Great Plains, although this spatial variation is not reflected in the way the policy works.

Cap-and-trade is an approach that seems applicable to global warming: set national caps for carbon dioxide and other greenhouse gases, and then permit trading of the emissions rights. Because greenhouse gases diffuse throughout the planet's atmosphere, hot spots are not a problem. A version has been implemented already in the European Union, and cap-and-trade is a leading contender for international agreements beyond the Kyoto Protocol. Cap-and-trade may be more difficult to adopt in the United States, however, after a proposal to institute this policy was attacked during the 2010 congressional elections.

Nothing is simple in governing commons. The European Union cap-and-trade program has been criticized because the cap was set so high that firms did not have to alter their emissions much, and the initial cap was raised rather than lowered, in response to political pressures. These are substantial problems. Having enlarged the scale of competence, political authorities must still govern competently. Cap-and-trade is a useful approach, but it still requires political will.

1. Paul L. Joskow, Richard Schmalensee, and Elizabeth M. Bailey, "The Market for Sulfur Dioxide Emissions," *American Economic Review* 88 (1998): 669–85.

2. Ibid.

to respond to changes in price. For this to happen, government is needed to secure property rights and enforce contracts, and to assure that buyers have access to information. The very success of markets, in turn, propels social changes that have implications for governing.

In the democracies of the rich nations of Europe, North America, and Japan, a political system of governance has emerged. In Figure 13.6 (page 370), this is labeled *polis*, using the Greek word that originally described the self-governing city-states of the ancient world. Holding their governments to account through competitive elections, the interests of citizens—particularly those who are organized effectively—have come to be reflected in the goals pursued by governments. Many of the policies adopted by governments affect actors in markets, protecting

FIGURE 13.6
Two ways in
which individual
competence
is extended
(imperfectly).

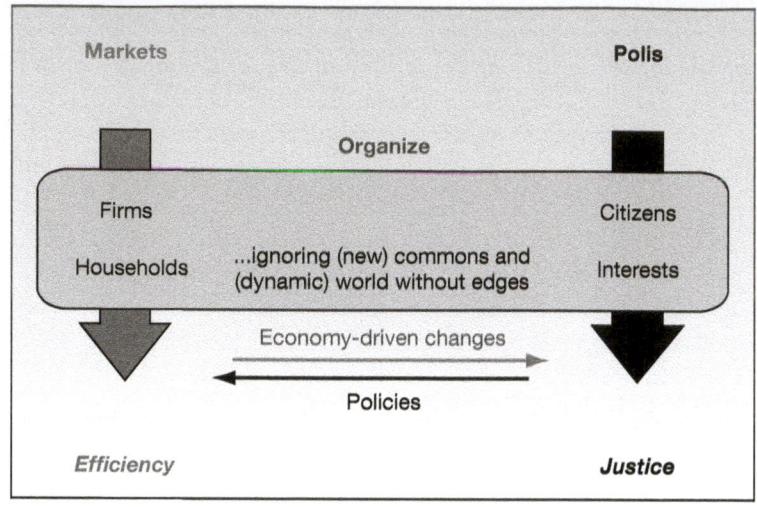

them from abuses such as pollution, but also interfering with the choices that the actors formerly could make. The resulting political responses lead to further changes and policy, as we discussed in Chapter 12. As we have seen in Chapter 4 and Part II, however, the world without edges drives change in ways that often ignore commons problems and social injustice. Governance struggles to keep up with the impacts of a dynamic economy.

The large arrows in Figure 13.6 identify the overall goals that theory says are the purposes for which people participate in markets and polities. Of course, this is an academic simplification. Most of us do not decide to join a community or society, exactly. We are born into a society, and if we choose to join a different one (usually by moving to another place), that choice usually is not made with fine calibration of costs, benefits, and risks.

Still, the rough approximation that markets lead toward efficient allocations of resources and that politics strives toward justice is a useful one. The dynamism of economies since the Renaissance has propelled waves of changes that have created and still drive the world without edges. The force here is self-interest in various forms—profit, competition, and the desire for a better life are some of its names.

But the power of markets also opens up paths to problems such as monopoly power. In the settlement of the American West, a semiarid landscape, control of water was of central importance to agriculture. Those who managed to acquire land in watersheds diverted and impounded the water in flowing streams, often to the disadvantage of their downstream neighbors; the dams and diversions also deprived fish and plants of their habitat.

The existing institutional structure can also ignore social inefficiencies, including environmental injustice and wasteful energy use. By imposing pollution costs on poor people, public health overall is harmed, with costs such as lost schooling and visits to emergency rooms that appear to be social problems but are really traceable to environmental causes. Yet poor communities have lacked the economic and political resources to correct environmental injustice on their own.

Analyses such as the one summarized in Figure 7.13 have shown that the largest source of greenhouse gas reductions are actually advantageous to energy users, who can save enough money on their energy costs that investments in energy efficiency pay for themselves within a year or two. Yet despite substantial efforts by utility firms and governments, many of those improvements have not been carried out. In short, even when firms and government combine forces, many energy efficiency gains are elusive, even though they are economically advantageous to all. This frustrating experience suggests that the competence needed for sustainable development remains beyond the reach of the institutions that currently exist.

All this human activity is rooted in nature. Those natural roots fade from view, however, as we move toward institutions that seem to be human in character. Remember the isolation of the Mayan kings from the growing vulnerability of their city-states' agricultural system discussed in Chapter 3. Human institutions that focus only on efficiency and justice may lose sight of environmental vulnerabilities. As we saw in Chapter 10, environmental awareness has led to a wider understanding of sustainable development, even though there is much left to do. Part of that work lies in improving technology and consumption, the subject to which we turn next.

FURTHER READING

Gowdy, John, and Sabine O'Hara. *Economic Theory for Environmentalists.* Delray Beach, FL: St. Lucie Press, 1995.

KEY TERMS

capital	externality	Pareto optimality	third-party effect
discounting	land	price	transaction cost
efficient allocation	market failure	social cost	

Chapter Fourteen

ADDRESSING AFFLUENCE

THE DILEMMA OF AFFLUENCE

No question is more troublesome for an environmentalist than this: What can *I* do to move myself and the world around me toward greater environmental responsibility and sustainable development? As we have seen in the discussion of grand challenges in Part II, and in the last two chapters, moving the world toward sustainability goes far beyond individual choices. What needs to change are the institutions that create and maintain the unsustainable practices of the material economy.

In Chapter 11 we encountered the dilemma of having "the wolf by the ears"—that is, because the material culture of the world is built on human domination of ecosystems, we cannot simply let go of industrialization and large-scale agriculture, even though it is also apparent that we cannot go on for long with the present trajectory of ever-increasing impacts. In this chapter we explore two avenues for dealing with the wolf of human domination: technology and consumption. The goal is not for you to learn about engineering or marketing, but rather, to develop ways of thinking about technology and consumption that enable you to understand the connection between the human domination of nature and the affluence that has become routine in the developed world. With this understanding, you will see how that connection can be modified to move toward greater sustainability.

We live in a world reshaped by technology, as we saw in Chapter 6. For the most part, technological innovation has been aimed at exploiting the resources and forces of nature more effectively so as to meet the needs of the present. This process has also expanded the ability of each successive generation to meet their needs, by building the stock of ideas that humans can apply to the transformation of landscapes. But as the world without edges expands, the very success of humans in dominating the natural world is now undermining the future of both the natural world and the human inhabitants of that world. Applying the inventiveness of people to recognize and counter that trend is vitally important going forward.

This inventiveness includes changes in our conceptions of need—that is, changes in the character of consumption and its relationship to well-being and sustainability. The increase in human domination of the natural world has been accompanied by an expanding set of material needs. In a meaningful sense, we have come to need automobiles even though they did not exist for most of human history. If we are to move toward a sustainable economy, we must shift the design of our cities and industries in such a way that we can meet our needs through a different profile of consumption. This seems like an impossibility, until one realizes that this shift is already under way. Much more needs to be done, to be sure, but the oil industry has come to accept that sales of gasoline in the United States are likely to decline from now on.

Learning Objectives
When you have finished studying this chapter, you should be able to

- recognize "biological nutrients" and "technical assets" in the consumer products that you use in your everyday life;

- identify nonmaterial needs in your own life, and the costs (material and nonmaterial) that you are willing to pay to meet those needs;

- look at the waste leaving your own household and consider whether it may be possible to build markets for some of the molecules being discarded;

- search for evidence to back up claims that a product is produced in an environmentally responsible way;

- find out whether your college or university votes on shareholder resolutions, and whether students can be involved in those decisions.

TECHNOLOGY, CONSUMPTION, AND ENVIRONMENTAL IMPACT

In the 1970s, a way to think about the sources of environmental impacts was popularized by two prominent environmental scientists: John Holdren, who later became science adviser to President Barack Obama, and Paul Ehrlich, the outspoken ecologist we met in Chapter 6. Together, they came up with the following formula:

$$\text{Impact} = \text{population} \times (\text{GDP per capita}) \times (\text{impact per unit of per capita GDP}) = \text{population} \times \text{affluence} \times \text{technology},$$
$$\text{or I} = \text{PAT.}$$

The IPAT formula, as it is known, is a definition rather than a law of nature—it is an **accounting identity**, a formula that is true because it defines "impact" in a certain way. The substantive questions are how population, affluence, and technology change, and in what units we estimate impacts, affluence, and technology. Through these questions we gain perspective on whether this definition of "impact" is a persuasive one. As the growth of population slows in the United States and elsewhere, many environmental impacts have continued to mount. The implication of this equation is that affluence and technology are the driving forces of intensifying environmental harm.

Even if the United States were to stabilize its environmental footprint, the continued growth of affluence in China, Brazil, and other parts of the world would still increase environmental damage. A quick look at the world beyond American borders shows how much technology matters in the material economy. The second column of Table 14.1 shows U.S. consumption of basic industrial commodities. The fifth column lists the factor by which consumption would need to increase for everyone in the world to consume at the U.S. level. These ratios provide specific numbers to illustrate the assertion by environmentalists that it would take "more than one planet" to meet the needs of a world consuming at the levels of today's developed countries.

Consider the ratios in Table 14.1 as crude indicators of the state of technology in the United States. The life-style we enjoy is obtained by using materials this way. If the rest of world attempts to match our standard of living, it may simply be impossible if they rely only on current technologies. We would need to improve technology by a factor of 7 or more merely to stay at the current levels of pollution from making plastics. This table is a companion to Figure 6.5, which shows the current impacts of humans on the global environment. Both tell us that the economy, as currently structured, is pressing against environmental limits and that economic growth, on the current trajectory, will press all the harder.

TABLE 14.1 U.S. CONSUMPTION PATTERNS PROJECTED TO THE WHOLE PLANET.

Commodity	1990 U.S. apparent consumption	1990 world production	Necessary world production for world per capita consumption to equal U.S. per capita	Factor increase
Plastic	25.0	78.3	530.0	6.8
Synthetic fibers	3.9	13.2	82.7	6.3
Aluminum	5.3	17.8	111.5	6.3
Copper	2.2	8.8	46.0	5.2
Salt	40.6	202.3	860.7	4.3
Fertilizer (potash)	5.5	28.3	115.5	4.1
Industrial sand and gravel	24.8	133.1	525.3	4.0
Iron and steel	99.9	593.7	2,117.9	3.6
Nitrogen	18.0	107.9	381.0	3.5
Cement	81.3	1,251.1	1,723.1	1.4

Production and consumption in millions of tons, based on U.S. Bureau of Mines data; calculations based on 1990 U.S. population of 249 million and world population of 5.3 billion.

Table 14.1 looks at getting the rest of *today's* world up to American levels of consumption. Of course, populations and economies will continue to grow. What does the IPAT formula suggest? We know that population is rising to somewhat more than 9 billion. We think affluence will continue its rise; certainly, billions of people around the world are striving to make it happen. So the environmental impact per person and per unit of affluence must go way down, and do so fairly rapidly—which is to say, during the working careers of the readers of this book—if sustainability is to be achieved. But instead of a sevenfold improvement, what is needed is more in the range of lowering environmental impact by 20–100 *times* today's levels. Is that feasible? Not if all the smart students go into finance or running environmental nongovernmental organizations, because a lot of new technology must be invented and developed.

The IPAT formula is, in an important sense, an oversimplification. Japan and Sweden are about as affluent as the United States as measured by income per person or in the Human Development Index described in Chapter 10. But in each of those nations, people use much less energy than do Americans, and in that respect they contribute much less, per person, to global warming. The differences are not simply in technology, as we normally use the term, but also in life-style and the way people use the technologies they have. Japanese and Swedish commuters use mass transportation to a far greater degree than American commuters, and their homes are smaller and thus require less energy to heat. These differences lead to different ways of living but reflect similar overall affluence.

Similarly, the disproportionality of some impacts suggests that very large changes in environmental damage can sometimes be obtained with small changes in technology or affluence. Recall the Magnesium Corporation of America plant in Rowley, Utah, that we discussed in Chapter 4, which accounts for more than 95 percent of the pollution in its industrial category in the United States. It could be cleaned up or closed down with minimal economic impact to the national economy, but with environmental benefits that affect the whole country's environmental performance. This example sounds a warning we often meet in looking for ways to describe large, complicated systems: be careful of averages and be careful of the way estimates are made. IPAT implies that lowering environmental impact can be done by lowering population or affluence or increasing the efficiency of technology. Technological change has already lowered environmental impact substantially from what it would otherwise have been, and much more can be done. This is a hope often overlooked by both environmentalists and their critics.

TECHNOLOGY AND MAGIC

British writer Arthur C. Clarke observed that "Any sufficiently advanced technology is indistinguishable from magic."[1] We live in a world in which magic has become routine: all of us, nearly all of the time, rely on powerful technologies that we do not understand and usually cannot fix when they go awry. Nearly all of us are happy to surf the Web with a computer whose inner workings are mysterious to us, but we are frustrated when a glitch appears that we do not know how to work around. We seek help from people who are much like sorcerers, who set things right by means we do not understand. This is equally true of medical technology or transportation, of sewers or electricity. And it is also true of *social* technologies, notably bureaucracy and accounting. When they work, we are happy to rely on them. And when they do not, we are powerless to remedy things without expert help.

The growth in human technological capability over the millennia, described in Chapters 6 and 10, is the keystone of our claim to be the most successful species. (For an interesting historical perspective, see Box 14.1: Technological Change and the Judeo-Christian Tradition.) With the rise of transportation technology, the world without edges came into being. The great majority of the environmental

TECHNOLOGICAL CHANGE AND THE JUDEO-CHRISTIAN TRADITION

Naïve faith in progress has played a central role in much of the environmental destructiveness of industrialization. From its beginnings in the Transcendentalist movement discussed in Chapter 2, environmentalism has wrestled with the dream of progress. When should humans respect the integrity of the natural world, and when should we enlarge the scale of human competence to reshape nature? These are questions that are both unavoidable and difficult to answer—and they are linked to cultural and religious beliefs. In 1967, as classical environmentalism was gathering steam, historian Lynn White published an influential and controversial essay, "The Historical Roots of Our Ecologic Crisis," in *Science*, the same journal in which Garrett Hardin laid out his view of the tragedy of the commons the following year.

White observed that one system of belief, the "religions of The Book" (as Judaism, Christianity, and Islam are known), has proved extraordinarily powerful. The Judeo-Christian religions raise profound questions about what nature *is* and enable a transformative change in the way people relate to nature. For this reason, he argued, the Judeo-Christian religions play a fundamental role in the way humans have become the most successful species. If this is the case, environmentalists need to rethink humans' relationship to the natural world in religious terms, as well as in material ones. Aldo Leopold's land ethic is an example of this kind of rethinking. White's analysis has three parts—theology, a concept of history, and science.

The Judeo-Christian theologies, White argued, constitute "the most anthropocentric religion."[1] Genesis, the first book of the Bible, contains this passage describing the sixth and final day of creation: "God said to them, 'Be fruitful and multiply, and fill the earth and subdue it; and have dominion over the fish of the sea and over the birds of the air and over every living thing that moveth upon the earth.'" (Genesis 1:28)

This declares a scale of competence—the competence to subdue. "Man shares . . . God's transcendence of nature" in this theology, White pointed out, and therefore, "Christianity made it possible to exploit nature in a mood of indifference to the feelings of natural objects."[2] By contrast, many cultures, including the Greeks and Romans, made religious observances in particular places, praying to the deities of the place; for example, the Greek god Poseidon (or Neptune, as the Romans called him) ruled the seas.

It is important to bear in mind that the original Hebrew word (*radah*) used in Genesis and translated as "dominion" includes the concept of stewardship—an idea that encompasses environmental responsibility as we understand the term today.[3] Some Christian evangelicals are questioning the mood of indifference that White identified, suggesting that dominion includes responsibility for climate change or other human impacts on landscapes. For example, the book of Leviticus includes this passage, in which God says, "Wherefore ye shall . . . keep mine ordinances and do them; and ye shall dwell in the land in safety. . . . And the land shall not be sold in perpetuity; for the land is mine: for ye are strangers and sojourners with me. And in all the land of your possession ye shall grant a redemption for the land." (Leviticus 25:18–25) Leviticus is one of the five books at the beginning of the Old Testament that form the foundation of the Judeo-Christian tradition.

The second leg of White's analysis focuses on the Judeo-Christian conception of history, the idea of linear time. Creation and the expulsion from the Garden of Eden imply that history is not cyclical, as the seasons are, but linear—that events and human actions can produce irreversible changes in the world. In linear time, progress is possible, though it is of course not assured.

Progress fueled by science is the third element of White's argument. By the Middle Ages in western Europe, science "was becoming the effort to understand God's mind by discovering how his creation operates."[4] The optimism that science could indeed fathom the basic laws of nature gathered force in Gilbert White's time, the period we call the Enlightenment (see Chapter 2). It was an optimism fueled by the physics of Sir Isaac Newton and the sense that humans could direct their own destiny—an idea reflected in the American Revolution and its startling notion that citizens should choose their government.

This faith in science had begun to challenge religion at the beginning of the seventeenth century, during the time of Galileo. That movement gathered force as the world without edges expanded. Optimism about science in turn fed the belief that "scientific knowledge means technological power over nature."[5] This promise took concrete form in the Industrial Revolution. Since 1850, explosive growth in science-based technology has led to fossil-energy machines and manufacturing, a chemical industry capable of turning crude oil into both fertilizer for the lawn and the furniture sitting on the grass, as well as biotechnology and information technology. Science-based technology, White argued, is "the greatest event in human history since the invention of agriculture."[6]

But environmental pressures originating (according to White) in the Judeo-Christian view of the world pose a different kind of challenge: "More science and more technology are not going to get us out of the present ecologic crisis until we find a new religion, or rethink our old one. . . . We shall continue to have a worsening ecologic crisis until we reject the Christian axiom that nature has no reason for existence save to serve man."[7]

What is surprising is that this environmentalist critique of an industrial economy "to serve man" has gotten as far as it has. By 1991, conservative commentator Charles Krauthammer felt impelled to defend the subordination of nature this way: "A sane environmentalism does not . . . ask people to sacrifice in the name of other creatures."[8]

Leopold's land ethic says that humans should find ways to be plain citizens, and thoughtful environmentalists struggle to do that. This includes a religious struggle, a search for the divine in a world that science says is shaped by random forces and impersonal laws. To many, like Krauthammer, the divine is found in human domination. This is an answer that makes environmentalists uncomfortable, but as a description, at least, one cannot disagree: human domination *has* reshaped reality, and in doing so has reshaped our sense of the divine. The discomfort felt by those witnessing this transformation is the seed of the land ethic and the germ of sustainable development.

1. Lynn White Jr., "The Historical Roots of Our Ecologic Crisis," *Science* 155 (March 1967): 1205.

2. White, "The Historical Roots of Our Ecologic Crisis," 1205.

3. Lee Canipe, "Rethinking Dominion in Genesis 1:27–28," *Christian Ethics Today*, 2010, www.christianethicstoday.com/cetart/index.cfm?fuseaction=Articles.main&ArtID=1172.

4. White, "The Historical Roots of Our Ecologic Crisis," 1206.

5. Ibid., 1203.

6. Ibid.

7. Ibid., 1206.

8. Charles Krauthammer, "Saving Nature, but Only for Man," *Time*, June 17, 1991, 82.

impacts from human domination of ecosystems are related to the technologies we use to exploit ecosystems, such as irrigation, automobiles, or cargo ships. In short, technology has reshaped human experience and the institutions that guide human choices, including those with significant environmental consequences.

Over the past generation, significant advances in environmental protection have come from lawyers and scientists, mostly acting within the classical environmentalist scenario. From the outset, businesses reacted by resisting change or adjusting reluctantly in response to legal mandates. But over the past generation, important contributions have come from engineers and designers operating with the mindset of businesspeople rather than bureaucrats. There is much interest in environmental responsibility in the business world. Windmills are being built. Organic produce is a growing segment of the grocery and restaurant industries. Many architects and builders have embraced green design and found it a profitable way to do business. Design, technology, and business are arenas in which some of today's students must make substantial contributions in the coming generation if the grand challenges of sustainability are to be met. Technology and business are channels for influence left in the background by classical environmentalism, but they are vital.

In addition, engineering and business choices are influenced by institutions, so what is needed are large improvements in the governance of commons. Those improvements are as essential to business and technology as they are to consumers and citizens. The world without edges has created a global web of material interdependences: coffee farmers in Indonesia depend on the customers of Starbucks; truck drivers in Mexico are using diesel fuel refined from tar sands in Canada; the soft drink Mountain Dew contains gum arabic from the edge of the Sahara in Africa. In nearly every case, however, the social element of that dependence is limited to the price signals that travel through markets. As we saw in Chapter 13, without the radical simplifications made possible by pricing, the material economy would be unmanageably complex. But this does not mean that the social interdependence can be reduced to price alone. The fact, discussed in Chapter 7, that American drivers finance many nations run by oppressive, nondemocratic governments is a grim example.

Engineers and businesspeople are often blamed for their roles in our unsustainable economy. What consumers and citizens are saying in that criticism is that technology and commerce should do a better job of addressing the wider environmental and social dimensions of the material interdependencies from which they profit. Fair enough. But citizens and consumers, through their participation in government and their purchasing choices, also need to grapple with their responsibilities to provide designers and firms with incentives that can move us all toward sustainability. These are moral burdens of the commons.

RECYCLING AND COMPOSTING

To architect William McDonough and Michael Braungart, critics of the existing industrial economy, a television set is a troublesome artifact.[2] A video monitor contains toxins and nonreusable materials in its circuitry and screen. It is not a set of hazards one would want to put inches away from a child, they note, but that is what we do.[3] The monitor delivers a service we want, although no consumer wants to be exposed to toxic compounds. McDonough's point is similar to something we saw with ecosystem services. We don't want the sewage treatment plant, but we do want the water-cleaning service it provides. Automobiles can now be rented by the hour in cities; their drivers pay for the service of transportation, but they do not need to own the vehicle when they want to drive.

How might we rethink our relationship to technology? McDonough proposes that we think of material artifacts such as video monitors as composed of two components, which should be kept separate:

Material artifacts = **biological nutrients** (deliberately cycled) + **technical assets** (permanently reused and remanufactured)

Biological nutrients include wood, fibers made from cotton or other plants, leather, and other materials that were once living things. These can be put back into the cycle of life through composting. A wooden cabinet housing a television might be handled this way. The nonbiological materials, which McDonough and Braungart call the "technical assets," including the toxic compounds in the electronic circuits, should not be put into a *natural* cycle at all, but retained in a technical economy. They are assets, like a valuable violin or a ring that has been kept in the family for generations—something to be handed down and reused rather than discarded. In Japan, discarded electronic gadgets are being melted down to recover precious metals.[4] The problem that McDonough is identifying is that biological nutrients and technical assets are now intertwined by the way things like cars or running shoes are made, so that sorting out the biological from the technical materials is almost always so costly as to be impractical.

It is also important to bear in mind that artifacts are connected to ecosystems affected by the extraction, use, or disposal of the artifact. Carefully managing the separate streams of biological nutrients and technical assets is important, but even carefully managed systems have impacts. Recycling paper, for example, is a significant way to manage the pressure to harvest forests and to expand landfills, but pollution and logging impacts continue even though they may be reduced. Intensive agriculture and the burning of fossil fuels have put more and more nitrogen compounds into waterways and the air. Although these nitrogen compounds are also found in nature and are biological nutrients, the human additions are overwhelming some ecosystems, as we saw in Box 6.1: Agricultural Ecosystems in Chapter 6, page 132. A more sustainable economy is one that is embedded intelligently within the ecosystems from which it draws, and it does not disrupt them. Keeping biological nutrients separated from technical assets is a useful tool in moving today's economy in more sustainable directions.

Are there practical ways to meet McDonough's challenge? One way is for technical assets not to be owned by individuals at all, as with rental cars. The Interface Corporation, a firm headquartered in Atlanta, Georgia, that supplies carpeting to offices and commercial businesses, does just that. Interface does not sell carpeting but rather sells the service of having a carpet on the floor. When an office is redecorated, Interface takes out the old carpet and installs a new one. If a portion of the carpet wears out because it is in a heavily used pathway, Interface will replace that portion of the carpet. Back at the plant, Interface remanufactures its carpeting, recovering and reusing its materials in an environmentally sophisticated process. The company is profitable, which is strong evidence that the idea of managing technical assets separately can make sense under some circumstances.

In many communities, residents now have three kinds of trash. Glass, paper, metal, and some plastics are recycled. Lawn and yard materials, and sometimes discarded food, are composted or collected to be composted in a central facility. And the rest is garbage, which is put into a landfill. In some cities, such as Seattle, garbage collection fees are set so as to encourage recycling—collection of unsorted trash is priced at a high rate, whereas recycled materials and compost are collected for a low fee or at no charge. The intent is to minimize the volume of material destined for permanent disposal in a landfill. Notice that recyclables include technical assets such as glass and metals, compostable materials are biological nutrients, and the remaining garbage is a stream of waste that is being reduced over time by the way waste collection is priced.

The garbage stream probably cannot be reduced all the way to zero. Pizza boxes usually cannot be recycled because the bits of food remaining attract insects and rodents. Paint is made from durable toxic materials and thus should be kept in the technical economy by renting it to a user and then taking it back, although how this could be done is not clear. One answer may be to make paint out of biological nutrients—thus moving it from the technical to the biological economy—though the preservative properties of paint come in large measure from its toxicity.

REDESIGN

To make it possible to separate biological nutrients from technical assets, products must be designed differently. Instead of simply minimizing manufacturing cost or maximizing performance, artifacts from clothing to phones to the packaging for hamburgers all need to be made so that their parts can be kept in two separate cycles of the industrial ecosystem. This can be done, sometimes, by making an entire product compostable, as is now done with some food packaging and clothing. Or it can be done by designing the product so that it is possible to disassemble the various technical assets easily. The European Union has adopted a policy to regulate automobiles that are being discarded, a step meant to encourage manufacturers to adopt designs that will lower the cost of recovering and recycling technical assets.[5] But thinking about manufactured products this way is still an unfamiliar idea, and thousands of products must be redesigned. This is fertile ground for students with an interest in art, engineering, architecture, and business.

Although Interface Corporation's business model shows that an ecological perspective can work in some instances, there is a long way to go. One hope is to rearrange the industrial economy so that everything that comes out of a factory can be sold profitably, including what is now considered waste. This is what two thoughtful engineers, Thomas Graedel and Brad Allenby, have called an *industrial ecology*.[6] They think of the material economy as a kind of composting, in which wastes are moved around so that they become valuable materials again.

To be sure, getting someone to buy one's wastes is not easy. Think of the person who collects the dirty dishes in the dining halls. How would you create a situation in which that person would pay you? (Your tuition and fees enable that person to be paid for taking the dirty dishes from you without charging you on the spot.)

What this view implies for engineering and design is discussed in Box 14.2: Industrial Ecology, page 384. Notice that changes in *both* technology and institutions are needed to move the material economy toward recognizing that biological nutrients and technical assets should be kept separate. Things need to be designed and made differently, and the way that people take responsibility for them needs to change at each stage, from manufacturing through use to disposal. The dream of industrial ecology is to make it profitable for businesses to do this by creating a market for every molecule. This is an objective that lies largely beyond the reach of classical environmentalism.

Think about what design means in this context. This sort of design is not primarily aesthetic; rather, it is a functional pursuit of trying to situate a technology within an industrial economy and a consuming society. This view of design is ecological in spirit, an attempt to translate, into the language of economics and engineering, Aldo Leopold's idea that a thing is right when it tends to preserve the integrity, stability, and beauty of the biotic community. A thing is right, in this context, when technical assets can easily by separated from biological nutrients, when technical assets are carefully reused, and when biological nutrients are integrated back into ecological cycles in a way that does not disrupt them.

Design can make a large difference in at least three different ways. First, major improvements can be made in both the products being manufactured and in the processes by which they are produced. Production processes tend to require expensive equipment such as oil-well drilling platforms, machinery, or fleets of trucks. Innovations in processes tend to be slower to take hold, and they are powerfully driven by the need to keep reducing costs, which is a constant theme of business management. Finding markets for materials that are by-products or wastes of a manufacturing process can improve the profitability of a business—and in this way induce technological changes that move toward sustainability.

Second, design can gain insights from **life-cycle analysis**—thinking about the impact of a product from the time its components are harvested from nature until it is used up or discarded. Some products, such as clothing, are meant to be

BOX 14.2

INDUSTRIAL ECOLOGY

How does a firm implement the radical changes of industrial ecology? A simple, ingenious, and practical way is by using the target diagram invented by Thomas Gradel and Brad Allenby. They begin with the matrix shown below. The idea is to rely on knowledge already available to those on the manufacturing and marketing teams of a business, for whom complying with environmental regulations has long been normal business practice. Instead of detailed technical studies, a simple 5-point ranking system is used: 0 means that an element of the matrix is being done by the firm as sustainably as anyone knows how to do it currently; 4 means that this is the only company left that does this process so inefficiently or wastefully; and rankings 1 through 3 are for intermediate levels of environmental competence.

These judgment are assembled systematically by putting all of the phases of production and use in the rows of the matrix, and taking account of the different kinds of environmental impact in the columns.

	Materials	Energy	Solid residues	Liquid residues	Gaseous residues
Pre-manufacture					
Production					
Packaging and delivery					
Use					
Disposal and recycle					

SOURCE: T. E. GRAEDEL AND B. R. ALLENBY.

Matrix to organize discussion of industrial ecology for a product.

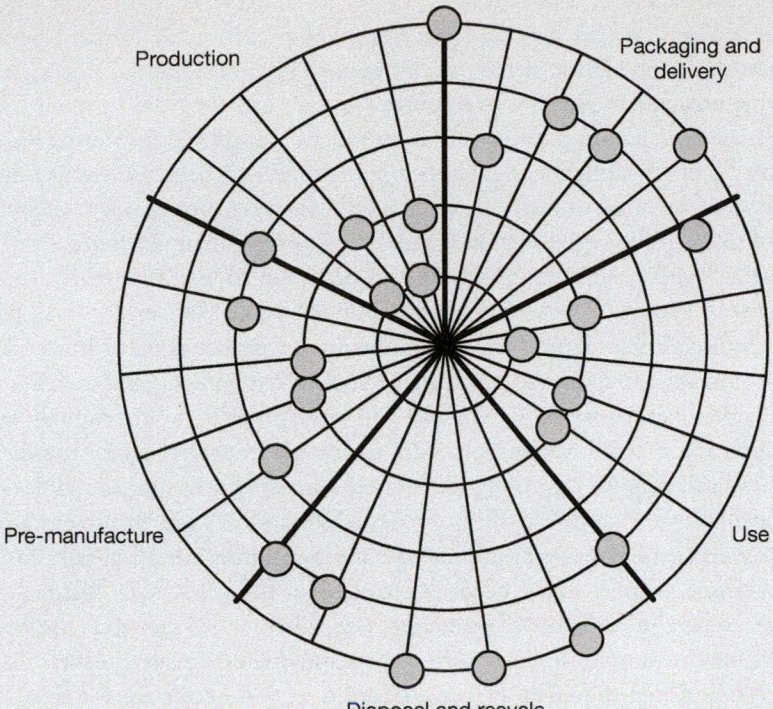

Production

Packaging and
delivery

Pre-manufacture

Use

Disposal and recycle

SOURCE: T. E. GRAEDEL AND B. R. ALLENBY.

Target diagram for industrial ecology.

The scores in the matrix grid can then be plotted on the target-shaped diagram. Scores of 0 are at the center of the target, 4s are dots on the outer ring, and the spokes of the wheel are the twenty-five cells of the matrix, organized by its rows. The firm can identify actions and priorities by seeing which points on the matrix are farthest from the target's center bull's-eye. Some of these points may be difficult to move, others easy. There maybe disputes over which actions to take, but these arguments are the seeds of change.

Consulting firms and academics have come up with much more elaborate methods for gathering and analyzing data to improve the sustainability of business today. But, as with the IPAT formula, the target diagram provides a useful way to see how to break down the intimidating complexity of the material economy into something tractable for debate and change.

discarded quickly, so that making an evening gown compostable might be more important than using fabric made out of organically grown fibers. Others, such as cars and appliances, require more resources in use than they did in manufacture. Walmart, which sells tens of thousands of products, is gradually instituting requirements on all of its suppliers to identify opportunities to decrease environmental impacts up to the point of sale. This is not a full life-cycle approach, but the market power of the world's largest retailer is already propelling many changes.

Third, designers can move products and processes toward sustainable supplies by using solar energy to power machines and electronic devices, for example, or by substituting biological products such as bamboo for plastics made from oil. The general principle is to design so that a technological system is aligned with available ecosystem services. Some civil engineers and governments are increasingly trying to conserve and nourish wetlands in order to complement the levees, seawalls, and other measures designed to protect human settlements against storms and flooding.

Some of these changes in the way firms do business are profitable already. Increasing energy efficiency in production and transportation often falls into this category. Some changes could become profitable if the rules were changed. Putting a price on the emission of pollutants (see Chapter 13) can shift the behavior of businesses through policies such as cap-and-trade. However, other changes will be costly. When the environmental benefits of these costs are worthwhile to society as a whole, the goal of governance should be to impose those costs fairly. Effective enforcement of rules is essential to forestall a tragedy of the commons. Technological improvements fit within the logic of governance we developed in Chapter 3. When they do so, shifts in technology and design can enable rapid and widespread progress.

Within the idea of renting the services of technical assets, one can hear an echo of the American Indian concept of property discussed in Chapter 2. A video monitor should not be private property, but a kind of window to the commons of the Internet or network broadcasting, a service that one can pay a fee to make use of, as we now pay for access to cable television. And the emphasis on services also reflects skepticism about material goods for their own sake. It is not the device but the experience that matters.

DEMATERIALIZATION AND DECARBONIZATION—CAN IT WORK?

Recycling of materials and energy efficiency have a long way to go. Figure 7.13 shows that a large reduction of greenhouse gas emissions is already economically advantageous to businesses and consumers. Many of these savings have been known for a long time but have been difficult to obtain. The U.S. Environmental Protection Agency reported that in 2008, Americans recycled about one-third of the 250 million tons of total waste generated.[7] Here, too, practice is far from the levels that are theoretically sensible from an economic point of view.

Yet, if one takes a longer view, hopeful signs can be found in the history of industrialism. Consider paper and wood products. Wood was once a primary fuel for heating and industrial use. Paper and paper-based packaging are still ubiquitous. But if one looks at the statistical trends, an interesting phenomenon can be seen. It is called dematerialization by students of industrial ecology. In one study, a group at Rockefeller University in New York reported that, between 1900 and 1993, population and affluence grew sixteenfold, yet the consumption of wood products increased by only 70 percent.[8] This is, as you will recognize, disproportionality: although two of the IPAT variables, population and affluence, grew, the impact, measured as consumption of wood, did not grow nearly as much. What happened?

Technological change brought greater efficiencies and led people to substitute other materials. Fossil fuels displaced wood for heating, which was not good news for the climate, but it did decrease pressure on forests. Paper use slowed, until by the 1990s it began to fall in absolute terms—somewhat similar to the way the demographic transition's slowing growth rate is leading to absolute declines in the populations of some nations in the next several decades. And the efficiency of forestry and processing increased. These shifts contributed to the long-term trend toward reforestation in North America.

Notice that dematerialization in a time of rapid economic and population growth does not necessarily lower human domination of landscapes. But the decoupling of human well-being from material impacts on the natural world is discernible. The contrast between a two-thirds increase in wood demand and a sixteenfold increase in population and affluence suggests that technological change over the course of the past century made a difference. It suggests that the hurdles of sustainability can be surmounted.

A similar, though not so optimistic, story can be told about energy and climate. Here, the process is called decarbonization, and it describes a change in the amount of carbon dioxide being emitted. As industrialization spreads around the world and grows nearly everywhere, the total quantity of carbon dioxide emitted is growing. Yet analysts have found an interesting pattern hidden within the overall growth (Fig. 14.1, page 388).

Between the mid-nineteenth century and today, the composition of the fuels used in the world economy has shifted. In 1850, wood and coal provided the major sources of fuel. They have largely been overtaken and displaced by oil and gas. In the latter part of the twentieth century, hydroelectric power from dams, nuclear power, and smaller amounts of solar and wind power came to play roles, too. This is

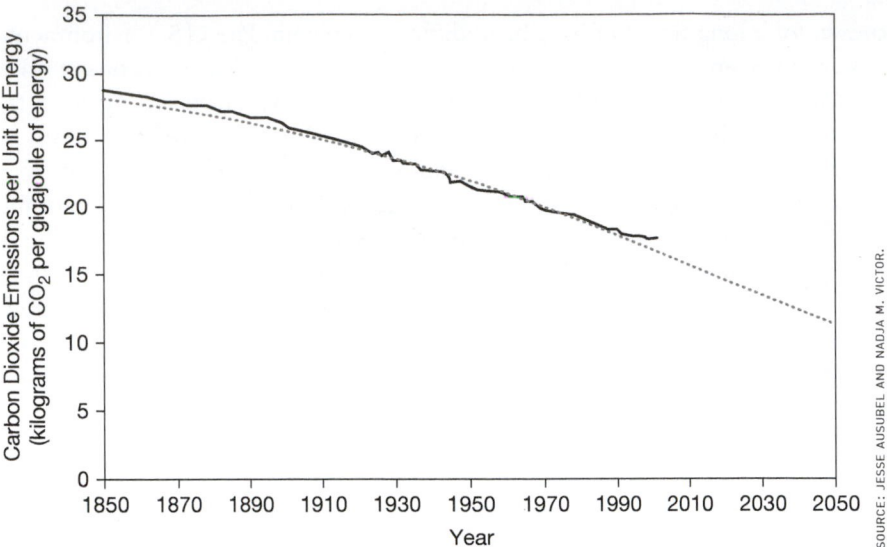

FIGURE 14.1
Decarbonization, 1850–2050.

SOURCE: JESSE AUSUBEL AND NADJA M. VICTOR.

how the carbon dioxide emissions per unit of energy declined by about 40 percent over a century and a half.

Of course, as the carbon per unit of energy declined slowly, the total amount of energy climbed very rapidly (Fig. 14.2). Thus, greenhouse gas emissions rose rapidly, and decarbonization is a slow and small trend hidden within the exponential growth in climate-changing emissions.

A further, hopeful point, however, is the idea of **energy intensity**: the amount of energy needed to produce a dollar of GDP. Within a generation, the energy intensity of the American economy fell by nearly 50 percent (Fig. 14.3). This means that energy use could stay roughly stable even as the economy doubled in size. As noted for paper use, this pattern is like the slowing growth rate of population, even as human numbers continue to grow. Since the oil embargoes of the 1970s raised oil prices substantially, one can see a long-term trend suggesting that continued steady progress is already underway, even though there has been little change in public policy.

The decarbonization curve in Figure 14.1 is slow compared with the changes needed to deal with climate. But computers have gained in speed much faster than this in a couple of decades. So feasibility is an open question. Is there a lesson here? One can imagine a world in which many people can telecommute to work, providing high-value services with little expenditure of resources, because what they are doing is moving information. In fact, this is already beginning to happen, as data-entry tasks in financial services are moved to India and as corporations staff

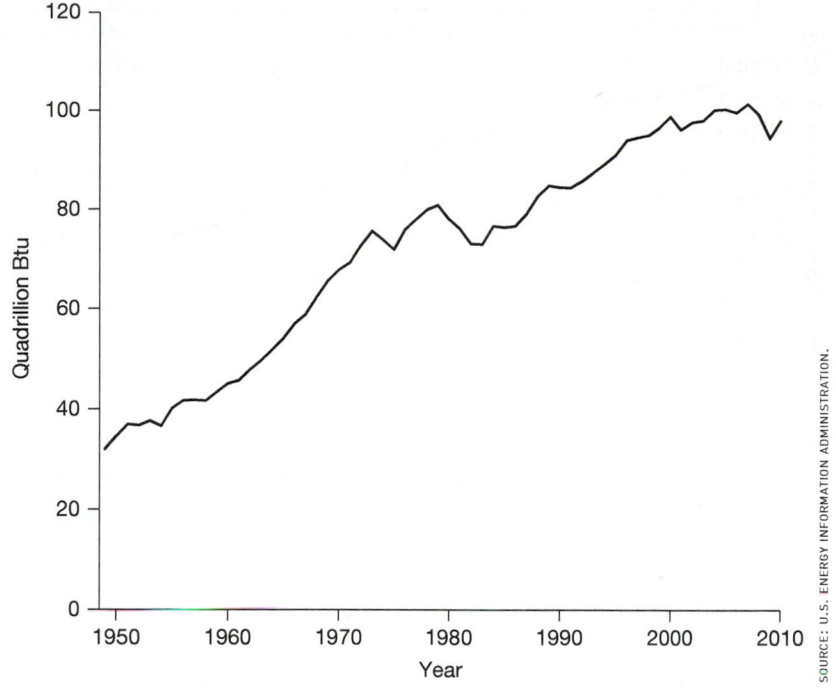

FIGURE 14.2
U.S. energy
consumption,
1949–2010.

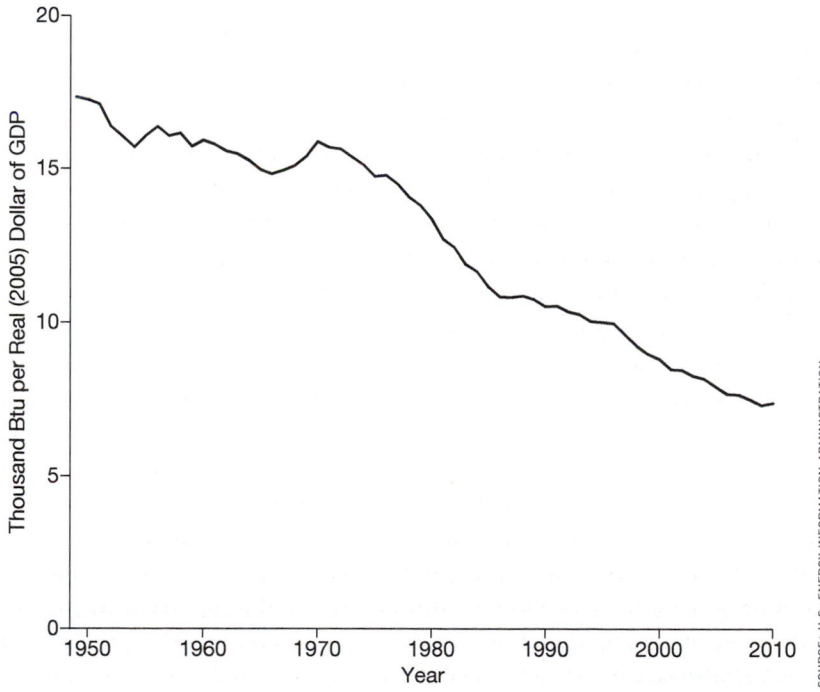

FIGURE 14.3
Energy intensity in
the U.S. economy,
1949–2010.

their customer help desks from offices in the Philippines. In both of these cases, the availability of a large, educated population that speaks English is crucial. These developments also move jobs out of the United States, putting pressure on U.S. workers to upgrade their skills or accept lower wages. In short, we are substituting information for energy to a degree, but the social changes that lie ahead are likely to lead to conflict and dislocation.

The example of fossil fuels also illuminates something important about energy. Why can't energy be recycled? Because the service we want from energy is the ability to do physical work, such as energy to turn motors or to drive the logical engine of a computer. A fundamental law of physics (the second law of thermodynamics) says that doing work creates heat, which is disordered energy. Compare heat to electric current, a highly ordered form of energy. Once electric current is converted into heat, as it is in a lightbulb, one still has energy but one cannot do as much work with it; in particular, one cannot use the heat to light another bulb. So energy is a flow that goes one way: from order to disorder. The fact that energy cannot be recycled, but that large-scale use of energy is central to industrial production, means that climate change poses a deep challenge. Energy use is fundamental to the industrial path to development; thus, cheap energy is essential to a nation's ability to meet the needs of the present. So far, this has meant fossil fuels.

"WE HAVE MET THE ENEMY . . ."

Consumption, the manifestation of affluence in the IPAT equation, is a major driving force in every economy. Even the most self-reliant Americans are drawn into the consumer economy. Here, we take up a question that has troubled every environmentalist: How much is enough and how do I know what I want? To most economists, the answer is simple: people are assumed to be better off if they have more, so they always want more, even though the intensity of the want diminishes as they get more. This does not feel right to many, but looking for alternative responses carries us in surprising directions—to economics, to philosophical reflections, and to innovations slowly making their way into a society where responsible consumption is gaining a foothold.

Consumption is a complicated subject, and it is one that has not been formulated into a tidy conceptual discussion the way that environmental politics could be summarized in Chapter 12. We will look at several different perspectives: how consumption is linked to environmental impact; the questions of need and frugality; and the complicated relationships among consumption, social justice, and political ideology.

The cartoon in Figure 14.4 appeared in 1971, when environmentalism was gaining a secure place in American society. It articulated a theme that classical environmentalism had put into the background: guilt. Whereas environmental policy activists had focused on industrial pollution and hazardous products such as pesticides in suburban gardens, citizen environmentalists doing recycling and energy conservation realized that consumers were part of environmental despoliation—a big part.

The idea that affluence produces a surfeit of stuff, and that we are all responsible because we all have and buy too much stuff—that we consume too much— became so deeply entwined with environmentalism in the popular mind that a

FIGURE 14.4
Walt Kelly, "We have met the enemy and he is us," *Pogo*, 1971.

reflexive sense of guilt is still the standard reaction when a conversation turns to the environment. When we look at the wolf we have by the ears, we see ourselves. But the success of cell phones shows that a sense of guilt does not halt growing consumption. Restraint can be overrun by the social and psychological forces set in motion by marketing and effective technology.

Moreover, as Methodist minister James Nash has emphasized, consumption is of fundamental importance to the U.S. economy.[9] More than two-thirds of the GDP is made up of purchases by consumers. This means frugality, once thought to be a virtue, also has an economic drawback: if people are frugal and refrain from consuming, this decreases the total activity of the American economy, unless that decrease is offset by new investment or government expenditure. Put another way, business firms use advertising to erode frugality, luring people in a multitude of ways to spend their money. One indication that businesses have succeeded is seen in Figure 14.5, which plots the fraction of total income that is saved. From the 1970s until the recession of 2008–11, the long-term trend was a steady decline.

The dilemma here is that a lower savings rate is dangerous for individuals as well as for the wider economy (because savings provide the investment capital essential to prosperity), yet lower consumption causes people to lose their jobs in the short run. Consumers buy cars and soap and haircuts. Consumer spending leads businesses to spend money to produce goods and services for sale to consumers. All of this accounts for about seven cents of every dime spent in the American economy. (The rest is investment and taxes. Investment includes building a new shopping center or planting a forest for eventual timber harvest. Taxes are required, unlike consumption. Tax payments are not the result of choices

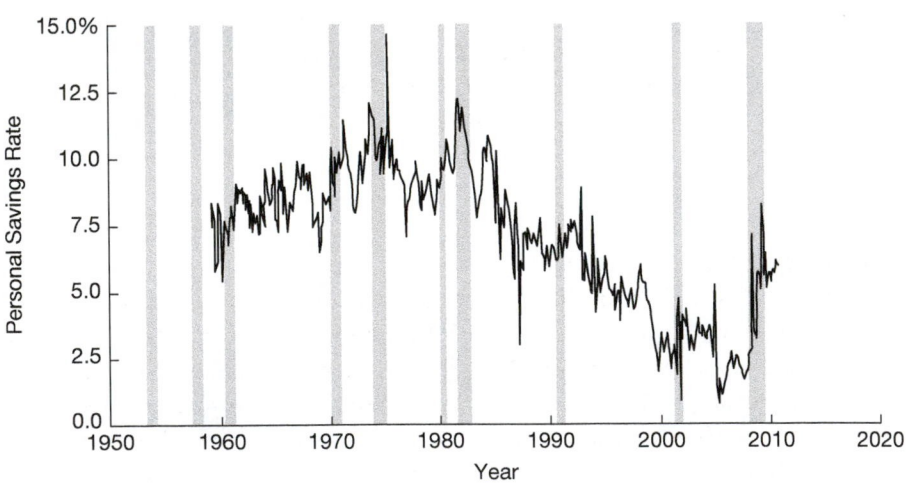

FIGURE 14.5
Personal savings rate in the United States, 1959–2010. (The shaded areas indicate periods of recession.)

SOURCE: FEDERAL RESERVE ECONOMIC DATA.

made by individual consumers.) Here again, we have the wolf by the ears: we rely on consumption as an engine of the economy, but consumption also drives unsustainable pressures.

NONMATERIAL NEEDS

How can we safely deal with the wolf of consumption? In the last three sections of this chapter, three pathways are described: thinking about the needs we meet through consumption, turning consumers toward responsibly produced products, and shifting the profile of businesses' investment toward greater sustainability.

The first approach is to consider the nonmaterial needs we meet through material consumption. Humans need food, but the desire for a meal in an upscale restaurant is a different matter: few restaurant diners suffer from malnourishment. What this example shows is that some of our material consumption satisfies nonmaterial wants, such as the cultivation of taste for fine food, or making a favorable impression on one's dinner guests.

It occurs to all of us that we might choose to make a charitable donation instead of going out to dinner. In some families, holiday gifts are partially converted to such ends. Instead of showing love for a relative by giving her something material, we can make a contribution to a cause that she considers important. How far can one take this idea? And, more important, how far can one expect the consumers of Brazil or the United States to embrace this practice and similar ones?

It is important, first, to realize that we have nonmaterials needs, including affection, justice, and the sense that one is leading a worthwhile life. Our ability to meet those needs is entangled with material consumption. This is a cultural matter—something that one can affect through individual choice but that one cannot escape, because other people's expectations are involved. If your relative is insulted when she finds that you have given her a year's membership to her public radio station, then your nonmaterial desire to show your affection will have been frustrated. (Box 14.3: Consumerism, page 394, considers the question of nonmaterial needs from a provocative direction—asking whether the American emphasis on consumption may have avoided or blunted the social conflicts inherent in the rapid changes of the past century.)

Environmental writer Bill McKibben considered these matters in an essay on Henry David Thoreau, the Transcendentalist author (see Chapter 2). Reflecting on the challenge of global warming, McKibben wrote that Thoreau "posed the two intensely practical questions that must come to dominate this age . . . : How much is enough? And, How do I know what I want?"[10] These are difficult questions to grapple with in an affluent society. "In the hundred and fifty years since *Walden*,

BOX 14.3

CONSUMERISM

In a study entitled *An All-Consuming Century*, the historian Gary Cross provided a provocative interpretation of the twentieth century. It was time of rapid social change—in race relations, urbanization, and the status of women—and saw enormous economic growth, the emergence of America as a world superpower, and the rise in environmental and consumer consciousness. Cross argued that none of the ideologies of the twentieth century—conservatism, liberalism, socialism, or libertarianism—could account for the ability of American society to handle these huge changes relatively peaceably. Instead, he suggested that the ideology that held the country together was consumerism: the right to buy with the income one earned, and to define oneself by what one bought. Such an ideology obviously directed people's energy toward participating in the economy rather than rebelling against it. The triumph of consumerism has significant implications for the environmental pressures exerted, as you can see in the 35 pounds of waste discussed in Chapter 4.

Consumerism redefined democracy, creating . . . opportunities for participation that transcended suffrage rights or political ideologies. A vision of a world of goods . . . in large part replaced the old idea of a republic of producers and challenged class, religion, and ethnicity as principles of political solidarity. Consumerism . . . reflected real social needs and . . . often fulfilled these needs with less conflict than did other, more substantial forms of social solidarity.

In the context of consumerism, liberty is not an abstract right to participate in public discourse. . . . It means expressing oneself and realizing personal pleasure in and through goods. In other ways, however, consumerism has been a threat to the kind of individual responsibilities and social solidarities that made political democracy work in the past. The fixation on personal goods has denied the necessity of sacrifice beyond the family. It has allowed little space for social conscience and confined aspiration to the personal realm.

When people display themselves through their goods, they are not required to reveal fragile egos and awkward manners. . . . Consumerism does not demand self-denial, . . . and it allows people to distinguish themselves without denying the rights or existence of others.

. . . Consumer culture is democracy's highest achievement, giving meaning and dignity to people when workplace participation, ethnic solidarity, and even representative democracy have failed. Of course, consumerism has done this without challenging manipulative power and inherited money.[1]

1. Gary Cross, *An All-Consuming Century: Why Commercialism Won in Modern America* (New York: Columbia University Press, 2000).

Thoreau has become ever more celebrated in theory, and ever more ignored in practice."[11] The reason for this, McKibben argued, is that Thoreau has become less a guide for how to live and more a figure of nostalgia, someone whose flinty frugality we wish for, but not very much. Our way of life has abandoned frugality, but like Pogo in Figure 14.4, we feel uneasy about this.

In a famous passage in *Walden*, Thoreau urged his reader to "Simplify! Simplify!"and do without the material comforts that he found nonessential.[12] By the early twenty-first century, Walden Pond had a gift shop next to it where the mugs shown in Figure 14.6 were being sold. The most nonmaterial of desires had become an item of consumption.

The question to face, as individuals and as a society, is how to satisfy our needs in nonmaterial ways. More practically, how do we shift our pursuit of nonmaterial wants into more nonmaterial forms. The rise of nongovernmental organizations

FIGURE 14.6
Mugs are sold at the Walden Pond State Reservation in Concord, Massachusetts exhorting one to "Simplify, simplify." A material way to fulfill a nonmaterial desire turns out to be declaring oneself in favor of . . . Thoreau's antimaterialism!

described in Chapter 12 is an indication that this is happening: people are putting more of their time and resources into community and charitable purposes; strengthening commons with important nonmaterial dimensions, such as museums or help for the disadvantaged; or the environment.

There is much farther to go, but one can begin by asking what one's needs are and how they might be satisfied in nonmaterial ways. Dining at a restaurant that serves locally grown produce may be a useful exploration to undertake, for example.

RESPONSIBLE CONSUMPTION

Is there such a thing as responsible consumption? Perhaps the most visible example is organic food, which is grown on land that has been certified free of pesticides and is grown using methods aimed at nurturing and rebuilding the productivity of the soil ecosystem. Organic food has been a rapidly growing segment of the grocery business, and major chain supermarkets now carry organic fruits and vegetables as a matter of course.

Other kinds of products certified as environmentally responsible have emerged. Lumber from sustainably harvested trees began to be certified by the Forest Stewardship Council (FSC) in the mid-1990s. The FSC is a nongovernmental organization governed by an assembly that seeks to represent three interests: environmentalists, forest communities, and the forest products industry. Perhaps it is surprising that, despite their many disagreements, these groups find enough common ground to have made a significant impact on forests and in the marketplace.

The major furniture retailer Ikea uses FSC-certified wood in its products. Timber companies not certified by the FSC worried that FSC wood would gain significant market share, so they developed processes to sustainably harvest wood. In the United States, the Sustainable Forestry Initiative (SFI) coordinates these efforts. One might expect that the industry-sponsored standards would hinder progress, but the resulting competition has led to an impressive adoption of responsible forestry practices among the firms using the SFI label. This does not mean that a large fraction of forests worldwide are being managed well. Institutional controls are weak in many countries, particularly in the tropics where rain forest is found. Yet in Bolivia, the FSC standards for timber harvesting have in effect become the national forest policy, providing both environmental and social benefits in the Bolivian corner of the Amazon.

Sustainably caught seafood is also a rapidly growing segment of the market. This is largely due to the global efforts of the Marine Stewardship Council (MSC). Like the FSC, the MSC is organized around multiple stakeholder groups. The

MSC certifies fisheries rather than specific harvesters (as is done with organic farmers). About one-eighth of the seafood traded in global commerce is now from MSC-certified fisheries. This makes a significant contribution to the battle against overfishing, but the struggle is far from won. MSC-certified fisheries, like organic farmers, tend to be the well-managed ones, and it seems unlikely that a voluntary approach such as certification can control those guilty of overfishing. Until that is done, the commons remains at risk.

Fair-trade and shade-grown coffee, as well as buildings built according to plans approved by the U.S. Green Buildings Council's Leadership in Energy and Environmental Design (LEED) program, provide other examples of products that have gone through an approval process to assure sustainable production. Seeing the advantages of marketing their products or services as green, many businesses now make environmental claims. However, many of these claims are exaggerated or undocumented, which is a problem for those, like the MSC and FSC, that certify their products in a rigorous fashion. Environmentalists call the exaggerated claims "greenwashing." Whatever it is called, this is another commons problem: specious claims undermine all claims. The search is on for an institutional way to manage this informational commons.

These certification approaches all rely on consumer pressure: encouraging people to make purchasing decisions based on whether a product is labeled as sustainably produced. The Monterey Bay Aquarium in California pioneered the Seafood Watch buying guide that rates the sustainability of fish using a three-color scale (Fig. 14.7). Red (Avoid) means that people should not purchase these severely overharvested species (the bluefin tuna is on this list). Green (Best Choices) means that the fish species are well managed and not overharvested (such as farm-grown

FIGURE 14.7
Seafood Watch is a pocket-size card designed to guide consumer choices in restaurants and fish markets. (The asterisk indicates fish that should be limited due to concerns about mercury or other contaminants.)

SOURCE: MONTEREY BAY AQUARIUM.

MONTEREY BAY AQUARIUM

Seafood WATCH®

RAINBOW TROUT

Central US
Sustainable
Seafood Guide
January 2012

BEST CHOICES	GOOD ALTERNATIVES	AVOID
Arctic Char (farmed)	Basa/Pangasius/Swai (farmed)	Caviar, Sturgeon* (imported wild)
Barramundi (US farmed)	Caviar, Sturgeon (US farmed)	Chilean Seabass/Toothfish*
Catfish (US farmed)	Clams, Oysters (wild)	Cod: Atlantic (Canada & US)
Clams, Mussels, Oysters (farmed)	Cod: Pacific (US trawled)	Crab: King (imported)
Cod: Pacific (US non-trawled)	Crab: Blue*, King (US), Snow	Flounders, Halibut, Soles (US Atlantic,
Crab: Dungeness, Stone	Flounders, Soles (Pacific)	except Summer Flounder)
Halibut: Pacific (US)	Flounder: Summer (US Atlantic)	Groupers (US Atlantic)*
Lobster: California Spiny (US)	Grouper: Black, Red (US Gulf of Mexico)*	Lobster: Spiny (Brazil)
Perch: Yellow (Lake Erie)*	Herring: Atlantic, Lake	Mahi Mahi (imported longline)
Sablefish/Black Cod (Alaska & Canada)	Lobster: American/Maine	Monkfish
Salmon (Alaska wild)	Mahi Mahi (US)	Orange Roughy*
Sardines: Pacific (US)	Perch: Yellow (Lake Ontario & Huron)	Salmon (farmed, including Atlantic)*
Scallops (farmed)	Pollock: Alaska (US)	Sharks*
Shrimp: Pink (Oregon)	Salmon (CA, OR, WA*, wild)	Shrimp (imported)
Striped Bass (farmed & wild*)	Scallops (wild)	Snapper: Red
Tilapia (US farmed)	Shrimp (US, Canada)	Swordfish (imported)*
Trout: Rainbow (US farmed)	Smelt: Rainbow	Tilapia (Asia farmed)
Tuna: Albacore (Canada & US Pacific,	Swordfish (US)*	Trout: Lake (Lake Michigan)*
troll/pole)	Tilapia (Central & South America farmed)	Tuna: Albacore*, Bigeye*, Skipjack,
Tuna: Skipjack, Yellowfin (US troll/pole)	Trout: Lake (Lake Huron & Superior)*	Tongol, Yellowfin* (except troll/pole)
Whitefish: Lake (Lake Huron & Superior)*	Tuna: Bigeye, Tongol, yellowfin (troll/pole)	Tuna: Bluefin*
Whitefish: Lake (Lake Michigan, trap-net)*	Walleye*	Tuna: Canned (except troll/pole)
	Whitefish: Lake (Lake Michigan, gillnet)*	

tilapia from the United States). Species on the yellow list (Good Alternatives), such as lake whitefish, fall in between.

Although sustainable harvesting began as a movement targeted at consumers, retailers and manufacturers are increasingly important to its success. In 2008, Walmart committed to selling only seafood certified by the MSC. This decision reverberated through seafood markets and fisheries around the world. It was a smart business move for Walmart as it worked to build a reputation for environmental responsibility. In 2009, Mars, one of the largest candy makers in the world, committed to purchasing only sustainably harvested cocoa for its chocolate within a decade. This is a major step because cocoa is grown in tropical countries, where management and governmental structures to assure environmentally responsible agriculture have been frail in the past. Mars changed the game for its competitors. Note that these firms are making these decisions because they believe it provides a competitive advantage—that moving toward sustainably produced goods will enable them to sell more or more effectively. In this way, greater consumption of some products may induce more sustainable management and production. If it works, this is a way to tame the wolf we have by the ears. As the continued decline in tropical forests and fish stocks shows, however, it is far from clear whether responsible consumption can be scaled up enough to underwrite sustainable development.

How might environmentalists further enlarge the scale of consumers' competence? There are two useful questions to ask. The first is, Which responsible practices can I help to make as common as recycling? Recycling is not widespread enough: biological nutrients are still mixed with technical assets, and the volume of garbage has not decreased. But recycling is probably the most environmentally responsible discretionary activity that people have integrated into their habits. Finding other ways to build habits that are more sustainable is an essential task. The popularity of farmers markets in many cities as a way to sell organic produce may be an example of the way that sustainable production can be woven into a new form of social capital, in this case the local-food movement, whose adherents call themselves "locovores."

The second question is, How can I help to create *institutional* arrangements such as the hybrid car, which help consumers make responsible choices *without* having to think about their choices at every moment? Because hybrids operate just like traditional gasoline-powered cars, one does not need to learn how to drive all over again. If new technologies such as all-electric vehicles become popular, people will need to adjust their behavior as they get into the habit of plugging in their vehicles and watching for pedestrians who cannot hear their quiet cars approach. The hope is that consumers will find these adjustments tolerable. People already make such changes, as when they switch to using mass transit instead of driving their cars to and from work. Institutional arrangements combine technology and the reorganization of human behavior. We are all so busy today that we need habits and institutions to change if we are to enlarge the scale of our competence for sustainability.

ENVIRONMENTALLY RESPONSIBLE BUSINESS

William McNeill said the market was planted like a chronic infection in imperial bureaucracies. The fever of markets has now become the normal state. Don Gifford described the world without edges as a web of interdependence that gives us a sense of independence (Chapter 3). These are products of the routine operation of markets. When they work, we are able to achieve the independence we value. And when markets fail abruptly, for example after a hurricane, government steps in temporarily, but replacing the smooth coordination of the market is usually not possible. In this way, the fever of the market has become an elevated body temperature in market economies—analogous, in a way, to the way warm-blooded mammals gained physiological advantages and became such a successful group of species, out-competing the cold-blooded ones.

At first it sounds impossible to create environmentally responsible businesses. But a lot is happening already, for four reasons. First, businesses want to avoid problems. Remember, the government makes rules that firms must deal with, and firms and governments both face public criticism and pressures. An example of a nonregulatory government requirement is the Toxics Release Inventory administered by the Environmental Protection Agency. Companies that routinely use toxic substances in their production processes must disclose their annual releases of specific toxins. The requirement is to disclose, not to decrease. Yet the availability of the information has spurred significant reductions by companies. Moving toward sustainability avoids trouble with the public.

Second, businesses can sometimes strengthen their competitive position by acting in an environmentally responsible fashion. Some buyers care about whether fruit is contaminated with pesticides. Home Depot carries FSC-certified lumber. Some buyers want it, and others think better of the retailer when they see plywood with the FSC stamp on the shelves.

Third, wastes can sometimes be sold or managed profitably. A large share of the aluminum produced in the United States is now made from recycled metal. It is a lot less costly to produce than aluminum from ore. Aluminum molecules have a market, and it is profitable to reuse them rather than to discard them.

Fourth, businesses compete for capital, and environmentally responsible practices attract some investors who demand that firms act in an environmentally responsible way. Although one would expect investors to be interested only in maximizing near-term profit, a small field of social investing has begun to take root.

Social investors have as yet little direct economic power through ownership, but they have exerted considerable influence through organizing shareholder campaigns. Most economically important businesses are publicly traded, with large numbers of shareholders. Once a year, the board of directors, who represent the shareholders, are required to meet in public. Among their tasks is the consideration of shareholder resolutions—petitions from some group of owners of the company. Environmentalists and social activists have used these resolutions to urge firms to change their behavior on a wide range of matters, from use of toxic materials in manufacturing to mining practices in foreign countries to labor practices that expose workers to high risk. Although few resolutions gain a majority vote among all the owners of a company, even a showing as low as 10 percent often persuades management to work with activists to make changes and avoid the adverse publicity of further resolutions in the future. Universities are major investors in the financial markets and hold many stocks in their endowments. Some schools now have committees to advise their trustees about how to vote on shareholder resolutions. In these ways, an institutional mechanism borrowed from democratic politics influences economic decision making. Such actions can move some firms in the direction of sustainability, and large corporations have in some cases embraced environmental responsibility seriously, although much remains to be done.

Consumption is significant in both the environmental impact of human activities and the human economy. Consumption is linked to broader social questions. Consumption is tied to social values and institutions: social justice, management of the economy, the rights of marketers to advertise, and social integration of a diverse population. For all these reasons, consumption cannot be easily managed in a market economy, but neither can it be ignored as a powerful factor in environmental studies and sustainable development.

People are continuing to work toward economic growth all over the world. How should we spend that higher GDP per capita? Greater affluence can go into increased consumption or increased investment, and individuals generally pursue both paths in tandem.

Today, as in the past, consumption is high—it accounts for more than two-thirds of the U.S. economy. Investments directed toward environmentally sound alternatives for providing well-being and better ways of meeting human needs are critical to a sustainable future. And technology is central in all three aspects of investment: investment in human resources, which increases the economically relevant skills of the workforce; investment in knowledge itself, through research; and investment in more effective or efficient technologies. As a consequence, an environmental education must be a technologically aware education. The search for sustainable development lies, to a significant degree, with those whose talents might lie in technology, design, and marketing. There are now environmentally responsible ways to pursue those careers, a recent development that should make environmentalists take heart.

FURTHER READING

Cross, Gary. *An All-Consuming Century: Why Commercialism Won in Modern America.* New York: Columbia University Press, 2000.

McDonough, William, and Michael Braungart. *Cradle to Cradle: Remaking the Way We Make Things.* New York: North Point Press, 2002.

Nash, James A. "On the Subversive Virtue: Frugality." Chap. 22 in *Ethics of Consumption: The Good Life, Justice, and Global Stewardship.* Eds. David A. Crocker and Toby Linden, 416–36. New York: Rowman & Littlefield, 1998.

KEY TERMS

accounting identity	energy intensity	life-cycle analysis

Chapter Fifteen

LEARNING

WHERE WE HAVE TRAVELED, HOW WE NEED TO PRESS ON

The world without edges is a daunting place. In this final chapter, we reflect on some of the paths we have explored and some of the things we have seen. Our purpose is not to draw conclusions. Instead, it is to emphasize the importance of keeping conclusions provisional—of pressing on with the search for an environmentally responsible way of living with an open mind and a willingness to learn. Learning is essential if we are to contribute constructively to the grand challenges facing people and the natural world.

A PARABLE: "HAVASU"

Edward Abbey was, like Henry David Thoreau, a writer whose encounters with nature became occasions for literary reflection. Abbey's 1968 memoir *Desert Solitaire* recounted his life in the Four Corners region of the American Southwest.[1] As the title implies, Abbey sought solitude: "I generally prefer to go into places where no one else wants to go." One of Abbey's adventures, vividly recounted in the story "Havasu," gives us a story to think about.

One day, about 25 miles south of the Grand Canyon, in the Havasupai Indian Reservation, Abbey climbed out of a side canyon of the Havasu River onto high ground. As the day wore on, he saw that he would still be far from his camp by

Learning Objectives
When you have finished studying this chapter, you should be able to

⊿ recognize attempts to create civic science in organizations, governments, and societies. For example, you should be able to analyze a discussion of whether a school should build a green, energy-efficient building;

⊿ analyze the shortcomings and barriers to civic science in those attempts, and to offer advice on how learning might be improved in specific instances;

⊿ identify elements of a sustainability transition in your community, and to devise ways to move such transitions forward more quickly or in a way that will be more readily accepted by those whose behavior needs to change;

⊿ answer the questions below, which were used in a final examination for a course based on this book.

1. Discuss this statement: The most successful species faces the challenge of matching the scale of its competence to the habitat that it dominates. In doing so, markets are both contributors to grave problems and tools for their management.

   Comment on how elements of this book, particularly the idea of sustainable development, connect to human striving, changes in gender roles, and the place of local, small-scale institutions in a world without edges undergoing rapid urbanization.

2. Using materials from this book, explain how you *agree* or *disagree* (or both) with this statement: New technologies such as fertilizers and pesticides have greatly increased the human carrying capacity over the past two centuries. Where human ingenuity is allowed to work, environmental problems ranging from polluted water to saving endangered species to providing organic food to supermarkets have been solved. Instead of trying to restrain businesses and innovators, environmentalists should be facilitating rapid changes in both humans and their relationship to landscapes.

3. Garrett Hardin argues that ethics cannot overcome the tragedy of the commons, whereas Aldo Leopold argues that the land ethic is central to environmentalism. Who is right (if either) and why?

4. Poor people, in America and elsewhere, may choose a profoundly different course with regard to natural resources and environment than the rich. What should rich environmentalists say to the poor? (Remember that your answer may depend on whom you seek to address, so be specific about how your response might vary for different audiences.) Are there poor environmentalists?

5. Environmental studies raises more questions than it answers. Identify the two most important and durable environmental questions that you now see on the agenda of your generation, and discuss how better answers might be sought in your lifetime. Develop and

amplify your discussion with materials from this book.

6. The three poems quoted below reflect contrasting perspectives on humans in landscapes. What are these perspectives, and how would you relate them to ideas and materials in this course? (Hint: How does religion enter into these poems? The discussion of the Hudson River School artists in Chapter 2 should be helpful.)

Corn planted us; tamed cattle made us tame.
Thence hut and citadel and kingdom came.[2]

　　　　　—*Richard Wilbur, 2000*

If no one sends you
messages to read, none
you can read
(so you have no
replies to shape)
still you may irrelevantly read

messages sent to
no one, light shaking
off a poplar leaf
(like seen wind chipped free)
or breaking into
threads

of bright-backed water
in brookstone shallows:
these
messages, though
not sent to you and
requiring no response

may nevertheless be
taken
down in strict

observances (like studied regard)
as if to be nearly adequate
messages to no one.[3]

　　　　　—*A. R. Ammons, 1983*

Nature is what we see,
The Hill, the Afternoon—
Squirrel, Eclipse, the Bumble-bee,
Nay—Nature is Heaven.

Nature is what we hear,
The Bobolink, the Sea—
Thunder, the Cricket—
Nay, —Nature is Harmony.

Nature is what we know
But have no art to say,
So impotent our wisdom is
To Her simplicity.[4]

　　　　　—*Emily Dickinson, c. 1863*

7. Your aunt is a multimillionaire with the sharp eye of a keen investor. You convinced her over Thanksgiving dinner that it is difficult but possible to achieve a sustainable future in the developing world. You've stressed, however, that many uncertainties lie ahead, and many of them cannot be addressed simply by markets, new technologies, or government. Which two or three experiments or pilot programs should her philanthropic foundation support, in order to explore ways to overcome the barriers to an environmentally sound future in the poor countries of the world? You should also identify things that citizens and consumers in rich countries should do to help these ventures succeed.

8. What is an environmentalist? How might one induce those living in the

dark, and he looked for a shortcut. He found a side canyon that seemed to lead back to the river.

It was a steep, shadowy, extremely narrow defile with the usual meandering course and overhanging walls; from where I stood, near its head, I could not tell if the route was feasible all the way down to the floor of the main canyon. I had no rope with me—only my walking stick. But I was hungry and thirsty, as always. I started down.

At first, walking in the narrow trench of the canyon, he made rapid progress. Then the canyon ended in a lip with a twelve-foot drop. Below was a stagnant pool with "a little water left over from the last flood, warm and fetid water under an oily-looking scum, condensed by prolonged evaporation to a sort of broth, rich in dead and dying organisms."

Here, Abbey did something that made it clear how far he was from Thoreau's thrift and planning at Walden Pond:

There was no way to continue except by dropping into the pool. I hesitated. Beyond this point there could hardly be any returning, yet the main canyon was still not visible below. Obviously the only sensible thing to do was to turn back. I edged over the lip of stone and dropped feet first into the water.

The green muck lay in a bowl-like depression. He swam through the unappetizing fluid to the other side. There was another overhang, also about a dozen feet high, with another pool below. The drop he had already made left him no path except forward. So he slid once more and swam to the far edge of the stagnant pond at the bottom, "to see what my fate would be."

Fatal. Death by starvation, slow and tedious. For I was looking straight down an overhanging cliff to a rubble pile of broken rocks eighty feet below.

Trapped in a narrow canyon where he knew he would die, he reversed course. It seemed impossible to scale the drops he had descended, but a desperate person has an unusual motivation to try. He crawled up one drop with the aid of his walking stick and climbed up a corner of eroded sandstone to make his way back up the upper chute. Before he emerged from the canyon, night fell and a rainstorm drove him under a ledge where he "suffered through the long long night, wet, cold, aching, hungry, wretched, dreaming claustrophobic nightmares. It was one of the happiest nights of my life."

Abbey would have disdained literary analysis of his story. And yet, his account provides an apt parable for this book. "Havasu" is a story of recklessness. Abbey slid into a trap set millions of years ago, when the strata of the Colorado plateau tilted, forming a sequence of dry falls on that streambed that rise higher and higher the deeper one drops down into the canyon. If he had understood the geology of the bedrock, he might not have tried to find a shortcut this way. After a moment of hesitation, when he saw the risk he was taking, he slid down the chutes, making seemingly irreversible choices. In the end, it turned out that he *did* know enough about the sandstone to see that the rough edges of the canyon's sheer walls might afford enough traction to a climber. From that knowledge, he fashioned a narrow escape from skill, the materials at hand, and desperate determination.

Having gotten this far into this book, you now know more than most people about the academic view of the planet and the humans that inhabit its landscapes. You have seen how human action is producing irreversible changes in the fundamental architecture of the biosphere by modifying climate and biodiversity. We face additional grand challenges in the urban landscapes being fashioned by humans and in the broader search for a sustainable economy. A reasonable person might say that we are plunging ahead without knowing how or when Earth's structure will put unforeseen, fatal drops ahead of us. As in the red rock canyons of the Havasu, the Earth does not care; it will wait for us to slide over a lip we cannot scale.

Is there a way that humanity can, like Edward Abbey, scramble to safety?

SUSTAINABILITY TRANSITION

Where does safety lie? In Abbey's story, he will survive only if he can go back—and in this respect the story of humans in the landscapes of the twenty-first century is different. We have no choice but to go forward. Where might that be? Sustainable development is one vision of the future, though moving toward sustainability has proved to be a slow process so far. Here is how the idea was originally described in 1987 by a United Nations committee called the World Commission on Environment and Development:

> Sustainable development is development that meets the needs of the present without compromising the ability of future generations to meet their own needs.
>
> It contains within it two key concepts:
>
> - the concept of "needs," in particular the essential needs of the world's poor, to which overriding priority should be given; and
> - the idea of limitations imposed by the state of technology and social organization on the environment's ability to meet present and future needs.[5]

The authors of this concept were international negotiators, who understood their duty to find words that would be accepted by the governments of a world deeply divided by differences in wealth, power, and ideology.

A generation has passed since the idea of sustainable development was first articulated. Looking back, we sense something is wrong. If sustainable development is the goal, why have we not achieved it? Progress has been made, as we have seen in the chapters of Part II. Yet those chapters describe grand challenges, none of which seems close to being surmounted yet.

The answer, we believe, is that sustainable development is not a goal in the usual sense. It is a direction toward which societies may move, similar to the way societies pursue justice or science pursues truth. Some kinds of injustice can be rooted out, but justice cannot be achieved for all and permanently. Science has investigated many phenomena, confirming some theories and disproving many others, but a complete understanding of the natural world is not at hand. Indeed, the range of important scientific questions seems to expand rather than contract, even as more and more is learned.

Progress remains uneven and the grand challenges persist. What is needed is a *transition* toward sustainability, a long-term shift similar to urbanization or the spread of women's rights that has a momentum larger than the decisions of presidents or corporate boards of directors, and a scale far larger than individuals or nongovernmental organizations. Of course, any such transition is a tapestry of individual, organizational, and collective choices. For humanity to move toward sustainability, and to continue to do so in the decades to come, billions of choices will need to preserve the ability of future generations to meet their own needs as we continue to improve our ability to meet the needs of the present. A sustainability transition can sound like wishful thinking, until one realizes that a transition toward sustainability can already be seen in the long-term trends of major variables.

The most evident sign is the demographic transition described in Chapter 8. Human numbers are still growing, but the rate of growth is slowing, and it has been since the 1960s. This is no guarantee that the trend will continue until the population levels off in the latter part of the twenty-first century. But this *is* the trend, and population scientists have documented the forces propelling families toward fewer children per woman. Slowing population growth makes it easier to meet the needs of the present, if other factors remain the same.

Of course, other factors will not remain the same. The global economy is continuing the pattern of growth that began with the Industrial Revolution, as we saw in Chapter 10. Thus, consumption is also continuing to rise, and major technological innovations and shifts in consumption are needed to counterbalance a rising impact on the world's ecosystems. As we saw in Chapter 14, the signs of progress are significant, such as transitions in the way paper is used, suggesting a leveling off of resource demands similar to that in human population.

The one truly fundamental element of industrial civilization is energy use, as discussed in Chapter 7. Humans have relied on fossil fuels for more than two hundred years to create material wealth and underwrite consumption. Energy use is still rising, but sources other than fossil fuels are beginning to play a role in the supply of energy for industrial civilization. Moreover, there are many inefficiencies in the way energy is used today. In many of these instances, energy use can be *reduced* in a way that is economically beneficial. This is what happens when a home is insulated—not only does the house feel more comfortable in winter because the insulation makes it warm and less drafty, but the cost to heat it also falls.

In 2008, the U.S. Energy Information Administration released an analysis that shows a sustainability transition in progress (Fig. 15.1). Since 1980, the use of energy per person has been declining, and so has the energy needed to generate a dollar of gross domestic product. Bear in mind that *total* energy use has continued to climb, as the population and the economy have grown. But, as with the rate of global population growth, energy use is stabilizing, and the trend is expected to continue. In 2009, the *Wall Street Journal* reported that executives at Exxon Mobil believed gasoline consumption in the United States had passed its all-time peak in 2007 and would continue to decline for the foreseeable future.[6]

The material world is dazzlingly complex, and there is no simple way to assess whether the U.S. economy or others worldwide are moving toward sustainability. In 1999, a committee of the National Academy of Sciences observed that by looking at time scales of 50 to 75 years, it was possible to map out a feasible course:

> For a successful transition to sustainability, the world must provide the energy, materials, and information to feed, house, nurture, educate, and

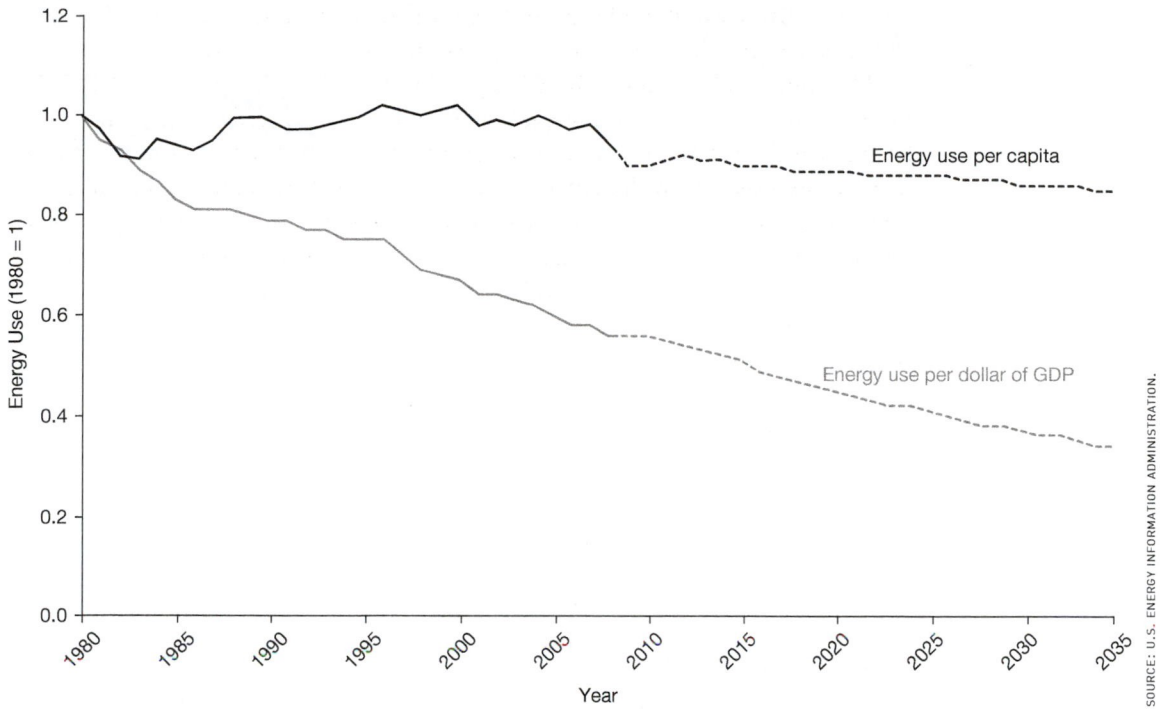

FIGURE 15.1
Energy use per capita and per dollar of gross domestic product (GDP), 1980–2035.

employ many more people than are alive today—while preserving the basic life support systems of the planet and reducing hunger and poverty.

. . . It is over the next two generations that many of the stresses between human development and the environment will become acute. It is over this period that serious progress in a transition toward sustainability will need to take place if interactions between the earth's human population and life support systems are not to significantly damage both.[7]

What needs to be managed is suggested by Table 15.1 (page 410). The projections, made by an independent team of analysts called the Global Scenario Group, are not predictions, but they provide a sensible starting point for thinking about what might be coming. By 2050, today's students will be well into their careers; most will be taking part in the ongoing transition, and some will be leading the search for a sustainable future.

What does Table 15.1 show?

↘ The world economy is likely to continue on a path of growth.

↘ Population growth is projected to be substantial, although the rate of growth will slow in comparison to the second half of the twentieth century (the time during which the grandparents of today's students saw their careers unfold).

TABLE 15.1 ACTUAL AND PROJECTED CHANGES IN GLOBAL VARIABLES.*

	Historical multiple, 1950–1993	"Conventional worlds" scenarios, 2005–2050
Economy (GDP)	5.1	5.8–5.9
Population	2.2	1.4
Energy	4.4	1.4–2.1

*Note: The second column shows the multiples for the 43-year period 1950–1993 (e.g. world population grew by a factor of 2.2). The third column shows projected multiples for the 45-year period 2005–2050.

Sources: National Research Council, Our Common Journey: A Transition toward Sustainability (Washington, DC: National Academies Press, 1999), 70; Christi Electris, Paul Raskin, Rich Rosen, and John Stutz, "The Century Ahead: Four Global Scenarios," technical documentation (Boston: Tellus Institute, 2010).

Energy growth may increase more slowly than in the past. But with widening prosperity in poor countries and growing economies in rich ones, energy growth is still likely to be faster than the growth in population or the demand for food. Climate change is clearly the central issue here: Will a doubling of energy use during the careers of today's students be accomplished by burning fossil fuels, or will there be a major shift toward noncarbon sources such as nuclear energy or higher energy efficiency?

With energy and economic growth come other grand challenges, notably urbanization and loss of biodiversity. Food needs, in particular, are expected to increase, though more slowly than in the past half century. Remember that much of the land that can be brought under cultivation in the coming decades is in tropical countries, which harbor higher biodiversity that is imperiled by the transformation of land use. So the problems of biodiversity conservation are of major importance. Since 2008, there have been several price spikes in basic grains and oils used for cooking. These increases appear to be the result of rising demand for grain used to feed animals raised for meat in China, India, Brazil, and other emerging economies. These price spikes have affected poor people in poor countries, contributing to political unrest in the Arab world beginning in 2011.

Two things are evident in these projections: the multiples going forward are smaller than in the past half century; but because the starting point in the late twentieth century is so much higher than in 1950, the absolute increases in population, food, energy, and GDP are all much larger than in the past two generations. This is the pattern we saw in the demographic transition, too.

BOX 15.1

ENVIRONMENTAL KUZNETS CURVES

The figure below draws together data from large sets of observations collected by the World Bank. The curves show six statistical trends that compare environmental indicators against income per capita in different countries. The vertical scales are all different. The upper pair examines water pollution from two perspectives; the middle two look at polluted air; and the bottom two pertain to garbage and greenhouse gas emissions.

The top pair shows that water and sanitation improve as people get richer. This makes sense (see Chapter 8). The bottom panels show that garbage and greenhouse gas emissions rise as people get richer, which is reasonable (see Chapter 7). These four correlations suggest that some environmental problems get better with economic development, whereas others get worse.

But what about the middle panels? It is surprising to discover that air quality responds to growing wealth. Air moves across boundaries, so it cannot be privately owned. How can pollutants dispersed by the winds be regulated by communities that may be far from where the pollution was generated? Air quality seems to be an obvious example of a hard-to-govern commons.

Consider a specific example. Mexico City, which was struggling with rapidly increasing air pollution and the rising health consequences for children and workers, saw several factors converge to improve air quality. First, technological improvements became available and affordable. Catalytic converters were mandated for motor vehicles, and the propane widely used in homes for cooking and heating water was reformulated so that leaking propane would be less polluting. An old and dirty oil refinery was closed, due in part to public pressure. Imports of cleaner gasoline from the United States also helped, a welcome environmental benefit of the world without edges. And in 1997, a municipal government was elected that was more willing to regulate than its predecessors. By 2001, the air in Mexico City had improved measurably.

The bending-over pattern in the middle panels of the figure is called a Kuznets curve after economist and Nobel laureate Simon Kuznets, who observed that income inequality in a developing economy would first rise and then moderate as the education needed to participate in an advanced economy became more widespread. Therefore, a measure of inequality should first rise and then decrease as income rises. Something like this was seen in Europe, Japan, and the United States in the years preceding 1970, although widening inequality has now been observed in the United States over the past thirty years. The environmental pattern is called a Kuznets curve by analogy, thinking of pollution as a transient phase of

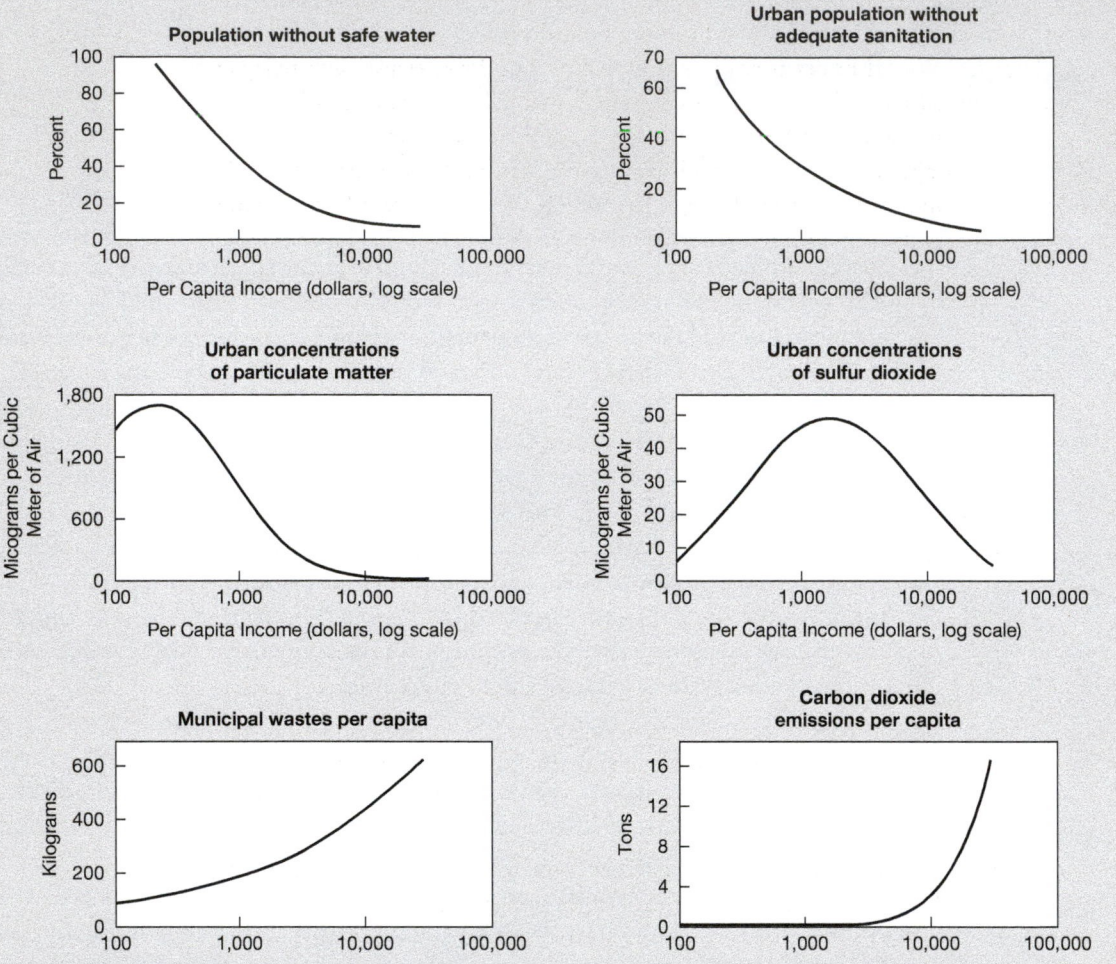

Patterns in the relationship between average income and different forms of environmental harm.

industrial development. That is, pollution rises initially as an economy develops, but then falls as public health and politics lead to more effective controls and more efficient technologies.

The curves in the middle panels have led some, particularly in the business world, to argue that economic growth *by itself* will solve environmental problems. The reasoning goes that when people have the money, they will take care of their environments. Yet the story from Mexico City suggests that markets alone do not seem to be the answer: government regulation (mutual coercion to protect a commons that cannot be easily cleaned except by preventing pollution) is also needed, along with technology and trade (modernization and globalization). Progressives will tell you that governmental action is essential. Conservatives will tell you that technology and trade are essential. Both seem to have evidence to support their views.

These are fascinating clues. They tell us that the natural history of sustainable development contains unexpected turns, for which we do not yet have a good understanding. In this, environmental protection is strikingly like the demographic transition: the signs are hopeful, even though we don't know how to turn them into a reliable outcome yet.

Is there a basis for hope? Box 15.1: Environmental Kuznets Curves, page 411, describes some curious statistical patterns discovered by economists studying economic development.

An environmentalist might think that growing prosperity would lead to worsening environmental conditions, and in some cases, this is what seems to be happening. But in other cases, the reverse is observed, as one sees in the top panels of the figures in the "Environmental Kuznets Curves" box. Because water can carry disease, one way to interpret these correlations is that a population living with poor sanitation and unsafe drinking water is too often sick to advance economically.

In still other cases, rising income has been accompanied by increasing governing capability, which has led to better control of air pollution. Even in the bottom panels, there is reason to think that rising prosperity may lead to more efficient use of resources and recycling of materials, as we discussed in Chapter 14. Therefore, further economic growth might be accompanied by the garbage and greenhouse gas curves bending over as well.

These are signs that a sustainability transition is not just wishful thinking. The National Academy of Sciences panel agreed:

Based on our analysis of persistent trends and plausible futures, the Board believes that a successful transition toward sustainability is possible over the next two generations. This transition could be achieved without miraculous technologies or drastic transformations of human societies. What will be required, however, are significant advances in basic knowledge, in the social capacity and technological capabilities to utilize it, and in the political will to turn this knowledge and know-how into action.[8]

DESPERATE PEOPLE CAN LEARN

The scientists concluded that a sustainability transition is likely to be technically feasible. But what about social capacity and political will? We learned in Chapter 3 that environmental problems are caused by people but cannot be solved by individuals alone:

➘ In an unregulated *commons*, rational self-interested behavior can make environmental problems worse.

➘ *New technologies and markets* have sometimes created or worsened mismatches between human responsibility and natural rhythms and flows. This has led to commons dilemmas, such as climate change, that have no historical precedent.

➘ In a world without edges, self-interest is often ambiguous because *causation becomes complex*. Many environmental changes are hidden in the invisible present, unfolding too slowly to be noticed by people who move every few years and who drive and fly through landscapes that they do not have time to observe. We also live within layers of infrastructure and technology that deliver the goods and services we demand, but through means that obscure their natural origins. Ours is a culture in which the magic of technology is routine and our *dependence* on one another and the natural world has come to feel like *independence* when everything works correctly, which is most of the time.

➘ We believe that outright *environmental disasters befall other people*. Even though we donate to relief organizations and imagine the pain and suffering of others, the lives of most of us who live in rich countries are rarely disrupted. We experience environmental destruction remotely rather than living through the hardships it inflicts on natural and social communities. The Haitian earthquake of 2010 and Hurricane Katrina in 2005 were seen as "natural" disasters, even though the destruction of Port au Prince and New Orleans was the product of human mismanagement and social frailties as much as it was the result of a big natural disturbance. In both cases, moreover, scientists had foreseen the cataclysm to come and sounded warnings, which were not heeded.

➘ In an *information economy*, people are overwhelmed by the information they can easily obtain, and yet it is often unclear how to turn that information into knowledge to guide effective action. The market prices of water, food, and other essential services do not reflect the environmental costs of producing them. Pollution-control regulations tailored to specific industries have to be so complicated that even those directly responsible for drafting and enforcing them have only limited knowledge of what the rules do or fail to do.

Although many historical examples of well-governed commons are known, the conditions in italics in the list above are more recent. The world without edges grew out of low transportation and energy costs—that is, from the Industrial Revolution. The information revolution is still under way, as information processing and storage costs continue to fall, though its roots go back to the invention of movable type at the beginning of the Renaissance.

These historically recent changes have unsettled our social institutions, particularly because human ties to landscapes persist. People remain dependent on ecosystem services, some of which cannot be imported. In that respect, some edges remain. And people remain emotionally and spiritually attached to places, by virtue of their innate and socially learned responses. In that sense, people choose edges through their loyalty to places and all the living organisms that inhabit them.

The challenges that became apparent in the twentieth century, which spawned the environmental movement of the past half century, are part of a long-term historical process of recognizing and addressing the implications of the immense influence that humans now exercise over nature. Classical environmentalism assumes that it is sufficient to bring this influence to light, and that once scientists have warned of environmental threats, the normal responses of politics, markets, and technological innovation will bring about reform. That is, classical environmentalism assumes environmental quality does not require fundamental changes in the institutional architecture of industrial and industrializing societies. The grand challenges suggest otherwise.

Does fundamental change require social convulsion, though? The political upheavals of fascism and communism in the twentieth century provide grim benchmarks. Even they did not, in the end, overturn industrialization or the untidy compromises of parliamentary democracy. So perhaps revolution is *not* a precondition of the fundamental changes implied by sustainable development.

Here is Edward Abbey facing the narrow canyon that threatened to entomb him:

The walls that rose on either side of the drop-off were literally perpendicular. Eroded by weathering, however, and not by the corrosion of rushing floodwater, they had a rough surface, chipped, broken, cracked. Where the walls joined the face of the overhang they formed almost a square corner, with a number of minute crevices and inch-wide shelves on either side....

When I had regained some measure of nerve and steadiness I got up off my back and tried the wall beside the pond, clinging to the rock with bare toes and fingertips and inching my way crabwise toward the corner. The watersoaked, heavy boots dangling from my neck, swinging back and forth with my every movement, threw me off balance and I fell into the pool. I swam out to the bank, unslung the boots and threw them up over the drop-off, out of sight. They'd be there if I ever needed them again. Once more I attached myself to the wall, tenderly, sensitively, like a limpet, and very slowly, very cautiously, worked my way into the corner. Here I was able to climb upward, a few centimeters at a time, by bracing myself against the opposite sides and finding sufficient niches for fingers and toes. As I neared the top and the overhang became noticeable I prepared for a slip, planning to push myself away from the rock so as to fall into the center of the pool where the water was deepest. But it wasn't necessary. Somehow, with a skill and tenacity I could never have found in myself under ordinary circumstances, I managed to creep straight up that gloomy cliff and over the brink of the drop-off and into the flower of safety. My boots were floating under the surface of the little puddle above.

He had done something that seemed impossible a few minutes before.

CIVIC SCIENCE

Abbey's is a story of learning. It suggests that another way to respond to the challenges of sustainability is to take learning seriously. The scientists of the National Academy of Sciences made a crucial observation about a sustainability transition—no one knows quite what it will look like:

> Because the pathway to sustainability cannot be charted in advance, it will have to be navigated through trial and error and conscious experimentation. The urgent need is to design strategies and institutions that can better integrate incomplete knowledge with experimental action into programs of adaptive management and social learning. A capacity for long-term, intelligent investment in the production of relevant knowledge, know-how, and the use of both must be a component of any strategy for the transition to sustainability. In short, this strategy must be one not just of thinking but also of doing.[9]

That is, learning is needed, not only in an academic setting but as a task of public life.

Taking learning seriously is the task of **civic science**, the attempt to *integrate rigorous science with practical politics*. This is what we called an integrative approach to problem solving at the beginning of this book. When civic science works, two things happen: action and learning. With the knowledge available, people act on important problems. Second, they learn from the experience of acting. Often the actions do not solve problems completely. Remember the Channel Island fox: all actions have unanticipated consequences.

Both rigorous science and practical politics are crucial. Practical politics matters because the search for an environmentally sustainable way of life almost certainly implies deep changes in industrial societies. We are looking for ways to move beyond fossil fuels while remaining the most successful species in a material sense. Achieving such transformations is inevitably a matter of society-wide changes. This raises difficult political questions: Should taxpayers subsidize a revival of nuclear energy? Can we raise taxes on gasoline? Should we foster compact urban development rather than sprawl patterns that demand lots of driving? Each of these issues has been politically contentious in the past; all of them need to be faced again as part of a transition toward sustainability. One would despair of our ability to handle the changes required for a prolonged shift toward sustainability, except for the fact that the pace of environmental reform was so rapid during the twentieth century.

Most of this change was accomplished by classical environmentalism, which made progress when the science was clear and the problem matched the scale of governmental powers. We are now facing problems, such as automobile-dependent urban landscapes, loss of biodiversity, and global warming, for which the scale and reach of conventional government is apparently inadequate. With the election of Ronald Reagan in 1980, an influential alignment of political forces—businesses resisting regulation, labor unions fearing for their members' jobs in polluting industries, and intellectual critics of social movements they dismissed as politically correct—created a strong backlash to the government regulation implied by classical environmentalism. Skepticism of "big government" resonates in American culture; the nation was born in rebellion, after all. Yet public opinion polls and community-level activism demonstrated the breadth and persistence of environmental concerns in American consciousness.

The struggles of environmental politics since 1980 have been focused to a large degree on defending regulations, including those in the Clean Air Act and the Endangered Species Act. Environmentalists often sounded alarmist in this struggle. Yet those seeking to weaken regulation also failed to persuade citizens that rapid development of public resources or reliance on free markets was a sensible alternative. Perhaps more important, the intense but esoteric battles over specific provisions of complicated laws often obscured the reality that conflict is one way that a *society* learns—it is difficult learning, usually, but sometimes there is no easier way.

In addition to practical politics, rigorous science is crucial to civic science because we know so little and need to learn so much so quickly. Yet rigorous science, as we understand how to do it now, depends on a functioning governmental framework. In 2011, political controversy over the U.S. budget delayed funding for a weather satellite, impairing climate observations and weather forecasting for more than a year into the future. What this shows is that rigorous science is not a substitute for the governing capacities needed to deal with environmental problems. What science provides is crucial understanding of the governance challenges involved in addressing commons dilemmas, and social science can illuminate the capabilities and limitations of the tools of governance available.

To put this simply, technology expands and reorganizes the commons that people can exploit, most dramatically in the world without edges, as distant ecosystems are coupled by trade. Natural and social science can illuminate the character of those commons, such as the atmosphere where we dump greenhouse gases. Science can help to identify the institutional tools that are needed to govern commons, as cap-and-trade may yet prove to do for greenhouse gas emissions. But, as this example demonstrates, science cannot impose or even design governance on its own. Science is necessary, not sufficient. But it *is* necessary.

MODES OF LEARNING

Civic science is a strategy for societies to learn. They learn what does not work and, haltingly, how to put together solutions that work better. Because nearly all environmental problems connect people of different interests and different knowledge bases, the solutions that work need a broad understanding of what the problem is and how it emerged. This requires the integration of science and social understanding. Then, to create a durable solution, they need to enlist human motivations of different kinds: the passion of the environmentalist and the procedural craft of the attorney; the inventiveness of the engineer and the profit motive of the businessperson. This requires an integrated grasp of people and institutions. (Box 15.2: Linking Knowledge with Action, page 420, describes an approach aimed at facilitating this integrated understanding in the real world of environmental decision making.) It is rare for the right combination of understanding and capabilities to come together. Learning is therefore essential.

The fastest known way to learn is by using experimental science. This does not always mean immediate results; a new cancer medication, for instance, may require a thirty-year investment in molecular biology and genetics. This is an example of the underlying premise of all modern medicine and technology: to control nature, we must first understand it.

The power of laboratory science derives from the ability of the experimenter to create an isolated world in a test tube or other apparatus. Within that literally bounded reality, a scientist can vary one factor at a time, holding the others constant. All key factors must be controlled if a reliable result is to be produced. From observations made as the key factors are identified, the way they relate to each other becomes scientific theory, a body of knowledge that enables other laboratories to repeat the findings. In this way, a scientific theory is like a recipe: bring together ingredients and process them as directed, and the result will be predictable in the same way a cake is the predictable result of a good recipe in the kitchen of a competent cook, even when he has not baked from that recipe before. A theory is sometimes a recipe in just that sense: the understanding it embodies not only enables prediction but also control and new technology. The theories of solid-state physics, for example, have led to the development of computers, flat-panel displays, and cell phones. Although solid-state physics shows how solar cells can be designed, so far it has proved difficult to bring down the cost of photovoltaic cells so that they are competitive with electrical generation by burning fossil fuels. This example illustrates the way that theories can be useful, but their range of application may be more limited than one might wish (Table 15.2).

For many of the most important components of public life, theories are incomplete or absent. For this reason, the bottom row of Table 15.2 provides a more plausible account of much of the learning that societies seem able to do. In areas from the conduct of diplomacy and war to domestic arenas such as poverty, education, and social issues like gay marriage, decision makers and voters ignore the theories that have been developed and engage in debate using facts and history

TABLE 15.2 MODES OF LEARNING.

Each mode of learning . . .	makes observations . . .	and combines them . . .	to inform activities . . .	. . . that accumulate into usable knowledge.	Example
Laboratory experimentation	Controlled observation to infer cause . . .	replicated to assure reliable knowledge . . .	enabling prediction, design, control.	Theory (It works, but range of applicability may be narrow.)	*Semiconductor physics and computer chips*
Adaptive management (quasi-experiments in field situations)	Systematic monitoring to detect surprise . . .	integrated assessment to build system knowledge . . .	informing model building to structure debate.	Strong inference (But learning may not produce timely prediction or control.)	*Green Revolution agriculture*
Trial and error	Problem-oriented observation . . .	extended to analogous instances . . .	to solve or mitigate particular problems.	Empirical knowledge (It works but may be inconsistent and surprising.)	*Learning by doing in mass production*
Unmonitored experience	Casual observation . . .	applied anecdotally . . .	to identify plausible solutions to intractable problems.	Models of reality (Test is political feasibility, not feasibility in practice.)	*Most statutory policies*

BOX 15.2

LINKING KNOWLEDGE WITH ACTION

Even though science has been recognized as an important element of decision making and policy since the mid-twentieth century, the effective linkage of scientific knowledge to the policy choices of governments and businesses remains an elusive goal. Scientists relying on the model of classical environmentalism are frustrated when problems such as loss of biodiversity or climate change are ignored or distorted in political debates, thwarting informed decision making. Politicians and resource managers, in turn, complain that scientists bring up many problems but offer few practical or timely solutions. Scientific knowledge, it seems, is too often not useful.

Why is this? Decision making is driven by different considerations than those that shape scientific discovery.[1] As a result, what counts as "good" information and "good" use of information is not the same to scientists and those in the decision process.

Science is the pursuit of *reliable* knowledge. Publication of results in a peer-reviewed journal is the basic guarantor of science good enough for other scientists to rely on in their own investigations. Gathering reliable knowledge is a meticulous process, with rigorous standards of evidence and disclosure of methods. Often, acceptance of new knowledge takes years of expert debate, replication of observations, and refinement of methods, as happened with the steadily improving understanding of the toxic properties of some components of industrial waste, such as mercury.

Decisions by policy makers are typically made under deadlines and amid controversy. Although decision makers seek knowledge to justify and guide their choices, often the knowledge that is usable is the knowledge that is available, however frail its basis in science. To be useful, knowledge should be scientifically credible, but it must be timely and relevant in the eyes of those in the decision-making process. The long-persisting public debate over global warming, for example, contrasts strikingly with the consensus among scientists about the basic dynamics of the global climate system. Scientists are convinced that changing the concentration of greenhouse gases in the atmosphere will change the climate—that this is reliable knowledge. Yet political discussion continues over whether climate change is a legitimate and urgent problem, with arguments about how warnings from scientists should be balanced against economic development or an unusually cold winter.

Such disconnects reflect deep differences between science and action. Reliable knowledge is essential in a complex and dynamic world. Decisions need to be made responsibly, which means that decision makers cannot wait for definitive proof or full understanding but need knowledge that is "good enough." Linking knowledge with action thus requires continuing management of the tensions between science and decision making, so as to honor both the demands of action and the rigor of science. A conceptual framework devised by scholars in the early twenty-first century provides a way to diagnose these tensions and suggests ways to manage the tensions of this intrinsically difficult relationship.[2]

The goal of Linking Knowledge with Action (LKwA) is knowledge that is useful and used to advance conservation and sustainability. As shown in the figure, LKwA is built around *joint* production of knowledge by users from the world of decision making and by scientists who bring specialized knowledge of geographic and disciplinary domains and mastery of analytical and other technical skills. Joint production means that research should aim to answer the questions facing decision makers, and that decision makers should grasp both the content and limitations of the scientific knowledge available to inform decisions.

Usable knowledge has three attributes, which often cannot be optimized at the same time. Decision makers need *credible* information—knowledge that has passed the tests of academic validation where it is available. In many decision-making situations, however, it is equally or more important that the information be *salient*—that is, relevant and timely. Decisions also need to withstand challenge, which means the information on which the decisions are based needs to be *legitimate*—gathered in ways that assure that the information is correct, complete,

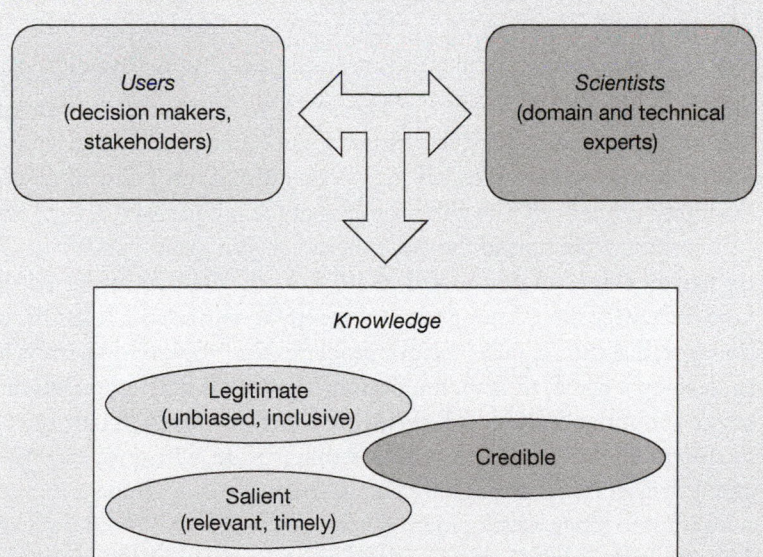

The goal of Linking Knowledge with Action is to produce knowledge that is legitimate, salient, and credible.

and unbiased. Legitimacy can often be strengthened by exposing research to peer review, by involving a wide range of stakeholders as research questions are defined, by relying on syntheses of knowledge carried out by independent scientists who have no political or financial interest in arriving at a preordained result, and by funding such syntheses from similarly independent sources. Credibility, salience, and legitimacy can often reinforce one another, but they are also often pulling in different directions, so that a central task of LKwA is making knowledge good enough to inform and improve decision making.

LKwA is now being used on a limited scale. One of its best-known successes is the Intergovernmental Panel on Climate Change (IPCC), a process begun more than twenty years ago to assemble scientific understanding on climate change for the members of the United Nations. The IPCC has worked closely with representatives of the United Nations to craft summary statements that are legitimate as well as credible for use by governments worldwide. This has facilitated informed policy choices, although it has not diminished the political difficulty of a challenge that demands fundamental changes in the way that societies use energy. The IPCC was recognized with the Nobel Peace Prize in 2007 for its contributions to linking knowledge with action.

1. Don K. Price, Chap. 5 in *The Scientific Estate* (New York: Oxford University Press, 1965).

2. William C. Clark, Ronald B. Mitchell, and David W. Cash, "Evaluating the Influence of Global Environmental Assessments," Chap. 1 in *Global Environmental Assessments: Information and Influence*, eds. R. B. Mitchell, W. C. Clark, D. W. Cash, and N. M. Dickson (Cambridge, MA: MIT Press, 2006).

to score points instead of illuminating wrenching trade-offs. Spectacular instances, such as a struggle over the fate of a person lying in a persistent coma, motivate legislation and public policy. The debates focus on competing stories and are decided on the basis of which side has the votes—that is, on political feasibility. Political feasibility is often not the same thing as the feasibility of the resulting policy.

Learning from unmonitored experience can seem foolish in hindsight. The Endangered Species Act, passed in 1973, enunciated an important principle, that human activity should not drive any species to extinction. It has become apparent, however, that the act raises serious problems for property owners. When they want to develop a parcel of land, the permits they need from government include language about the presence of endangered species. Development is widely accepted in American business as a legitimate thing to do with property. But the Endangered Species Act may put a barrier in the way—something exploited by activists who want to block development for reasons that have little to do with biodiversity. The resulting political struggle, both by champions of development and by conservationists, has led to a stalemate for more than a generation. Progress toward harmonizing the goals of economic investment and protecting habitat has been slow, even though they are often not necessarily incompatible. A major advance came in the 1990s, when the federal government agreed to use ecosystem-scale habitat conservation plans. These plans said that development conforming to the plans would meet the requirements of the Endangered Species Act. This gives developers certainty about the regulations they would face, while assuring reasonable habitat to endangered species.

Can't we do better? Yes, often we can. Two alternatives are identified in Table 15.2. One is **trial and error**, a deliberate process of solving problems by trying things out. It works, although surprises typically arise because the knowledge one gains from this problem-driven approach does not get to the bottom of the phenomena. That is, trial and error usually does not lead to scientific theories, because there is insufficient time or capability to set up the controlled observations that characterize laboratory experiments. Consider how we learn to use personal computers. Rarely do we grasp and apply the principles of semiconductor physics or digital logic or software architecture; instead, we trade gossip with others, bumble along, and discover ways to work around problems. Experience as a student is often similar. Students often do not understand how to put the material in a course into a coherent whole until after the course has ended. People try and sometimes fail, piecing together ways of doing problem sets and writing papers, and doing the best they know how. The result, for those who keep at it, is learning. Trial and error works, especially in the protected setting of academia, where successes and failures can both be part of the learning process.

Another approach, **adaptive management**, is built around the idea that policies are experiments—whether or not the policy makers acknowledge that fact. Policy is action and may have unanticipated consequences. Equally important, every policy is adopted because it is believed to have *anticipated* consequences. A policy is therefore a set of hypotheses about human behavior and the natural world. Consider:

> If property rights were assigned in the United States to greenhouse gas emissions, and a market in those rights were made legal, then
>
> ↘ greenhouse gas emissions, and the concentration of greenhouse gases would decline;
> ↘ climate change would follow a different, slower course;
> ↘ some nations and firms and people would gain economically from trade in emissions rights and others would find themselves poorer or unable to carry out economic activities as planned; and
> ↘ technologies would be developed and sold in response to a world in which emissions of greenhouse gases have a cost.

None of these consequences is assured, however. All can be monitored and compared against the experience of other nations and compared to benchmark predictions made by political actors and analysts when the policy was under debate. As we saw in Box 13.5: Cap-and-Trade in Chapter 13, page 368, the predicted price for emissions rights was uncertain when the Clean Air Act Amendments were adopted in 1990. The actual market prices that emerged turned out to be lower than predicted.

By treating policies as experimental tests of hypotheses and then measuring the effects, people can learn systematically over time. Adaptive management is a difficult approach to implement, however, because those who advocate policies do not want to admit they might have been wrong to do so, particularly when they are working in the public arena. The policy area in which adaptive learning has been most successful is medicine, as we sometimes learn that treatments thought to work are actually not safe or effective, or they need to be used differently. So far, adaptive management has languished in environmental policy, even though learning to deal systematically with the uncertainties of nature would help to inform large-scale decisions in resource management, such as energy policy.

When the consequences matter, as they always do in environmental policy, the ability to learn is critical. This is a point that is not often appreciated in the classical model of environmentalism.

LEARNING AND SOCIAL LEARNING

It has long been a cliché to say that liberal education leads to lifelong learning. This evokes an image of cultured leisure and of knowing some art, music, literature, history, and science so that one can enjoy museums and concerts and travel to Europe. Environmental historian William Cronon proposed something strikingly different.

> Learning is, to me, a way of emphasizing the questions we face in dealing with humans in the landscape.
>
> A liberal education is not something any of us ever achieve; it is not a state. Rather, it is a way of living in the face of our own ignorance, a way of groping toward wisdom in full recognition of our own folly, a way of educating ourselves without any illusions that our education will ever be complete.[10]

To Cronon, the learning that is desperately needed to deal with what we are calling grand challenges includes a liberal education, what some might consider the most "useless" kind. (See Box 15.3: Only Connect, page 426.) This was a theme announced in the opening chapter of this book. You are now in a position to judge whether we have made our case.

Why is the open-minded but also open-ended character of knowledge in the humanities, social sciences, and natural sciences—the heart of a liberal education—important? Because the grand challenges of sustainability entail large-scale social change. Not only must individual people come to see that they should act in environmentally responsible ways, but they must live in institutional settings where acting in responsible ways is feasible and advantageous. Think about crime. Criminals usually get what they want, in the short run. But they live in an institutional setting that inhibits criminality through policing and other forms of monitoring, backed by punishments, fines, and other forms of enforcement. Where monitoring and enforcement work, everyone can rely on a low level of criminality. People refrain from bad behavior even if there is little likelihood of getting caught. The benefits of that social atmosphere include going to the assistance of people who call out for help, because one believes they really need it. These are collective benefits, ones that individuals acting alone cannot achieve. Social scientists call these collective benefits social capital.

Building and maintaining social capital is essential to tackling the grand challenges of sustainability, because, for all of them, the simple approaches of enacting rules or setting the right prices seem to be necessary but insufficient. Instead, humanity is engaged in the messier, error-prone modes of **social learning**, a process that will continue to unfold over the lifetimes of today's students. It is helpful for individuals, particularly those who lead and shape the organizations at the center of this historical process, to understand that it is a process of *social* change, not only a matter of making a single right choice of policy. The broader, historical perspective is helpful in one practical sense: it enables one to evaluate the consequences of failure, and to decide how to press forward, or whether to do so. For this, the liberal education that Cronon described is useful indeed, because setbacks and errors have been plentiful in the search for environmental responsibility. That risky state of affairs seems likely to continue.

Over the course of a career, one can contribute to surprisingly large changes. Which changes turn out to be feasible, and how to achieve them, is not often clear until after the fact. But the wide scope of the idea of sustainability means that things do not need to be clear yet when one is in college. What matters instead is to shape one's education in two somewhat contradictory ways. First, develop solid skills: clear writing with which to think and to persuade; quantitative and graphic skills to deal with a world with so much information that it must be dealt with, in part, through statistics and numbers and graphics. Second, maintain and broaden curiosity about the world and its complexities, such as the implications that lie within the Kuznets curve or the demographic transition, and the connection

"ONLY CONNECT": THE GOALS OF A LIBERAL EDUCATION

In a speech to graduating college students, the environmental historian William Cronon offered this description of someone who has gained the benefits of a liberal education.

1. *They listen and they hear.*
   This is so simple that it may not seem worth saying, but in our distracted and over-busy age, I think it's worth declaring that educated people know how to pay attention—to others and to the world around them. They work hard to hear what other people say. They can follow an argument, track logical reasoning, detect illogic, hear the emotions that lie behind both the logic and the illogic, and ultimately empathize with the person who is feeling those emotions.

2. *They read and they understand.*
   This, too, is ridiculously simple to say but very difficult to achieve, since there are so many ways of reading in our world. . . . Skilled readers know how to read far more than just words. They are moved by what they see in a great art museum and what they hear in a concert hall. They recognize extraordinary athletic achievements; they are engaged by classic and contemporary works of theater and cinema; they find in television a valuable window on popular culture. When they wander through a forest or a wetland or a desert, they can identify the wildlife and interpret the lay of the land. They can glance at a farmer's field and tell the difference between soy beans and alfalfa. They recognize fine craftsmanship, whether by a cabinetmaker or an auto mechanic. . . .

3. *They can talk with anyone.*
   Educated people know how to talk. They can give a speech, ask thoughtful questions, and make people laugh. They can hold a conversation with a high school dropout or a Nobel laureate, a child or a nursing-home resident, a factory worker or a corporate president. Moreover, they participate in such conversations not because they like to talk about themselves but because they are genuinely interested in others.

4. *They can write clearly and persuasively and movingly.*

What goes for talking goes for writing as well: Educated people know the craft of putting words on paper. I'm not talking about parsing a sentence or composing a paragraph, but about expressing what is in their minds and hearts so as to teach, persuade, and move the person who reads their words. I am talking about writing as a form of touching, akin to the touching that happens in an exhilarating conversation.

5. *They can solve a wide variety of puzzles and problems.*

The ability to solve puzzles requires many skills, including a basic comfort with numbers, a familiarity with computers, and the recognition that many problems that appear to turn on questions of quality can in fact be reinterpreted as subtle problems of quantity. These are the skills of the analyst, the manager, the engineer, the critic: the ability to look at a complicated reality, break it into pieces, and figure out how it works in order to do practical things in the real world. Part of the challenge in this, of course, is the ability to put reality back together again after having broken it into pieces—for only by so doing can we accomplish practical goals without violating the integrity of the world we are trying to change.

6. *They respect rigor not so much for its own sake but as a way of seeking truth.*

Truly educated people love learning, but they love wisdom more. They can appreciate a closely reasoned argument without being unduly impressed by mere logic. They understand that knowledge serves values, and they strive to put knowledge and values into constant dialogue with each other. The ability to recognize true rigor is one of the most important achievements in any education, but it is worthless, even dangerous, if it is not placed in the service of some larger vision that also renders it humane.

7. *They practice humility, tolerance, and self-criticism.*

This is another way of saying that they can understand the power of other people's dreams and nightmares as well as their own. They have the intellectual range and emotional generosity to step outside their own experiences and prejudices, thereby opening themselves to perspectives different from their own. From this commitment to tolerance flow all those aspects of a liberal education that oppose parochialism and celebrate the wider world: studying foreign languages, learning about the cultures of distant peoples, exploring the history of long-ago times, discovering the many ways in which men and women have known the sacred and given names to their gods. Without such encounters, we cannot learn how much people differ—and how much they have in common.

8. *They understand how to get things done in the world.*

...Learning how to get things done in the world in order to leave it a better place is surely one of the most practical and important lessons we can take from our education. It is fraught with peril because the power to act in the world can so easily be abused—but we fool ourselves if we think we can avoid acting, avoid exercising power....And so we study power and struggle to use it wisely and well.

9. *They nurture and empower the people around them.*

Nothing is more important in tempering the exercise of power and shaping right action than the recognition that no one ever acts alone. Liberally educated people understand that they belong to a community whose prosperity and well-being are crucial to their own, and they help that community flourish by making the success of others possible. If we speak of education for freedom, then one of the crucial insights of a liberal education must be that the freedom of the individual is possible only in a free community, and vice versa. It is the community that empowers the free individual, just as it is free individuals who lead and empower the community. The fulfillment of high talent, the just exercise of power, the celebration of human diversity: Nothing so redeems these things as the recognition that what seem like personal triumphs are in fact the achievements of our common humanity.

10. *They follow E. M. Forster's injunction from* Howards End: *"Only connect."*

More than anything else, being an educated person means being able to see connections that allow one to make sense of the world and act within it in creative ways. Every one of the qualities I have described here—listening, reading, talking, writing, puzzle solving, truth seeking, seeing through other people's eyes, leading, working in a community—is finally about connecting. A liberal education is about gaining the power and the wisdom, the generosity and the freedom to connect.

Source: "'Only connect ...': The Goals of a Liberal Education," by William Cronon. Reprinted from *The American Scholar*, Volume 67, No. 4, Autumn 1998. Copyright © 1998 by the author.

between their odd behavior and the history being made by humans who rarely think of themselves as making history at all.

If one can maintain that dual focus of specific expertise married to broad curiosity, then one is in a position to see large-scale, long-term patterns, such as the increasing economic effienciency of energy use or the emergence of nongovernmental organizations to form a robust civil society. And with that knowledge, one can contribute to the still bumbling search for a sustainability transition.

As we have seen throughout this book, social learning is an attempt to improve the ability of individuals and organizations to pursue collective goals by improving knowledge and restructuring rules and incentives. This happens in a variety of ways:

- ↘ By recognizing and responding to crises, as we see in endangered species protection or in the response to industrial threats, including pesticides, carcinogens, radioactivity, and other pollutants.
- ↘ By working to improve human well-being sustainably, from the individual household to the decades-long pursuit of economic development.
- ↘ By fostering responsibility in consumption and supporting the stabilizing trajectory of the demographic transition.
- ↘ By reforming large firms and governments, so as to create rules and practices that apply fairly across the economy and safeguard ecosystem services.
- ↘ By nurturing institutions that manage commons sustainably and advance the land ethic.
- ↘ By designing technologies to avoid environmental problems.
- ↘ By learning from experience (civic science).
- ↘ By accepting conflict as a mode of learning, and choosing the battles worth fighting.

Some of these methods of social learning involve regulation and government directly, but others, such as industrial ecology, do so only indirectly. Some of these modes are compatible with a market economy, such as ecotourism conducted in a sustainable fashion; others envision intervening in markets. This array of possibilities suggests that environmental responsibility does not cleanly fit along an ideological spectrum. Even though the governance of commons does require rules, the rules that endure are ones that come to be accepted as legitimate. All the listed approaches to social learning are necessary, but even together they may not be sufficient; the grand challenges remain grand.

Social learning builds on a base of amazingly rapid social change over the past generation, including the rise of environmental responsibility as a goal in every society. What does responsibility look like?

- ↘ It looks like recycling, the large-scale, routine acceptance of responsibility for sorting one's trash.
- ↘ It looks like pollution-control programs that use rules and enforcement mechanisms to organize and limit the use of commons in water and air, and

bring about the long-term isolation of the dangerous wastes of an industrial economy.

- ↘ It looks like conservation biology, the search for ways to salvage the world's hard-pressed biodiversity.
- ↘ It looks like industrial ecology, the pursuit of environmentally responsible ways of producing manufactured goods and using industrial technology while continuing to earn profits.
- ↘ It looks like mass-media coverage of weather and environmental news that alerts and informs, and reminds people that the world without edges is a world that can be affected by human activity.
- ↘ It looks like major corporations voluntarily adopting targets for reducing greenhouse gas emissions.
- ↘ It looks like social and economic development in poor countries when that development works together with policies that advance the down-turning pattern of the Kuznets curves.
- ↘ It looks like improvements in the life chances of poor people so as to slow population growth without coercion.
- ↘ It looks like communities combating sprawl.
- ↘ It also takes shape as contemplation of religion and ethics, which help us to articulate the place of humans in the landscape in ideas such as the land ethic.

Practice isn't perfect in any of these areas. Far from it. The point, rather, is that there has been unmistakable change, and in positive directions, within the lifetimes of today's students and their parents. The awakening of environmental consciousness has spread all over the world. It is an amazing qualitative change in only one human generation. What is not clear yet is whether that consciousness can be translated into action quickly enough. Surely for the endangered species being lost, corrective action will not come in time. A sensible response to climate change will take more wrangling. As India, China, and other once-poor societies surge toward wealth, the challenges of environmentally responsible growth have now spread beyond those nations that were already rich when Rachel Carson launched the contemporary environmental movement.

Many people are now aware that local and global scales are connected. This reality is being taken into account in the way institutions operate, although here especially a lot of work remains to be done. Environmental awareness is one legacy today's students inherit; continuing to turn that awareness into durable change is the hurdle the current generation faces. And if change is to be made durable, it is essential that environmental responsibility be perceived as equitable in a world of sharp differences in human well-being and natural resource endowments.

We end with a reminder that, however much humans dominate landscapes, we cannot transcend the inherent uncertainties and open-ended character of the world we share with other forms of life. A. R. Ammons, a celebrated twentieth-century American poet, put his meditations on this theme in a poem called "Corsons Inlet."[11] Ammons describes a walk along a beach in New Jersey, where he observed the shifting rhythms of shore and wildlife:

> a young mottled gull stood free on the shoals
> and ate
> to vomiting: another gull, squawking possession, cracked a crab,
> picked out the entrails, swallowed the soft-shelled legs, a ruddy
> turnstone running in to snatch leftover bits . . .

Watching the pulse of predator and prey, Ammons came to terms with his human longing for certainty:

> terror pervades but is not arranged, all possibilities
> of escape open: no route shut, except in
>   the sudden loss of all routes:
>   I see narrow orders, limited tightness, but will
> not run to that easy victory:
>   still around the looser, wider forces work:
>   I will try
>   to fasten into order enlarging grasps of disorder, widening
> scope, but enjoying the freedom that
> Scope eludes my grasp, that there is no finality of vision,
> that I have perceived nothing completely,
>   that tomorrow a new walk is a new walk.

The pursuit of sustainability is also an attempt to widen the scope of human competence. And we accept, with Ammons, that living responsibly requires, not finality of vision, but rather an appreciation of the wonders of an inhabited world.

FURTHER READING

Clark, William C., Ronald B. Mitchell, and David W. Cash. "Evaluating the Influence of Global Environmental Assessments." Chap. 1 in *Global Environmental*

Assessments: Information and Influence. Eds. R. B. Mitchell, W. C. Clark, D. W. Cash, and N. M. Dickson. Cambridge, MA: MIT Press, 2006.

Electris, Christi, Paul Raskin, Rich Rosen, and John Stutz. "The Century Ahead: Four Global Scenarios," technical documentation. Boston: Tellus Institute, 2010.

Kingsolver, Barbara. "High Tide in Tucson." Chap. 1 in *High Tide in Tucson: Essays from Now or Never.* New York: HarperCollins, 1995.

KEY TERMS

adaptive management	civic science	trial and error
	social learning	

NOTES

Chapter 1

1. Friends of the Island Fox, "Island Fox Population Update," November 15, 2011, www1.islandfox.org.
2. World Commission on Environment and Development, *Our Common Future* (New York: Oxford University Press, 1987), 43.
3. Ibid., 2.

Chapter 2

1. Barbara Novak, *Nature and Culture: American Landscape and Painting, 1825–1875* (New York: Oxford University Press, 1980), 160.
2. "Nature Is What We See" by Emily Dickinson. Reprinted by permission of the publishers and the Trustees of Amherst College from *The Poems of Emily Dickinson*, Thomas H. Johnson, ed., Cambridge, Mass.: The Belknap Press of Harvard University Press, Copyright © 1951, 1955, 1979, 1983 by the president and fellows of Harvard College.
3. Ralph Waldo Emerson, *Nature*, ed. Kenneth Walter Cameron (New York: Scholars' Facsimiles & Reprints, 1940), 12–13.
4. Bill Moyers, "America's First River—Bill Moyers on the Hudson," broadcast on Public Broadcasting Service, April 23–24, 2002.
5. John Magnuson, "Long-Term Ecological Research and the Invisible Present," *BioScience* 40 (1990): 495.
6. Gary Snyder, "The Place, the Region, and the Commons," in *The Practice of the Wild* (San Francisco: North Point Press, 1990), 25–47.
7. Ibid.
8. Joan Didion, "Los Angeles Notebook," in *Slouching toward Bethlehem* (New York: Noonday Press, 1968), 220–21.
9. Edward Abbey, *Desert Solitaire: A Season in the Wilderness* (1968; repr., New York: Ballantine, 1971), 23.
10. Ibid., 24.
11. Tony Hiss, *The Experience of Place* (New York: Vintage, 1991).
12. Olmsted, Vaux & Company, "Preliminary Report to the Commissioners for Laying Out a Park in Brooklyn, New York," quoted in Witold Rybczynski, *A Clearing in the Distance: Frederick Law Olmsted and America in the Nineteenth Century* (New York: Scribner, 1999), 274.

Chapter 3

1. World Commission on Environment and Development, *Our Common Future* (New York: Oxford University Press, 1987), 46.
2. Garrett Hardin, "The Tragedy of the Commons," *Science* 162, no. 3859 (1968): 1243–48.

3. Ibid., 1244.

4. Willard W. Cochrane, *The Development of American Agriculture: A Historical Analysis* (Minneapolis: University of Minnesota Press, 1979).

5. Peter Barnes, *Who Owns the Sky?* (Washington, DC: Island Press, 2001).

6. Gary Snyder, "The Place, the Region, and the Commons," in *The Practice of the Wild* (San Francisco: North Point Press, 1990), 30.

7. John Tierney, "A Tale of Two Fisheries," *New York Times*, August 27, 2000.

8. Hardin, "The Tragedy of the Commons," 1245.

Chapter 4

1. Robert U. Ayres, "Industrial Ecology: Wealth, Depreciation and Waste," in *Frontiers of Environmental Economics*, ed. Henk Folmer, H. Landis Gabel, Shelby Gerking, and Adam Rose (Northampton, MA: Edward Elgar, 2001), 214–49.

2. Steven Ashley, "When the Twin Towers Fell," *Scientific American*, October 9, 2001, www.sciam.com/article.cfm?id=when-the-twin-towers-fell.

3. William R. Freudenburg, "Privileged Access, Privileged Accounts: Toward a Socially Structured Theory of Resources and Discourses," *Social Forces* 84, no. 1 (2005): 89–114.

4. Natural Resources Defense Council, "Alfalfa: The Thirstiest Crop," Clean Water & Oceans: Water Conservation & Restoration: In Brief: Fact Sheet, 2001, http://web.archive.org/web/20010702041017/http://www.nrdc.org/water/conservation/fcawater.asp.

5. Peter H. Gleick, *California's "Economic Productivity" of Water Use: Jobs, Income, and Water Use in California* (Oakland, CA: Pacific Institute, 2004); Peter H. Gleick, *California's Past and Future Water Use: Future Increases in Demand Are Not Inevitable* (Oakland, CA: Pacific Institute, 2004); Marc Reisner and Sarah Bates, *Overtapped Oasis: Reform or Revolution for Western Water* (Washington, DC: Island Press, 1990).

6. The numbers for toxic releases by various industrial facilities come from U.S. Environmental Protection Agency, *1996 Toxics Release Inventory: Public Data Release—Ten Years of Right-to-Know* (Washington, DC: U.S. EPA, Office of Pollution Prevention and Toxics, 1998).

7. Ibid., 247.

8. Emissions rights auction price for 1993 from 58 Fed. Reg. 27563–67 (May 10, 1993).

9. Acid Rain Retirement Fund, "ARRF Has Retired over 2 Million Pounds," press release, April 1, 2011, http://acidrainretirementfund.org/anew.htm#press7.

10. Theodore Roosevelt, Welcoming Statement to a White House Conference on Conservation (1908), in *Major Problems in American Environmental History: Documents and Essays*, ed. Carolyn Merchant (Lexington, MA: Heath, 1993), 350–52.

11. Don Gifford, "How We Are Housed," chap. 6 in *The Farther Shore: A Natural History of Perception, 1798–1984* (New York: Vintage, 1991).

12. Richard Mabey, *Gilbert White: A Biography of the Author of* The Natural History of Selborne (London: Century, 1986), 2.

13. Letter XXXVIII to the Hon. Daines Barrington (February 12, 1778) excerpted in Gilbert White, *The Natural History and Antiquities of Selborne* (1789), in *The Norton Book of Nature Writing*, ed. Robert Finch and John Elder (New York: Norton, 1990).

14. Gilbert White, *The Natural History and Antiquities of Selborne* (London: B. White & Son, 1789), 225.

15. Gifford, *The Farther Shore*, 217–18.

16. White, *The Natural History*, 12.

17. Gifford, *The Farther Shore*.

18. Henry David Thoreau, *Walden; or, Life in the Woods* (1854), in *Walden and Other Writings by Henry David Thoreau*, ed. Joseph Wood Krutch (New York: Bantam Classic, 1981), 200–8.

19. Henry David Thoreau, *Walden; or, Life in the Woods* (1854; repr. New York: Macmillan, 1937), 149.

20. Gifford, *The Farther Shore*, 230.

21. Gifford, *The Farther Shore*, 237.

22. George L. Cowgill, "On Causes and Consequences of Ancient and Modern Population Changes," *American Anthropologist* 77 (1975): 505.

Chapter 5

1. Edward O. Wilson, *The Diversity of Life* (Cambridge, MA: Harvard University Press, 1993), 35.

2. J. B. L. Noel, *The Story of Everest* (New York: Blue Ribbon Books, 1931), as quoted by H. Rahn in A. S. Paintal and P. Gill-Kumar, eds., *Respiratory Adaptations, Capillary Exchange and Reflex Mechanisms* (Delhi: Vallabhbhai Patel Chest Institute, 1977), 94–104. Both are cited, in turn, in the supplemental online information of Raymond B. Huey and Peter D. Ward, "Hypoxia, Global Warming, and Terrestrial Late Permian Extinction," *Science* 308 (2005): 398–401, with supporting online material at www.sciencemag.org/cgi/content/full/308/5720/398/DC1.

3. U.S. Geological Survey, "Earth's Water Distribution," http://ga.water.usgs.gov/edu/waterdistribution.html.

4. PhysicalGeography.net, "Introduction to the Hydrosphere," chap. 8 in *Fundamentals of Physical Geography*, 2d ed. (ebook), www.physicalgeography.net/fundamentals/8b.html.

Chapter 6

1. Joel E. Cohen, *How Many People Can the Earth Support?* (New York: Norton, 1995), 34.

2. Jared Diamond, *Guns Germs, and Steel: The Fates of Human Societies* (New York: Norton, 1997); William H. McNeill, *The Human Condition: An Ecological and Historical View* (Princeton, NJ: Princeton University Press, 1980).

3. Diamond, *Guns Germs, and Steel*, 128.

4. William Cronon, Chap. 7 in *Changes in the Land: Indians, Colonists, and the Ecology of New England* (New York: Hill and Wang, 1983).

5. Cohen, *How Many People Can the Earth Support?*, 36.

6. Karl August Wittfogel, *Oriental Despotism: A Comparative Study of Total Power* (New Haven, CT: Yale University Press, 1957).

7. McNeill, *The Human Condition*, 7–8.

8. Paul R. Ehrlich, *The Population Bomb* (New York: Ballantine Books, 1968).

9. Julian L. Simon, *The Ultimate Resource* (Princeton, NJ: Princeton University Press, 1981).

10. Julian L. Simon, "Resources, Population, Environment: An Oversupply of False Bad News," *Science* 208 (1980): 1431–37.

11. John Tierney, "Betting on the Planet," *New York Times*, December 2, 1990, www.nytimes.com/1990/12/02/magazine/betting-on-the-planet.html?scp=4&sq=tierney+betting+planet&st=nyt&pagewanted=print

12. Katherine Kiel, Victor Matheson, and Kevin Golembiewski, "Luck or Skill? An Examination of the Ehrlich-Simon Bet," *Ecological Economics* 69 (2010): 1365–67.

13. Gretchen C. Daily et al., "Ecosystem Services: Benefits Supplied to Human Societies by Natural Ecosystems," *Issues in Ecology* 2 (1997): 1, www.esa.org/science_resources/issues/FileEnglish/issue2.pdf.

14. May R. Berenbaum, "Losing Their Buzz," *New York Times*, March 2, 2007, www.nytimes.com/2007/03/02/opinion/02berenbaum.html?sq=may%20berenbaum&st=nyt&scp=4&pagewanted=all.

15. Peter M. Vitousek, Harold A. Mooney, Jane Lubchenco, and Jerry M. Melillo, "Human Domination of Earth's Ecosystems," *Science* 277 (1997):494.

Chapter 7

1. Intergovernmental Panel on Climate Change, Chap. 2 in *Climate Change 2007* (Fourth Assessment Report), www.ipcc.ch/publications_and_data/ar4/syr/en/contents.html.

2. Intergovernmental Panel on Climate Change, *Climate Change 2007*. Many scientists think this projected sea-level rise is an underestimate, given the latest observations in the field.

3. Kenneth S. Deffeyes, *When Oil Peaked* (New York: Hill & Wang, 2010).

4. McKinsey & Company, *Pathways to a Low-Carbon Economy*, 2009, 6.

Chapter 8

1. U.S. Census Bureau, U.S. & World Population Clocks, www.census.gov/main/www/popclock.html.

2. T. H. Hollingsworth, *Historical Demography* (Ithaca, NY: Cornell University Press, 1969).

3. UN Department of Social and Economic Affairs, Population Division, *Long-Range World Population Projections: Based on the 1998 Revision* (ESA/P/WP.153), executive summary, 1999.

4. John Bongaarts, "Population Policy Options in the Developing World," *Science* 263 (1994): 771–76.

5. Joel Cohen, *How Many People Can the Earth Support?* (New York: Norton, 1995), chap. 3, 4.

Chapter 9

1. Millennium Ecosystem Assessment, *Ecosystems and Human Well-Being: Synthesis* (Washington, DC: Island Press, 2005), 2, www.millenniumassessment.org/documents/document.354.aspx.pdf.

2. UN Food and Agriculture Organization, "What Is Agrobiodiversity?" (fact sheet), 2004, www.fao.org/docrep/007/y5609e/y5609e02.htm.

3. D. J. Rapport and A. M. Friend, *Towards a Comprehensive Framework for Environmental Statistics: A Stress-Response Approach* (Ottawa: Statistics Canada, 1979).

4. Ruth H. Thurstan, Simon Brockington, and Callum M. Roberts, "The Effects of 118 Years of Industrial Fishing on UK Bottom Trawl Fisheries," *Nature Communications* 1 (May 2010), DOI: 10.1038/ncomms1013.

5. Millennium Ecosystem Assessment, *Ecosystems and Human Well-Being*, 3.

6. Martin Jenkins, "Prospects for Biodiversity," *Science* 302 (2003):1175–77.

7. Millennium Ecosystem Assessment, *Ecosystems and Human Well-Being*, 2.

8. D. C. Nepstad, C. M. Stickler, and O. T. Almeida, "Globalization of the Amazon Soy and Beef Industries: Opportunities for Conservation," *Conservation Biology* 20 (2006): 1595–1603.

9. UN Environment Programme, World Conservation Monitoring Centre, "Growth in nationally designated protected areas from 1872 to 2008," 2009, www.bipindicators.net/pacoverage.

10. S. Chape J. Harrison, M. Spalding, and I. Lysenko, "Measuring the Extent and Effectiveness of Protected Areas as an Indicator for Meeting Global Diversity Targets," *Philosophical Transactions of the Royal Society B* 360, no. 1454 (2005): 443–55, DOI: 10.1098/rstb.2004.1592.

11. Paulo Prada, "Poisonous Tree Frog Could Bring Wealth to Tribe in Brazilian Amazon," *New York Times*, May 30, 2006.

Chapter 10

1. World Commission on Environment and Development, *Our Common Future* (New York: Oxford University Press, 1987), 43.

2. William R. Catton Jr., *Overshoot: The Ecological Basis of Revolutionary Change* (Urbana: University of Illinois Press, 1980).

3. Wendell Berry, "The Futility of Global Thinking," in *Learning to Listen to the Land*, ed. Bill Willers (Washington, DC: Island Press, 1991), 150–56. Originally published in *Harper's*, September 1989.

Chapter 11

1. Aldo Leopold, "The Land Ethic," in *A Sand County Almanac* (1949; repr. New York: Oxford University Press, 1966), 217.

2. Leopold, "The Land Ethic," 219.

3. Ibid., 239.

4. Ibid., 220.

5. Aldo Leopold, *Round River: From the Journals of Aldo Leopold*, ed. Luna B. Leopold (1953; repr. New York: Oxford University Press, 1993), 147.

6. Leopold, "The Land Ethic," 219–20.

7. Thomas Jefferson, "Letter to John Holmes (1820)," in *The Founders' Constitution*, vol. 1, ed. Philip B. Kurland and Ralph Lerner (Chicago: University of Chicago Press, 1987), 575.

8. Leopold, "The Land Ethic," 240.

9. Rachel Carson, *Silent Spring* (1962; repr. Boston: Houghton Mifflin, 2002), 189.

10. Ibid., 189.

11. Ibid., 204.

12. Ibid., 205.

13. Ibid., 208.

14. Ibid., 187–88.

15. Ibid., 188.

16. Ibid., 181.

17. Ibid., 176.

18. Ibid., 181.

19. Ibid., 13.

19. Ibid., ???.

20. Linda J. Lear, *Rachel Carson: Witness for Nature* (New York: Henry Holt, 1997), 428.

21. Ibid., 409.

22. Alfred, Lord Tennyson, "Canto LV," in *In Memoriam* (London: Edward Moxon, 1851), 80.

Chapter 12

1. George Modelski and Gardner Perry III, "Democratization in Long Perspective," *Technological Forecasting and Social Change*, 39 (1991): 22, and "Democratization in Long Perspective: Revisited," 69 (2002): 359.

2. The Ocean Project, "Youth," *America, the Ocean, and Climate Change: New Research Insights for Conservation, Awareness, and Action*, 2008, http://theoceanproject.org/wp-content/uploads/2011/11/final_youth.pdf.

3. The Ocean Project, "Youth."

4. Tim Weiner, "Terrific News in Mexico City: Air Is Sometimes Breathable," *New York Times*, January 5, 2001, www.nytimes.com/2001/01/05/world/terrific-news-in-mexico-city-air-is-sometimes-breathable.html?pagewanted=all.

5. Jenny Gitlitz and Pat Franklin, *The 10-Cent Incentive to Recycle*, 4th ed. (Culver City, CA: Container Recycling Institute, 2006), www.container-recycling.org/download/index.php?itemID=1.

6. Thomas Pynchon, *Gravity's Rainbow* (New York: Penguin, 1995).

7. Resources for the Future, William A. Pizer, and Raymond Kopp, "Calculating the Cost of Environmental Regulation" (Discussion Paper 03-06), 2003, www.rff.org/Documents/RFF-DP-03-06.pdf.

8. U.S. Environmental Protection Agency, *Benefits and Costs of the Clean Air Act: Retrospective Study: 1970 to 1990*, October 15, 1997, www.epa.gov/air/sect812/design.html; and *Benefits and Costs of the Clean Air Act: Second Prospective Study: 1990 to 2020*, March 2011, www.epa.gov/air/sect812/prospective2.html.

9. Wendell Berry, "The Futility of Global Thinking," in *Learning to Listen to the Land*, ed. Bill Willers (Washington, DC: Island Press, 1991), 150–56. Originally published in *Harper's*, September 1989.

10. Kenneth Prewitt, "Foundations," in *The Nonprofit Sector*, ed. Walter W. Powell and Richard Steinberg (New Haven, CT: Yale University Press, 2006), 357.

11. Elizabeth T. Boris and C. Eugene Steuerle, "Scope and Dimensions of the Nonprofit Sector," in *The Nonprofit Sector*, ed. Walter W. Powell and Richard Steinberg (New Haven, CT: Yale University Press, 2006).

12. Boris and Steuerle, "Scope and Dimensions of the Nonprofit Sector."

13. Helmut K. Anheier and Lester M. Salamon, "The Nonprofit Sector in Comparative Perspective," in *The Nonprofit Sector*, ed. Walter W. Powell and Richard Steinberg (New Haven, CT: Yale University Press, 2006), 89.

14. The Foundation Center, *Foundation Yearbook*, 2010, http://foundationcenter.org/gainknowledge/research/pdf/fy2010_highlights.pdf.

15. Jeffrey Ball, "The Green Machine: Hal Harvey Spends His Environmental War Chest with One Guiding Principle: Winning," *Wall Street Journal*, February 12, 2007, R12.
Chapter 13

1. William H. McNeill, *The Human Condition: An Ecological and Historical View* (Princeton, NJ: Princeton University Press, 1980), 44.
Chapter 14

1. Arthur C. Clarke, "Hazards of Prophecy," in *Profiles of the Future* (New York: Harper & Row, 1973).

2. William McDonough and Michael Braungart, *Cradle to Cradle: Remaking the Way We Make Things* (New York: North Point Press, 2002).

3. William McDonough and Michael Braungart, "The NEXT Industrial Revolution," *Atlantic Monthly*, October 1998, 82ff, www.theatlantic.com/issues/98oct/industry.htm.

4. Hiroko Tabuchi, "Japan Recycles Minerals from Used Electronics," *New York Times*, October 4, 2010.

5. European Commission, "End of Life Vehicles," http://ec.europa.eu/environment/waste/elv_index.htm.

6. T. E. Graedel and B. R. Allenby, *Design for Environment* (Upper Saddle River, NJ: Prentice Hall, 1996).
NOTES   A-7

7. U.S. Environmental Protection Agency, "Municipal Solid Waste Generation, Recycling, and Disposal in the United States: Facts and Figures for 2008," www.epa.gov/epawaste/nonhaz/municipal/pubs/msw2008rpt.pdf.

8. Iddo K. Wernick, Paul E. Waggoner, and Jesse H. Ausubel, "Industrial Ecology and Wood Products," *Journal of Forestry* 98, no. 10 (2000): 8–14.

9. James A. Nash, "On the Subversive Virtue: Frugality," chap. 22 in *Ethics of Consumption: The Good Life, Justice, and Global Stewardship*, ed. David A. Crocker and Toby Linden (New York: Rowman & Littlefield, 1998), 416–36.

10. Bill McKibben, "Walden Revisited," *DoubleTake* (Spring 1997): 127.

11. McKibben, "Walden Revisited," 129.

12. Franklin Benjamin Sanborn and Bradford Torrey, eds., *The Writings of Henry David Thoreau*, vol. II: *Walden*. (Boston: Houghton Mifflin, 1906), 102.

Chapter 15

1. Edward Abbey, "Havasu," in *Desert Solitaire: A Season in the Wilderness* (1968; repr. New York: Ballantine, 1971). All quotations in this section are from the "Havasu" chapter of this book.

2. "A Short History," from *Collected Poems 1943–2004* copyright © 2004 by Richard Wilbur, reprinted by permission of Houghton Mifflin Harcourt Publishing Company. Originally published in *Mayflies*. All rights reserved.

3. "Nature Poetry," from *Lake Effect Country* by A. R. Ammons. Copyright © 1983 by A. R. Ammons. Used by permission of W. W. Norton & Company, Inc.

4. "Nature Is What We See," by Emily Dickinson. Reprinted by permission of the publishers and the Trustees of Amherst College from *The Poems of Emily Dickinson*, Thomas H. Johnson, ed., Cambridge, Mass.: The Belknap Press of Harvard University Press, Copyright © 1951, 1955, 1979, 1983 by the president and fellows of Harvard College.

5. World Commission on Environment and Development, *Our Common Future* (New York: Oxford University Press, 1987), 43.

6. Russell Gold and Ana Campoy, "Oil Industry Braces for Drop in U.S. Thirst for Gasoline," *Wall Street Journal*, April 13, 2009.

7. National Research Council, *Our Common Journey: A Transition toward Sustainability* (Washington, DC: National Academies Press, 1999), 3, 31.

8. National Research Council, *Our Common Journey*, 7.

9. Ibid., 10.

10. William Cronon, " 'Only Connect': The Goals of a Liberal Education," *The Key Reporter* 64, no. 2 (Winter 1999): 2–4.

11. "Corsons Inlet." Copyright © 1963 by A. R. Ammons, from *Collected Poems 1951–1971*, by A. R. Ammons. Used by permission of W. W. Norton & Company, Inc.

Glossary

accounting identity A mathematical relationship that defines a connection among variables. An accounting identity is not a scientific law but a definition. The IPAT equation in Chapter 14 is an accounting identity.

adaptation The pattern of life that results from natural selection. Species are adapted to the environmental conditions (including other living things) faced by their forebears.

adaptive management An approach to policy implementation that treats policies as experiments, to be designed so that the policy can be monitored to see if it is working and to detect surprises and unexpected outcomes.

age structure The distribution of people in a population, according to age.

agenda setting The process in which the major institutions of a society, including its government, decide which choices are important enough that institutional leaders must pay attention to them. Agenda setting is a chaotic and competitive process in which mass media and visible leaders such as a president have a disproportionate influence, but no single actor or sector can dominate for long.

agent A person or organization that acts on behalf of another person or group of persons. The latter are known as *principals*.

agriculture A set of human institutions organized to produce food by cultivation of plants and feeding of domesticated animals.

anthropogenic radiative forcing Changes in Earth's heat balance caused by human activities, including agriculture, industry, and consumptive activities like driving and heating buildings.

appropriation The influencing or control of natural processes by humans. Damming a river appropriates its flow; clearing land and growing crops on it appropriates the photosynthesis of the plants originally growing there.

biodiversity The range of variation in the species living in an ecosystem, and the genetic variability within each species.

biogeochemical cycle A concept used to track the flows of basic chemical constituents of Earth's ecosystems, including carbon, oxygen, nitrogen, water, and sulfur, as they move from one compartment to another. The compartments are often defined broadly as land, water, and atmosphere.

biogeography The study of the spatial distribution of living things and the

changes in that distribution over time.

biosphere The zone of life on Earth.

bureaucratic organization The human institution for implementing rules and most formal processes. Bureaucratic organization was necessary to the operation of imperial governments that ruled large territories. Bureaucratic organization is still needed for large enterprises such as corporations, as well as contemporary governments.

capital The physical resources—machinery, furniture, and buildings—of a business. Capital is measured in financial terms, such as dollars, but its economic significance lies in what the money can buy, rather than in the money itself.

carbon dioxide The best known of the greenhouse gases in Earth's atmosphere, it is produced by the respiration of living things and by the burning of fossil fuels.

carnivore An animal that eats animals.

cartogram A world map used to illustrate the differences among nations of the world in a particular property, such as population. Instead of using a display that represents the relative *areas* of the nations, the cartogram distorts the areas in proportion to the *property* under study. In a population cartogram,

China, with more than 1.3 billion people, appears about 60 times larger than lightly populated Australia, although the land areas of the two countries are about the same.

civic science The commitment to combining rigorous science with practical politics in a search for environmentally responsible governance.

civil society An umbrella term for social activities occurring outside of governmental institutions and the profit-making world of commerce. Nonprofit organizations play prominent roles in civil society.

classical environmentalism A pattern in which a group of concerned citizens and scientists work to identify environmental problems and put pressure on government agencies to solve them. Classical environmentalism has been effective in addressing many of the more clearly defined problems of the past but has been less successful in dealing with more subtle or complex challenges, some of which are no less significant.

climate The long-term pattern of weather in a place.

community governance One way to handle commons problems, through management by the community that has traditionally used the resources of the commons.

community-governance principles A set of generalized rules for communities managing commons, developed by the political scientist Elinor Ostrom.

concentrated interests Interest groups that have fewer members, greater ease of communication, and often a direct financial interest in governmental action. Such groups are often called "special interests."

demographic transition A scenario describing the historical trajectory of human population size. This pattern was first observed in Europe in the seventeenth century, and then later elsewhere in the world.

demography The science of population, including the collection of birth and death records, and their analysis using models incorporating estimates of disease and other causes of mortality, as well as changes in fertility and birthrate.

diffuse interests Commonly defined as the opposite of concentrated interests. Diffuse interests are shared by wide ranges of citizens who generally expect nonfinancial benefits and may therefore be less attentive, politically, to actions that undermine their interests. Environmental issues are classic examples of diffuse interests, although environmental nongovernmental organizations can be seen as developing at least some concentrated interest groups around environmental issues.

discounting An analytical process for estimating the present value of a service or good available in the future.

disproportionality The surprising fact that a large fraction of environmental damage usually comes from a small part of the economy, so that addressing this damage would not be costly to society, either as a whole or in terms of jobs lost.

division of labor A form of social organization in which economic and social roles are specialized. Women specialize in childcare in many societies, for example, and legal tasks such as police work are usually separate occupations in industrialized societies.

economic inequality A measure of the difference in economic welfare between rich and poor. Inequality may be estimated for individuals, nations, or regions.

economic specialization In a complex, industrialized economy, different firms do different things. Specialization is a driving force behind *interdependence* and *disproportionality*.

ecosystem A set of living organisms, together with the nonliving geological and aquatic environment they inhabit. Some ecosystems are nearly closed, meaning that their species do not usually move in and out of the system or interact with living things outside. Most ecosystems, however, are open to some degree.

ecosystem services Services desired by people that are provided by a managed or unmanaged ecosystem. These include the fertility of soil in which crops are grown, detoxification of wastes in flowing streams, and moderation of floods by vegetation growing on slopes.

ecotourism Businesses built on recreational uses of natural areas for viewing animals, plants, and places in a nondestructive way. Ecotourism is often based on visits to protected areas, with local people often acting as guides and hosts for visitors.

efficient allocation With a fixed set of resources (capital, labor, and land), a community of economic actors can use those resources in various ways. The allocation that produces the maximum satisfaction to households and the maximum profit to firms is the efficient allocation. Economic theory provides an explanation of how a market can lead to an efficient allocation of resources in the society using that market.

enclosure The historical transformation in Britain from a system of open fields, called commons, where all could have access, to a system of privately held pastures and fields. Enclosure displaced large numbers of landless peasants, whose ancestors had earned their livelihood in the commons.

endemic species A species whose spatial range is limited to a single place.

energy efficiency The ratio between the social value created by the use of an energy source and the amount of energy expended to achieve that result. Raising energy efficiency, or energy conservation, constitutes the largest source of energy that could be tapped while reducing greenhouse gas emissions.

energy intensity The energy input required to generate the output of an economy. Increases in energy efficiency lower energy intensity, whereas new products or activities that use energy without a corresponding gain in economic output raise the intensity.

environmental justice A field of policy and academic study to analyze the intersection of social justice and environmental problem solving. Environmental justice is a response to the fact that the environmental risks borne by minority populations are often much higher than in neighboring communities.

evolution The process in which natural selection produces new species.

experimental (theory-driven) science Science that advances understanding through experiments that test the predictions of theory. Virtually every scientific discipline draws upon both experimental and *observational* science.

exponential growth Growth in a variable over time that is characterized by a *proportional* change in the variable in each time period. An exponentially growing population will double in size in a specific time and will continue to do so.

externality Costs (or benefits) that are real but are not taken into account in market transactions. Pollution is an example of a negative externality of many consumption and production processes.

food web The relationship among organisms in which plants and phytoplankton are consumed by herbivores, which are in turn consumed by carnivores and organisms that eat both plants and animals (omnivores). Food webs include organisms that decompose dead organisms, and parasites that do not kill their prey but do consume their tissues.

fossil fuel Oil, natural gas, coal, or other mineral resources formed by the decay and transformation of once-living things. Fossil fuels have provided a potent source of concentrated energy to power industrial economies.

free rider. A free rider is a person (or organization) who enjoys a benefit, such as clean air or improved environmental conditions, while paying less than a fair share of the costs.

free rider problem The question of how to spread costs equitably, particularly in cases in which Hardin's "mutual coercion" isn't feasible.

gene A component of a living cell that contains information for producing the constituents of the cell or its descendants. The information is coded in deoxyribonucleic acid (DNA) molecules.

genome The genetic information that defines a species—usually the DNA sequences found in the cellular nuclei of an organism.

Gini coefficient A statistical measure of the degree of inequality in a set of measurements. Used to analyze disproportionality.

global warming The observed and predicted rise in Earth's average temperature. Although the magnitude and rate of increase of Earth's temperature in the future cannot be predicted with precision, the physics of the greenhouse effect are not in doubt.

government ownership One way to handle commons problems, through public management of a commons.

grand challenge A term this book applies to large, significant commons problems (including climate, biodiversity, urbanization, and sustainable development), for which solutions have not yet been found, even though the problems themselves threaten the continuation of human civilization in its present form.

greenhouse effect The process by which light energy enters a system

and is trapped there as heat, which cannot escape as easily in the form of infrared radiation.

greenhouse gas A gas in Earth's atmosphere that is transparent to visible light but opaque in the infrared. Although greenhouse gases such as water vapor, carbon dioxide, and methane are found in low concentrations in the atmosphere, they play a significant role in determining the steady-state average temperature of the planet. They produce the greenhouse effect, the mechanism of global warming.

gross domestic product (GDP) per capita GDP is a measure of the total economic activity of a country, which is often used to compare the average incomes of people in different societies. GDP per capita is the standard measure of economic welfare collected in national statistics annually by the World Bank.

herbivore An animal that eats only plants.

heritable trait A characteristic of an individual that can be inherited by its descendants, such as hair color.

Human Development Index A composite measure of well-being that combines national statistics on GDP per capita, life expectancy, and two measures of education. It is published annually in the UN Development Programme's *Human Development Report*.

hypothesis A plausible or potential interpretation or explanation for an observed pattern or phenomenon that is intended to be tested by more detailed, scientific investigation.

Industrial Revolution The harnessing of nonanimal sources of energy to manufacture goods, which began in the late eighteenth century in northwest Europe.

informal employment Economic activities that are "off the books" and unrecognized by government. The informal economy is dominant in most poor countries, accounting for about three-quarters of all employment.

informal housing Housing built without legal land title or, in most cases, access to utilities such as water and electricity. Slums, which house more than 1 billion people worldwide, are a prominent example of informal housing.

infrared radiation Light that is longer in wavelength than can be seen by the human eye. Earth's atmosphere is opaque to much infrared radiation, whereas it is transparent to the visible wavelengths.

infrastructure Permanent installations that enable human settlements to harness the resources of nature for human ends, such as water, sewer, transportation, communications, and energy distribution systems. Infrastructure is usually built as a network linking suppliers and a population of consumers.

institution A set of rules, expectations, rights, and duties that have been accepted by a human group, which shapes behavior among its members and is more enduring than the participation by individual members.

institutional solution A change in an institution's rules, expectations, and practices, or the creation of a new institution, to address an environmental problem. Examples include privatization, regulation, and modification of a community's governance of shared resources.

interdependence The economic and social relations of mutual dependence and exchange that grow within a specialized, industrialized economy.

interest group A group organized to pursue its political aims in a pluralist political system.

invisible present A name for environmental changes that occur too slowly for humans to perceive, which sometimes means that human institutions have difficulty recognizing those changes.

jet stream High-altitude, fast-flowing winds along the seams between the large cells of the atmosphere. Jet streams generally flow from west to east.

land The economic term used to describe resources taken from the natural world. Land includes ecosystem services such as soil fertility and natural resources such as fish.

land ethic The idea that the treatment of ecosystems (land) must be considered in ethical terms, as matters of right and wrong.

life-cycle analysis An accounting system that measures the environmental impacts of a product's manufacturing and use, so that the use of energy, water, and other materials, and the generation of pollution and waste, can be understood over the entire life of the product.

logarithmic A way to present quantitative information using the logarithm of the value of a variable. In a logarithmic presentation, exponential growth appears as a straight line.

LULU "Locally undesirable land uses," which may bring profits to proponents, but which tend to bring risks, damage, and/or stigma to almost any place they are located.

macroparasite Individuals and institutions that seize goods or compel services from people. According to William McNeill, governments in the preindustrial world may be understood as macroparasites.

market failure A set of actual or potential economic transactions in which externalities are ignored. As a result, decisions to produce or to consume a good or service are made in response to erroneous signals. Market failures can inhibit the provision of desirable things, such as research, as well as encourage over-use of other things, as with driving when there are equally effective mass transit options.

Maya A group of people who emerged in Central America and created impressive city-states and a civilization that lasted for more than a thousand years before collapsing rapidly in about AD 1000.

metabolism The biochemical processes of living organisms that are devoted to sustaining life but may not lead to growth or reproduction.

microhabitat The local environment in which a living thing spends much of its life.

mutation A change in a gene that may alter the characteristics of a cell or its descendants. Mutations arise spontaneously, and humans can also cause mutations using a variety of techniques.

mutual coercion mutually agreed to Restraints on individual choices, which are needed to address tragedies of the commons.

natural selection The process by which variations in the survival of members of a population, due to differences in the traits of the individuals, produces changes in the characteristics of that population over generations.

net primary productivity (NPP) A measure of the photosynthetic activity of an area; usually measured in the mass of carbon fixed by photosynthesis.

NIMBY "Not in my backyard," a label given (usually by a project's proponents) to the people who seek to avoid the location of an undesirable facility near their homes or communities.

nonprofit organization A civil society organization whose mission is not centered on earning profits to be distributed to owners. Examples include universities, churches, and community land trusts.

observational (field) science Science that advances understanding by seeking generalizations based on observation. Virtually every scientific discipline draws upon both observational and *experimental* science.

oil reserves The volumes of oil below ground that are known to exist but have not yet been extracted for human use.

omnivore A life-form whose diet includes both plants and animals.

open access The condition under which no rules are in place to limit the use of a commons.

optimal The level or quantity that best realizes the economic return possible under a given set of circumstances.

outliers Data points in a set of measurements that have extreme values, outside the range that seems to be defined by the majority of the data.

Pareto optimality An allocation of resources between two actors in which one actor cannot move to a different, better allocation without making the other actor worse off. This is an efficient allocation.

permanent settlement Year-round habitation of people in a community, with individuals usually having specialized roles.

philanthropic foundation A civil society organization that makes grants to nongovernmental organizations. Foundations are governed by independent boards of trustees who make decisions about distributing funds from the foundation's endowment.

photosynthesis The chemical process by which light is captured by living things and converted into chemicals used for energy and the growing of material parts of organisms, such as leaves. In a transformation called "fixing," the carbon from carbon dioxide in the atmosphere is transformed into sugars and other compounds.

phytoplankton Single-celled organisms that carry out photosynthesis. Phytoplankton play a vital role in freshwater and marine ecosystems, where they are responsible for a large share of the net primary production.

pluralism A theory of politics that interprets governmental behavior as the result of competition among interest groups.

precautionary principle The idea that environmental (or other) changes should not be presumed to be harmless until shown otherwise. Instead, the proponents of such changes should be required to analyze the predicted effects and disclose them to affected communities.

precipitation Water falling from the atmosphere to the ground in the form of rain, snow, hail, sleet, or mist deposited by fog.

pressure-state-response (PSR) The evaluation of readily available measures of stresses on an ecosystem (pressures), the condition of that ecosystem (state), and human responses to changes in state, so as to gain an understanding of the ecosystem useful enough to guide human actions in its care and use.

prevailing winds The average direction and speed of the wind in a place. Prevailing winds provide indications of the large-scale circulation of Earth's atmosphere.

price The amount exchanged between buyer and seller in a market transaction.

principal A person representing him- or herself, or the interests of a group. When the group is an organization, the principals may be the board of directors or leadership body. When the group is not formally organized, the members of that group may be principals. In all cases, principals may have actions taken in their name by agents, although whether the agents are faithful to the interests of the principals is often unclear.

privatization One way to handle commons problems, through divestiture of common resources into privately held assets that are managed in accord with their owners' priorities.

protected area A spatial zone, such as a national park, in which some human activities that harm species or ecosystems are restricted, such as agriculture or hunting. The restrictions on human use may not,

however, enable ecosystems under pressure to become more resilient over time.

public policy Governmental rules, usually based on legislation or other publicly adopted mandate for public action.

public trust A legal concept that defines commons as resources owned by the people and administered by government as a trustee of its citizens.

regulated industry Firms whose economic behavior is affected by government regulations, including environmental regulations.

regulation A public policy that prescribes or prohibits behaviors of economic and nongovernmental actors.

rent-seeking Activities directed toward changing public policy to favor a specific economic interest. Rent-seeking includes political contributions, lobbying, and funding scientists and other experts who make a case on behalf of the rent seekers before government and in the media.

resilience A property of an ecosystem that describes its ability to maintain its functions and structure as stresses are imposed upon it and as individual species within the system die out or are introduced. A resilient ecosystem can withstand shocks and rebuild itself when necessary.

risk analysis The study of the probabilities that govern environmental impacts and their anticipated consequences, which are often delayed and uncertain.

scenario A story used to explain a pattern of behavior. Scenarios are often helpful in translating the abstract projections of computer models into terms that relate to human experience. Scenarios are not predictions, however.

sea-level rise As climate change warms ocean water, causing it to expand, and as the volume of water increases due to runoff from melting snow and ice on land, the average level of ocean water slowly rises.

self-sufficient Being able to live independently, without ties to the interdependent economy of the contemporary world.

social capital The value of social relationships and networks, which is shared by the members of a network and by society beyond the members. Trust—in people and in the networks in which they participate—is an important attribute of social capital, as is the predictability of institutions and their rules.

social cost Costs inflicted by externalities, which are not taken into account in market transactions. If social costs are left out of market transactions, the resulting allocation of resources will not be efficient.

social learning Social change reflecting improved understanding of what works.

specialized roles A social order in which not every family is engaged full time in subsistence. Usually possible only with a productive agricultural system.

"stationary man" A term used by eighteenth-century English clergyman Gilbert White to describe himself and his willing attachment to a small, well-defined place.

steady state A condition in a dynamic system in which competing processes produce a stable level of a key variable. For example, the generation of heat and the loss of heat in the human body produces a steady-state body temperature that is useful as a measure of health.

sustainable development An approach that aims to reconcile improvement in the quality of human life, the achievement of equity (especially poverty alleviation), and protection of the environmental systems on which humans and other species depend.

thermodynamics The science of heat developed in the early nineteenth century that provided the scientific underpinning for designing machines such as the automobile, which harness the burning of fossil fuels to do mechanical work useful to humans and the economy.

third-party effect Intended or unintended impacts of economic decisions that fall on people who are not party to the transactions.

topography The shape of Earth's surface, including, for example, hills, valleys, plains, and mountain ranges.

Toxics Release Inventory (TRI). A government database that reports the annual releases of pollutants by large industrial facilities.

tradition A long-established value or process shared by a community. Traditions play an important role in community governance of commons.

tragedy of the commons A situation in which members of a group face incentives as individuals that produce behavior contrary to the interest of the group as a whole.

transaction cost A cost that arises in the course of making a transaction in a real (that is, imperfect) market. The cost of information is a common example of a transaction cost, as is the cost due to delays in completing the transaction.

treaty An agreement among two or more nations. In contrast to a law administered by a government, which usually binds all of its citizens, a nation can withdraw from a treaty without formal sanctions.

trial and error A method of learning by doing and by learning from mistakes.

urban penalty The difference in human mortality between urban and rural areas. The urban penalty is absent now—life span is longer in cities than in surrounding rural areas—but this is a historically recent development in poor countries.

urbanization The process of human settlement in cities. In many statistical summaries, an urban place is defined as having more than five thousand residents.

usufruct right The right to use a resource temporarily, usually on the understanding that the resource will not be depleted or harmed.

well-being A concept of human development broader than economic welfare. The Human Development Index is a widely used measure of well-being.

world without edges The name given to a global economy and worldwide information system in which the clearly defined sense of place that characterized older societies has been transformed into a world in which people's material ties and information sources extend far beyond their knowledge or understanding.

Credits

Chapter 1

Figure 1.1: Kevin Schafer/Peter Arnold/Getty Images; **Figure 1.2**: Friends of the Island Fox (www1.islandfox.org).

Chapter 2

Figure 2.1: Yale University Art Gallery/Art Resource, NY; **Figure 2.2**: © The Metropolitan Museum of Art/Art Resource, NY; **Figure 2.3**: Montclair Art Museum, Museum Purchase; Lang Acquisition Fund, 1945.8; **Figure 2.4**: Wikimedia Commons; **Figure 2.5**: © Christie's Images/The Bridgeman Art Library International; **Figure 2.6**: Marc Riboud; **Figure 2.7**: Apollo 17 Crew/NASA; **p. 37**: David Dethier, Williams College; **Figure 2.8**: Don Smith/The Image Bank/Getty Images; **Figure 2.9**: Frank Lynch.

Chapter 3

Page 56: A Less Mighty Mississippi by David Constantine and Joe Burgess from "Time to Move the Mississippi, Experts Say" by Cornelia Dean, *The New York Times*, September 19, 2006. Copyright © 2006 The New York Times. Reprinted by permission of the New York Times Company. All rights reserved.

Chapter 4

Figure 4.1: Figure 2b from "Privileged Access, Privileged Accounts: Toward a Socially Structured Theory of Resources and Discourses." Freudenburg. *Social Forces* 94 (1): 89–114. Reprinted by permission of Oxford University Press Journals; **Figure 4.2**: Figure 3b from "Privileged Access, Privileged Accounts: Toward a Socially Structured Theory of Resources and Discourses." Freudenburg. *Social Forces* 94 (1): 89–114. Reprinted by permission of Oxford University Press Journals; **Figure 4.3**: William Day, c1789/Derby City Council/Derby Museum & Art Gallery; **Figure 4.4**: Superstock/Everett Collection (1095-540); **Figure 4.5**: Arvind Garg/Corbis.

Chapter 5

Figure 5.1 (left): Kai Lee; **Figure 5.1 (right)**: Apollo 17 Crew/NASA; **Figure 5.2**: Figure 4 from: *Meteorology Today: An Introduction to Weather, Climate, and the Environment*, 1985, by C. Donald Ahrens. Copyright 1985 © Cengage Learning; **Figure 5.3**: C. Donald Ahrens, *Meteorology Today: An Introduction to Weather, Climate, and the Environment,* 1985; **Figure 5.4**: Figure "Net Precipitation," by J. M. Wallace, Y. T. Hwang, D. Frierson, University of Washington, 2011. Reprinted courtesy of Y. T. Huang, University of Washington; **Figure 5. 5**: NOAA; **Figure 5.6**: C. Donald Ahrens, *Meteorology Today: An Introduction to Weather, Climate, and the Environment,* 1985; **Figure 5.7**: C. Donald Ahrens, *Meteorology Today: An Introduction to Weather, Climate, and the Environment,* 1985; **Figure 5.8**: Mark Monmonier, *Air Apparent: How Meteorologists Learned to Map, Predict, and Dramatize Weather* (Chicago: University of Chicago Press, 1999), 11; **Figure 5.9**: NASA; **p. 115**: © The Metropolitan Museum of Art/Art Resource, NY; **p. 117**: Ursula Goodenough, *The Sacred Depths of Nature* (Oxford, UK: Oxford University Press, 1998); **Figure 5.10**: © Academy of Natural Sciences of Philadelphia/Corbis; **Figure 5.11**: David Peart, Dartmouth College; **5.12**: Bill Hatcher/National Geographic/Getty Images; **Figure 5.13**: Ralph A. Clevenger/Corbis; **p. 122**: Marc L. Imhoff et al., "Global Patterns in Net Primary Productivity (NPP)," 2004; data distributed by the Socioeconomic Data and Applications Center, http://sedac.ciesin.columbia.edu/es/hanpp.html.

Chapter 6

Figure 6.1: Figure 4-24: "Generalized pyramid of energy flow in the biosphere," from *Living in the Environment*, 8th Edition, by G. Tyler Miller Jr. Belmont, CA: Wadsworth, 1994. P. 92; **p. 132**: Source: U. S. Environmental Protection Agency, "Hypoxia in the Gulf of Mexico and Long Island Sound," December 2008, http://cfpub.epa.gov/eroe/index.cfm?fuseaction=detail .viewMidImg&ch=50&lShowInd=0&subtop=315&lv=list. listByChapter&r=201562; **Figure 6.2**: Georg Gerster/Photo Researchers, Inc.; **Figure 6.3**: InflationData.com; **p. 145**: GlobalFootprint.org: http://pthbb.org/natural/footprint. Copyright © Jerrad Pierce. Creative Commons Attribution. Reproduced with permission; **Figure 6.4**: Millennium Ecosystem Assessment, *Ecosystems and Human Well-Being: Synthesis* (Washington, DC: Island Press, 2005); **p. 150**: NASA; **p. 151**: NASA; **Figure 6.5**: From *Human Domination of Earth's Ecosystems* by Vitousek et al, 1997. Reprinted by permission of the American Association for the Advancement of Science.

Chapter 7

Figure 7.1: Figure 1.1 from *Geoengineering the Climate: Science, Governance and Uncertainty.* Royal Society Policy document, October 2009. London. Reprinted by permission of Royal Society (UK) 2009; **Figure 7.2**: Dr. Pieter Tans, NOAA/ESRL (www.esrl.noaa.gov/gmd/ccgg/trends) and Dr. Ralph Keeling, Scripps Institution of Oceanography (scrippsco2.ucsd.edu); **Figure 7.3**: Figure 1.3 from *Geoengineering the Climate: Science, Governance and Uncertainty.* Royal Society Policy document, October 2009. London. Reprinted by permission of Royal Society (UK) 2009; **Figure 7.4**: Climate Change 2001: The Scientific Basis. Contribution of Working Group I to the Third Assessment Report of the Intergovernmental Panel on Climate Change, SPM, Figure 5. Cambridge University Press; **p. 166**: Liu Yongqiu/Xinhua Press/Corbis; **Figure 7.5a and 7.5b**: NASA Image/Overlay by Sharron Macklin; **Figure 7.6**: National Snow and Ice Data Center, http://nsidc.org/images/ arcticseaicenews/20090917_Figure1.png; **Figure 7.7**: for 1949–2009, U. S. Energy Information Administration, Annual Energy Outlook, Total Energy, October 19, 2011, Table 1.2, Primary Energy Production by Source, http://www.eia.gov/ totalenergy/data/annual/showtext.cfm?t=ptb0102; for 1775– 2009, ibid., Table E.1, Estimated Primary Energy Consumption in the United States, 1645-1945. http://www.eia.gov/total energy/data/annual/showtext.cfm?t=ptb1601http://www.eia .gov/totalenergy/data/annual/showtext.cfm?t=ptb0102; **p. 173**: SASI Group (University of Sheffield), www.worldmapper.org/ display.php?selected=119; **Figure 7.8**: U.S. Energy Information Administration, *Annual Energy Review 2010*, Fig. 2.0, www.eia .gov/totalenergy/data/annual/pecss_diagram.cfm; **Figure 7.9**: U. S. Energy Information Administration, Real Prices Viewer, Annual Average Motor Gasoline Prices, January 2012; **Figure 7.10**: "Who Has The Oil, Who Uses The Oil" by Environmental Action. http://www.timescapemedia.com/ uptake/ whohasoilmap.pdf; **p. 182**: Lugar Energy Initiative, http://lugar.senate.gov/energy/ graphs/oilimport.html, citing data from U.S. Energy Information Administration; **Figure 7.11**: Data from U.S. Energy Information Administration and BP Statistical Review for 2007, www.eia.doe.gov/

emeu/international/reserves.html; **Figure 7.12**: Data from U.S. Energy Information Administration, http://tonto.eia.doe.gov/ dnav/pet/hist/mcrfpus1a.htm; **Figure 7.13**: From "Pathway to Low Carbon Economy Report," McKinsey & Co. Reprinted by permission of the publisher.

Chapter 8

Figure 8.2: From *Historical Demography* by T. H. Hollingsworth, p. 311. Reprinted by permission of Hodder Education; **Figure 8.4**: UN Department of Economic and Social Affairs, Population Division, *World Population Prospects: The 2008 Revision*; **Figure 8.5**: UN Department of Economic and Social Affairs, Population Division, *World Population Prospects: The 2008 Revision*; data from Population Reference Bureau; **p. 203**: Age structures of the rich and poor," World Population Data Sheet (briefing), August 2009, www.prb.org, citing data from UN Population Division, World Population Prospects, the 2008 Revision, page 2. Reprinted by permission of the Population Reference Bureau; **p. 207**: Figure: 1-7 Mexico City Urban Subsystem, 1995 by Gustavo Garza. National Research Council, *Cities Transformed* (2003). Reprinted by permission of Colegio de Mexico; **Figure 8.6**: Figure 1-1, *Cities Transformed* (2003), National Research Council. Reprinted by permission of the World Bank; **Figure 8.9**: Kai Lee; **Figure 8.10**: Figure 1-3, *Cities Transformed* (2003), National Research Council. Reprinted by permission of the World Bank; **Figure 8.11**: Kai Lee; **Figure 8.12**: Kai Lee; **Figure 8.13**: Kai Lee.

Chapter 9

Page 233: Jim Carlton, Williams-Mystic Maritime Studies Program; **Figure 9.2**: UN Environment Programme, World Conservation Monitoring Centre, Global Generalised "Original" Forest dataset (v 1.0), March 1998; and Global Coral Reef Distribution, 2010; **Figure 9.3**: Kai Lee; **Figure 9.4**: Reprinted by permission of the publisher from *The Diversity of Life* by Edward O. Wilson, p. 195, Cambridge, Mass.: The Belknap Press of Harvard University Press, Copyright © by Edward O. Wilson; **Figure 9.5**: Reprinted by permission of the publisher from *The Diversity of Life* by Edward O. Wilson, p. 191, Cambridge, Mass.: The Belknap Press of Harvard University Press, Copyright © by Edward O. Wilson; **Figure 9.6**: Michael Fay/National Geographic/Getty Images; **Figure 9.7**: Figure 2 on page 3 of "The Overall Synthesis Report." Reprinted by permission of the Millennium Ecosystem Assessment; **Figure 9.8**: "Figure: Biodiversity hotspots include many areas of rapid population growth." From Population Action International; **Figure 9.9**: KILIAN FICHOU/AFP/Getty Images/News

Chapter 10

Figure 10.1: NASA; **Figure 10.2**: Angus Maddison, *The World Economy: A Millennial Perspective* (Paris: OECD, 2003), 262; **Figure 10.3**: UN Development Programme, *Human Development Report 2005*; **Figure 10.4**: Paul Nevin/Photolibrary/ Getty Images; **p. 271**: Courtesy of Charles Benjamin;

p. 272 top, bottom: Courtesy of Charles Benjamin; **Figure 10.5**: Human Development Index (HDI) value: HDRO calculations based on data from UNDESA (2011), Barro and Lee (2010), UNESCO Institute for Statistics (2011), World Bank (2011a) and IMF (2011); **p. 275**: Gapminder.org; **p. 276**: Gapminder.org;

Chapter 11

Figure 11.1: © The Metropolitan Museum of Art/Art Resource, NY; **Figure 11.2**: © 2011 DigitalGlobe/Getty Images.

Chapter 12

Page 316: George Modelski and Gardner Perry III, "Democratization in Long Perspective," *Technological Forecasting and Social Change*, 39 (1991): 22, and "Democratization in Long Perspective: Revisited," 69 (2002): 359; **Figure 12.1**: From "Environment," Gallup.com, 2010. http://www.gallup.com/poll/1615/environment.aspx. Reprinted by permission of Gallup.

Chapter 13

Figure 13.1: Figure 1.1 from *Economic Theory for Environmentalists* by Gowdy and O'Hara, 1995. St. Lucie Press, Delray Beach, FL. Reprinted by permission of Taylor & Francis Group, LLC; **Figure 13.2**: Figure 1.2 from *Economic Theory for Environmentalists* by Gowdy and O'Hara, 1995. St. Lucie Press, Delray Beach, FL. Reprinted by permission of Taylor & Francis Group, LLC.

Chapter 14

Page 384: Drawn after Graedel, Thomas E. and Brad Allenby, 1998. *Design for Environment*, 1st ed., Englewood Cliffs, N.J.: Prentice-Hall. P. 109. Reprinted by permission of Pearson Education; **p. 385**: Drawn after Graedel, Thomas E. and Brad Allenby, 1998. *Design for Environment*, 1st ed., Englewood Cliffs, N.J.: Prentice-Hall. P. 110. Reprinted by permission of Pearson Education; **Figure 14.1**: Figure from "Is Richer Greener?" by Jesse Ausubel and Nadja M. Victor. *The New York Times,* April 22, 2009. Reprinted by permission of the author; **Figure 14.4**: © 2011 Okenfenokee Glee & Perloo Inc.; **Figure 14.5**: Data Source: FRED, Federal Reserve Economic Data, Federal Reserve of St. Louis: Personal Saving Rate [PSAVERT]; U.S. Department of Commerce: Bureau of Economic Research; http://research.stlouisfed.org/fred2/series/PSAVERT; **Figure 14.6**: Kim Yi; **Figure 14.7**: Courtesy of Monterey Bay Aquarium.

Chapter 15

Figure 15.1: U. S. Energy Information Administration, "Energy Demand," *Annual Energy Outlook 2010*, www.eia.doe.gov/oiaf/archive/aeo10/demand.html; **p. 412**: World Development Report 1992. *Development and the Environment.* New York: Oxford University Press; p. 11. Reproduced by permission of the World Bank; **p. 421**: From "Linking Knowledge with Action." Reprinted with permission of the David and Lucile Packard Foundation.

Index

Albania, well-being of people in, 273
Aleutian Islands, 261
Alidhoo Island, Maldives, *166*
All-Consuming Century, An (Cross), 394
Allenby, Brad, 383, 384
almond production, 144
al-Qaeda, 105
aluminum, recycled metal in, 399
Amazon River basin, 105, 108, 242
America
  carbon emissions in Japan *vs.* in, 174, 186
  Hudson River School and idealistic portrait of, 23–24
American Dream, 281
American economy, tragedy of the uncommon and, 86–87
American Fork River, Utah, cleanup of, 341
American Museum of Natural History, 241
American Progress (Gast), 29, *29, 30*, 61
American Revolution, 91, 316, 378
American Romanticism, descendants of, 31
Ammons, A. R., 404, 431
Amnesty International, 335
analytic geometry, 352
Andes Mountain, 106
Angola
  land suitable for conversion to agriculture in, 242
  oil production in, 182
animals, 111
  climate and, 156
  in crown groups on tree of life, *117*
  cultivated, settlement and, 135–36
  domestication of, 131, *133t*, 134, 261
Annex B countries, 188
Antarctica, 261

global warming and Larsen B ice shelf in, 169, *170*
  major changes in ice cover in, 249
  ozone hole above, 57–58
anthropogenic radiative forcing, 159
anthropology, 34, 35*t*
antibacterial products, from natural products, 255
antibiotics, 229–30
antipollution policies, classical environmentalism and, 13
antislavery activism, 92
Appalachian Mountains, 100, 312
Appalachian Trail, 21
apple skin, thickness of, *102, 103*
appropriation, 145
Arab uprisings of 2011, 316
Aral Sea, Kazakhstan
  dust storm over, *151*
  shrinkage of, 149, *150, 150*–51, 250, 332
Archea, *117*
Arches National Park, Utah, 40, *40*
Arctic National Park, Gates of, 253
Arctic National Wildlife Reserve, 300
Arctic region
  climate change and, 360
  imprint of human activity in, 11
  major ice cover changes in, 249
Argentina, land suitable for conversion to agriculture in, 242
aristocracy, 139
Arizona
  in rain shadow, 105
  "sky island" ecosystems in, 166
Arkwright, Richard, 92, 315
Army Corps of Engineers, 322
Arrhenius, Svante, 158, 159, 163

art, Romanticism in, 20, 21.
  see also Hudson River School paintings
artifacts, redesign of, 383–83, 386
Asia, 277
  age structure in, 202
  density of population in, 192
  economic growth in, 270
  gross domestic product per capita in, 1700-2000, 267
  new generation of nuclear power plants in, 180
  oil reserves in, *183*
  population estimate, growth rate, total fertility rate in, 193*t*
  population growth in, 201
  urbanization in, 213
  urban population by region, 1950, 1990, and 2005 in, *212*
  urban population growth in, 208
Asian carp, in Mississippi River, 247
asthma, 306, 366, 367
astronomy, 33, 35*t*, 43
Atlantic barnacle *(Amphibalanus improvisus)*, on San Francisco Bay seawall, *234*
Atlantic Ocean, 105, 232, 234
atmosphere, 103, 123
  amount of carbon in, 161–62
  Earth's carbon cycle and, *162*
  Jupiter, visible banding of, 111, *111*
  nitrogen in, 149
  public trust in, 63
  three-cell model of, 108, *110, 111*
  two-dimensional model of, *104*
  water in, 105
atmospheric commons, governing of, 155
atomic power, 179

auctions
  electromagnetic spectrum and, 60, 61
  emissions rights and, 63, 86
Audubon, John James, 116, 118, *119*
Australia, deserts in, 108
Australian tubeworm *(Ficopomatus enigmatius)*, on San Francisco Bay seawall, *234*
automobiles, 20, 34, 98, 282
  complex institutions built around, 12
  hybrid, 398
  low density of suburbs and, 73
  mileage standards for, 180–81, 328
  renting, 380, 381
  tragedy of the commons and, 53
averages, 80
avian flu virus, 147
Ayres, Robert, 78

bacteria, 117, *117*
bald eagles, 323
  DDT ban and resurgence of, 8
  golden eagles and, on Channel Islands, 9
bamboo, 386
Bangladesh, 72, 264
  population growth rates in, 213
  sea-level rise and, 165
barcode of life, 241
barnacles, on San Francisco Bay seawall, 233, *234*
Barnes, Peter, 63
Baxter State Park (Maine), 21
beans, domestication of, 131
beauty, Leopold's insistence on, 293
beaver pelt trade, New England colonists and, 52–53, 57, 88, 266
bees, pesticides and, 303
Beijing, China, 211*t*
belief, nature and, 24–32

DDT, 295
banning of, 8, 12, 18, 299–300
birds and, 7, 8, 12, 295, 302
Channel Island foxes and, 7–8
dead zones, 249
death rate, 193, 194, *194*
in cities, 218
falling, population growth and, 195–96
decarbonization, 386–88, *388,* 390
deciduous trees
diversity and height of, in rain forests, *119*
seasonal fluctuations and, 112
decomposers, *129*
Deepwater Horizon well oil spill (2010), 184
deer populations, elimination of wolves and increase in, 226–27, 290
Deffeyes, Kenneth, 185
deforestation, hot spots, *245*
Delhi, 277
dematerialization, 386–88, 390
democracy, 139, 282, 314, 315–17, 357
consumerism and, 394–95
reinvigorating, 281
rent-seeking and dilemma of, 54
Democratic Party, 317, 321
demographic projections, making, 200
demographic transition, 193–95, *194,* 196, 197, 202, 224, 262, 264, 408, 410, 425
accelerating conclusion of, 204
relative to broad phases of human history, *198*
demography, 192
Department of Agriculture, 322
deserts, 100, 107–8
Desert Solitaire (Abbey), 40, *40,* 402

design
ecological, 383, 386
environmentally responsible careers in, 401
sustainability and, 379–80
designing experience, 41–43
developed countries
age structures of, *203*
private property transactions and, 49
developing countries
age structures of, *203*
economic growth in, 270
legitimacy in, 272–73
living among people of, 269
urbanization in, 205, 208, 214, 223
Development of American Agriculture, The (Cochrane), 61
Diamond, Jared, 131, 134
Dickinson, Emily, 20, 24–25, 27–28, 94, 115, 404
Didion, Joan, 39, 40, 43
diffuse interests, 326
dinosaurs, 112, 240
diphtheria, 196
direct pressures
biodiversity and, 243–45, 251
response to, 252–53
discounting, 364, *364,* 365
disease organisms, human settlements and, 138
disproportionality, 124, 387
in American economy, 87
environmental ignorance and, 77–78
of environmental impact across set of polluting firms, 82–84, 98
environmental impacts and, 79
environmentalism and, 303
human domination of ecosystems and, 127
humans in the landscape and, 80
Mayan civilization and, 96–98
in population growth, 201

diversity, 111–12
biogeography and, 114, 116, 118–21
forests and fostering of, 121
division of labor, 136
DNA barcode, 241, 256
Dogon people, Songo, Mali, 271
dogs, 117
doldrums, *110*
dolphins, 243
domestication, of plants and animals, 133t, 134, 261
droughts, 157, 165, 207, 232, 271
drugs, from natural products, 255
dry climates
of land areas on Earth, *109*
in Western Hemisphere, *106*
Duke Power, 282
Dupont, 282
Durand, Asher, 24, *24,* 25–26, *26,* 42
dysentery, 219

Early Morning at Cold Spring (Durand), *24,* 24–25, 42
Earth, 33, 45, 100, 111
average temperature of, 161
axis of, 104
carbon cycle of, *161,* 161–62
climates of land areas on, *109*
countries with highest biodiversity on, 242
energy budget of, *159*
human dominance or alteration of several components of systems on, 149, *152*
as human-dominated planet, 225
at night, *263*
radius of, 103
total footprint of humans on, 146

two-dimensional model of atmosphere, *104*
Earth Day, first, 301
Earth from Space, 1972 (NASA Apollo 17 Mission), 30–31, *31*
earthquakes, 179
earth sciences, 11
Earth Summit, Rio de Janeiro (1992), 11, 12, 14
Earth Summit, Sweden (1972), 288
Eastern Europe, oil reserves in, *183*
ecological footprint, 145–46
of various countries, *145*
Ecological Society of America, 306
ecologists, 11, 301
ecology, 301–2
economic activity
across boundaries, 361, *362*
circular flow of, 350, *351*
environment and, 11
economic choices, reasoning graphically about, *352,* 352–53, *353*
economic development, 15
decrease in population growth rates and, 201
environmental protection and, 253
iron triangle and, 325
more sustainable profile of, 351
privatization and, 61
sustainable development vs., 14–15
economic growth, 282
in developing countries, 270
environmental impacts of, 15
environmental limits and, 374
grand challenges tied to, 410
human success and, 264
natural cycles and, 266
resource prices and, 141–42

environmental harms
(continued)
   patterns between average
      income and forms of,
      412
environmental impacts,
   technology, consumption, and, 374–76
environmental injustice, 79
environmentalism, 301, 304
   classical. see classical
      environmentalism
   disproportionality and,
      303
   dream of progress and, 377
   emergence of, 261,
      287–88
   opponents of, 325
   Pogo cartoon and, 391
   polarization of American
      politics and, 317
   Romanticism and rise
      of, 21
   as social movement, 44
environmentalists
   activism and influence of,
      300–301
   self-described, 339
environmental justice, 304–7
environmental Kuznets
   curves, 411–13, 412
environmental law, pioneering field of, 300
environmental legislation,
   U.S., 1964-82, 319t
environmentally responsible
   business, 87
environmental movement
   classical environmentalism
      and, 13–14
   launching of, 8, 294, 430
environmental nongovernmental organizations,
   306, 330, 334, 335
environmental nonprofits,
   338–39
environmental policy, 250,
   318–20, 322
   classical environmentalism
      and, 313
   effectiveness of, 323–24
   history behind, 318–20,
      322

precautionary principle
   and, 292
environmental politics,
   324–25, 327, 341
environmental pressures,
   rising economic activity
   and, 277
environmental problems, 43
   as commons problems, 75
   disparities in economic
      and political power tied
      to, 50
   frustrating realizations
      about, 279
   institutional responses
      and, 47
   integrative approach to,
      13, 416
   mismatches between
      human use, natural
      systems and, 71t
   origins of, 45–47
   responsibility, mismatch
      and, 70–73
   tragedy of the commons
      applied to, 52–53
environmental problem
   solving, 6
   classical environmentalism
      and, 14
   liberal education and, 18
environmental protection
   global, 282
   trends in public opinion
      about priority of economic growth and, 318
Environmental Protection
   Agency (EPA), 62, 79,
   80, 85, 86, 306, 320, 325,
   329, 330, 331, 342, 366,
   387
   formation of, 300
   risk analysis and, 302
   Toxics Release Inventory,
   79, 80, 85, 301, 309, 399
environmental quality, as
   legitimate component of
   social welfare, 14
environmental regulation
   economic impact of, 329
   social costs and, 366–67
environmental responsibility
   equitability and, 431

some goals of, 429–30
environmental science, contemporary, synthesis of
   traditions in, 34–35
environmental studies
   central skills of, 16
   moral deliberation and,
      74–75
   observation crucial to, 35
   property as idea and, 49
   search for responsible role
      for humans in landscape
      and, 17
epidemics, falling death rates
   from, 196
equator, 106, 108, 116
Erie Canal, 23
ESA. See Endangered
   Species Act
ethics
   biodiversity and, 229,
      256
   Hardin's analysis of the
      commons and role of,
      73–76
eukaryotes, 117
eukaryotic cells, 349
Europe
   falling birth rate in, 196,
      198
   grazing commons in, 54
   greenhouse gas emissions
      in, 179, 282
   gross domestic product
      per capita in, 1700-2000,
      267
   income distribution and,
      81
   oil reserves in, 183
   population estimate,
      growth rate, total fertility
      rate in, 193t
   urban population by
      region, 1950, 1990, and
      2005 in, 212
   urban transition in, 209
European Union
   automobiles discarded
      in, 382
   cap-and-trade in, 369
   fuel use in, 173
evaporation, net precipitation and, 107, 107

evergreen trees, diversity and
   height of, in rain forests,
   119
evolution, Darwin's theory
   of, 25, 89, 112, 113, 115,
   116
evolutionary biology, 112,
   302
experimental (theory-driven) science, 32–33,
   35, 35t, 418
exponential change, 266
externalities, 362–63, 367
   positive and negative, 363
   social cost and, 363
extinction
   endemism and risk of,
      235, 237
   as local phenomenon, 235
Exxon Mobil, 342, 408

fabrication, 34
Facebook, 282
fair trade, Pareto criterion
   and, 356
fair-trade coffee, 397
family planning, 201
family size, decline in,
   196–98, 200
farmers markets, 398
farming, 11
   endemic species and, 121
   transformations of natural
      systems by, 11
farms, ecosystems of, 133
fascism, 415
FCMA. See Fisheries Conservation and Management Act (FCMA)
FDA. See Food and Drug
   Administration (FDA)
Federal Reserve, 289
feral pigs, golden eagles,
   Channel Island fox and,
   8, 9, 10, 246
Ferre cell, 110
fertility levels, at or below
   replacement, 200
fertilizers, 133, 332, 343
   crop subsidies and, 322
   indirect pressure on ecosystems and, 248–49
   nitrogen-containing, 160

feudalism, end of, 347
field sciences, 33, 35, 35*t*, 103, 106, 111, 116
financial crisis of 2008, 265
fire, humans and use of, 131
First Nature, 221, 222*t*
fish, DDT and, 7
fisheries, 238–39, 256
  depletion of, 15, 16
  MSC-certified, 397, 398
Fisheries Conservation and Management Act (FCMA), 252
fishing, 229, 252
  decline in fish populations, 5
  natural systems transformed by, 11
floods, 157
Florida, 165
Florida Everglades, 287
Florida Keys, 240
flowering times, climate change and, 249
Food and Agriculture Organization (FAO), 230
Food and Drug Administration (FDA), 289, 298
food chain, 253
  DDT and, 7
  golden eagles on Channel Island and, 9
  ocean acidification and, 166
food production, falling death rate and, 196
food supply
  agriculture and control of, 131, 134–35
  price spikes and, 410
food web
  carnivores and, 128–29
  rising to top of, 130–31
Ford Foundation, 342
Ford Model T, 69
forest fires, 166
forestry, 387
  economic life and, 229
  sustainable, 396
forests, 147, 156
  clearings in, 243
  diversity fostered by, 121

microhabitats in, 120–21, 235
of northeastern U.S., replenished, 234
Forest Stewardship Council (FSC), 396
Forster, E. M., 428
fossil fuels, 15, 20, 45, 74, 130, 169, 172, 181, 229, 279, 387, 390, 408
  carbon flows and burning of, 162
  carbon offsets and, 358
  climate change and burning of, 15
  competitive pricing and, 178
  harnessing of, 261
  imports of, 182
  in industrialized economy, dilemma of, 155–57
  Industrial Revolution and harnessing of, 33–34, 189
  nitrous oxide formation and, 160
  U.S. primary energy consumption and, 176, *177*
  world without edges and, 186
fossil record, five mass extinctions in, *242*
foundations, governance of, 342
Fourier, Joseph, 158, 163
Framework Convention on Climate Change (1992), 188
France, nuclear energy development in, 179
free markets, 58–59, 350
free rider problem, 338
free riders, 334–35
free speech, 317
French Impressionists, 25
French Revolution, 91
freshwater, percentage of water on Earth as, 105
Friends of the Earth, 300
Friends of the Island Fox, 9
frugality, economic drawback to, 392

FSC. *See* Forest Stewardship Council (FSC)
fuel use cartogram, *173*
Fukushima Daiichi nuclear plant, accident after tsunami, 2011, 179, *299*
functional commons problems, 98
functional mismatch, human responsibility, natural world and, 72–73
"Futility of Global Thinking, The" (Berry), 279–80
future, valuing, 364–65
futures contracts, 142–43

Galileo, 378
Gallup, 317
Gandhi, Mohandas K., 92
garbage stream, reducing, 382
gasoline
  consumption, peak and decline in, 408
  sales of, 373
gasoline prices
  rising, 96
  since cars became popular, 178, *178*
Gast, John, 29, *29,* 30, 31
GDP. *See* gross domestic product (GDP)
General Electric, 282
genes, 117
  industrial chemicals and, 296
  mutations in, 112
Genesis, 377, 378
genetics, invisible present and, 117–18
genome, 227
genomics, 33
geography, climate patterns and, 105–7, *106*
geology, economics, politics and, 182–83
Germany, nuclear energy phaseout in, 179
Ghana, doing the laundry in Nima, *213*
Gifford, Don, 88, 90, 94–95, 148, 399

Gilded Age, 281
Gini, Corrado, 81
Gini coefficient, 80, *81*
  distribution of emissions by industrial polluters and, 82–84
  measuring inequality and, *81,* 81–82
  toxic releases from major U.S. industrial sectors and, *82*
glaciers, 119, 144, 156, 176, 239, 261
global circulation, climates and, 107–11
global climate change, 11, 34
  China's expanding use of coal and, 30, *30*
  human responsibility for, 14
  international treaties and, 188–89
  Maldives and, *166,* 166–68
  responsibility, greenhouse gas emissions and, 72
global economy
  ecological footprint and, 146
  emergence of, with planetary impacts, 47
  extraordinary complexity of, 77
  planetary change and, 76
globalization, 3, 413
global-local linkages, environmental effects of, 87
global population, urban and rural, 1950-2030, *209*
Global Scenario Group, 409
global system of production, 87
global variables, actual and projected changes in, 410*t*
global warming, 12, 35, 38, 58, 105, 130, 155, 156, 157, 160, 162–65, 169–71, 205, 228, 269, 280, 281, 314, 324, 417
  cap-and-trade approach to, 367, 369
  coal burning and, 185

individual competence, extending, *370*

Indonesia, old-growth rain forest in, *120*

industrial chemicals, holes in the ozone layer and, 57–58

industrial ecology, 383, 384–85, 387, 429, 430
matrix for a product, *384*
target diagram for, *385*

industrial economies, wastes generated by, 79

industrial gases, 159

industrialism, 30, *30*

industrialization, 20, 21, 43, 92, 175, 261, 262, 377
age structure and, 202
air pollution and, 270
enclosing of commons and labor for, 54

Industrial Revolution, 20–21, 29, 101, 123, 130, 139, 149, 160, 172, 188, 189, 261, 312, 408, 415
harnessing of fossil fuels and, 33–34
labor struggle and, 350
launching of, 92, 315
optimism about science and, 378

inequality, 45, 51, 137, 282
commons and, in Mississippi Delta, 56–57
Gini coefficient and measure of, *81*, 81–82
grand challenges and, 5
Kuznets curve and measure of, 411
sustainable development and dilemma of, 54
sustainable development and riddle of, 14–16
well-managed commons and, 70

infant mortality rate
decline in, 274
in urban *vs.* rural environments, 218, 218*t*

infectious diseases, 218–19, 297
falling death rate from, 196
Native American populations ravaged by, 49
urbanization and, 205

inflation, oil prices and, 178

influenza epidemic, 196

informal employment, 213

informal housing, rise of, 215

information economy, 414

infrared radiation, 158

infrastructure, 271
Second Nature and, 221
urbanization, ecosystem services and, 222–23

innovation, foundation funding and, 344

inputs, 350

insects, 7, 116

institutional solutions, 58–59

institutions
complicated and subtle language of, in landscape, 70
defined, 12
human connection to nature and, 44
shaping of access and, 50
social science and definition of, 47
speed of human alteration of ecosystems and, 123
sustainability and, 140
world without edges and, 98

insurance, global warming and, 165

integrative problem solving, 13, 16–17, 416

intelligent design, 31, 112
Darwinian evolution *vs.*, 25

interest groups, 321, *321,* 325, 345

Interface Corporation (Georgia), 381–82, 383

Intergovernmental Panel on Climate Change (IPCC), 58, 155, 165, 228, 422

Internal Revenue Code, Section 501(c)(3) of, 338

Internet, 34, 95, 337, 386

invader species, 247–48

investments, 392
carbon offsets and, 358
environmentally responsible businesses and, 399–400

philanthropic, 343–44
technology and, 400–401

"invisible hand," 59

invisible present, 35–38, 101, 124, 205, 237–40, 258, 345
climate change hidden in, 36, 104, 155, 165
economic activities across boundaries and, 361, *362*
ecosystem services and, 148, 149
enclosures, role of government and, 326
genetics and, 117–18
of New England winters, 37
shifting baseline and, 239

IPAT formula, 374, 375, 376, 385, 387, 390

IPCC. *See* Intergovernmental Panel on Climate Change (IPCC)

Iran, oil production in, 182

Iraq, 281
Hussein and civil society in, 337
oil production in, 182
site of Uruk in, *136*

iron triangle, 325
environmental policy and modification of, *327*
pluralism and, *321, 322*

irrigation systems, 136–37

Islam, rise of, 204

Istanbul, Turkey, 211*t*

ivory-billed woodpecker, 242

Jackson, Andrew, 28

Jackson, Lisa, 306

Jakarta, Indonesia, 211*t*

Japan
affluence and life-style in, 375
carbon emissions in America *vs.* in, 174, 186
democratic rule in, 316
falling birth rate in, 196, 198
fuel use in, *173*
Fukushima nuclear complex meltdowns in, 179

GDP and energy use in, 174
income distribution and, 81
Pearl Harbor attacked by, 183

jararaca, captopril derived from venom of, 255, 256

Java, Indonesia, gathering firewood in, *270*

Jefferson, Thomas, on slavery, 293

Jesus Christ, 261, 335

jet streams, 109, 111

jobs
environment *vs.*?, 84–87
offshoring, 390
urban migrants and, 218, 219

Johannesburg, 277

John, king of England, Magna Carta and, 318

Johns Hopkins University, 342

Jon Stewart Intelligence Agency, 336

Judeo-Christian-Islamic traditions, 261, 377–79

Jupiter, visible banding of atmosphere, 111, *111*

justice, individual competence and, *370*

Keats, John, 20

Kepler, Johannes, 33

Kindred Spirits (Durand), *26,* 26, 42

King, Martin Luther, Jr., 92

Krauthammer, Charles, 379

kudzu vines, 226, 246

Kuznets, Simon, 265, 411

Kuznets curves, 425, 430
environmental, 411–13, *412*

Kyoto climate treaty, 14

Kyoto Protocol, 64, 188, 343, 369

labor unions, 338

Labrador, number of bird species in, 118*t*

Lahore, Pakistan, 211*t*

lakes, 105

Lake Superior, 149

Natural Resources Defense Council, 85, 300, 333–34

natural sciences, 13, 75

natural selection, 33, 112, 113–14, 119–20, 123

natural systems, mismatches between human use and, 71t

natural theology
Darwin and, 115–16
in Hudson River School paintings, 24, 29, 31, 115

nature, 24
belief and, 24–32
climate, life, and provinces of, 100–101
energy and control of, 171–72, 175–81, 183–85
environmentalists and view of, 19
Hudson River School and, divine order in, 24
human defiance of, 12
human reliance on, 4–5
Judeo-Christian tradition and, 377–79
markets and, 359–62
rapid increase in human domination of, 282
roots underlying our visual aesthetic of, 31
salvaging wealth of, 225–28
services through, human need for, 144
sublime image of, in *Earth from Space*, 31, *31*
Transcendentalism and appreciation of, 28
transformation of and reliance on, 219, 221–22
view of, in Hudson River School and Transcendentalism, 29
wealth, power, and, 271–72

Nature Conservancy, 4, 8

Nature (Emerson), 25

Nature's Metropolis (Cronon), excerpt from, 220–21

nature writing
of Gilbert White, 89–92
of Henry David Thoreau, 92–94

Nazca Desert (Peru), 108

needs
changing our conceptions of, 373
wants *vs.,* 11–12

NEPA. *See* National Environmental Policy Act (NEPA)

net benefits, maximizing, household decisions and, 353, *353*

net precipitation, 107, *107*

net primary production, 161

net primary productivity (NPP), 122, *122,* 145, 146

Nevada, in rain shadow, 105

Newark, NJ, 210t

Newburyport Meadows (Heade), *291*

New England, 26, 266
colonists, private ownership of land and, 60, 61, 134
forces changing landscape of, 1620-1830, 135t
Pilgrims arrival in, 52

New England mussel *(Geukensia demissa),* on San Francisco Bay seawall, *234*

New England winters, invisible present of, 37

Newfoundland, number of bird species in, 118t

New Orleans, Hurricane Katrina in, 56–57, 165

Newton, Isaac, 33, 378

New World, 139, 261

New York City, 204, 205, 210t, 277

New Yorker magazine, 294, 295, 299, 333

New York state, number of bird species in, 118t

NGOs. *See* nongovernmental organizations (NGOs)

Nigeria
oil fields in, 312, 313
oil production in, 182

night, Earth at, *263*

Nima, Ghana, doing the laundry in, *213*

NIMBY. *See* not in my backyard (NIMBY)

nineteenth century
popular culture in, 20, 21
rapid change in America during, 23

nitrogen
atmospheric, 149
biogeochemical cycle and, 161

nitrogen fixation, human dominance or alteration of, *152*

nitrous oxide, 159, 160

Nixon, Richard, 320

nomadic way of life, 135, 136

nonagricultural employment, in sub-Saharan Africa, 213

nonferrous metals, toxic burden from, disproportionality and, 83, 84

nongovernmental organizations (NGOs), 311, 313, 339, 340, 341, 342, 395, 428–29
environmental, 300, 301, 306, 330, 334, 335
philanthropy and, 344

nonhuman world, invisible present and, 38

nonmaterial needs, 393, 395–96

nonmigratory animals, 111

nonprofit organizations, 334
characteristics of, 338
creating civil society and, 337–39
role and responsibilities of, 339

nonrenewable resources, depletion of, 143

North Adams (Massachusetts), example of invisible present in, 36, 104

North America, 267
falling birth rate in, 196, 198
oil reserves in, *183*

population estimate, growth rate, total fertility rate in, 193t
urban population by region, 1950, 1990, and 2005 in, *212*
urban transition in, 209

Northern Hemisphere, 108
average temperature of, over the past thousand years, *163*

northern spotted owl, timber industry and, 330

North Pole, global warming and retreat of ice caps at, 169

Norway, DDT banned in, 8

not in my backyard (NIMBY), 305

Novak, Barbara, 24

NPP. *See* net primary productivity (NPP)

nuclear electric power, U.S. primary consumption of, 2010, *177*

nuclear energy, 417
economic growth and, 410
waste disposal and, 179–80

nuclear fission process, waste disposal and, 179–80

nuclear power, 179
commercial, 175
political problems related to, 180

nutrition, falling death rate and, 196

Obama, Barack, 265, 374

Obama administration, 306

obesity, 279

observation, 111

observational (field) sciences, 33, 35, 35t, 103

Oceania
population estimate, growth rate, total fertility rate in, 193t
urban population by region, 1950, 1990, and 2005 in, *212*

oceanography, 34

oceans, 103, 104, 250
  acidification of, 166, 250
  Earth's carbon cycle and, *162*
  percentage of water on Earth in, 105
oil, 172, 181, 186, 387
  geology, economics, politics and, 182–83
  U.S. imports, monthly average, June 2005 to November 2006, *182*
  U.S. production and imports, 1920-2008, *184*
  world reserves of, by country, *181*
oil prices, 142, 178, 388
  world price of barrel of oil, *143*
oil production, decline in, 184, 185
oil reserves, 183, *183*
Old Testament, 137
Olmsted, Frederick Law, 42, 253
omnivores, humans as, 129
one-child policy, in China, 74
OPEC. *see* Organization of Petroleum Exporting Countries (OPEC)
open access, 65, 69, 70
  commons and conditions of, 51
  degradation of pasture lands and, 52
  institutional solutions and, 58–59
  technology and, 53
optimal level of livestock, 51
organic agriculture, 308
organic food, 301, 379, 396
organizational life, reasons for participation in, 340–41
organizational work, social change and, 333–35
Organization of Petroleum Exporting Countries (OPEC), 184, 185
organizations, ecosystem of, 341

Origin of Species, The (Darwin), 25, 113, 115
Osaka-Kobe, Japan, 210*t*, 211*t*
Ostrom, Elinor, community-governance principles for well-managed commons by, 65–69, 66*t*
outliers, 80
overfishing, 16, 55, 57, 239, 397
overharvesting, 243
"Oversupply of False Bad News, An" (Simon), 141
Oxbow, The (Cole), 22–23, *23,* 25, 26, 28, 134
oxygen, 146
  biogeochemical cycle and, 161
  generation of, photosynthesis and, 128
ozone layer, 12
  holes in, 44, 57, 58
  industrial gases and, 159
  protecting and healing, 13

Pacific Asia, economic transformation in, 213
Pacific Institute, 84
Pacific Northwest, timber industry of, 329–30
Pacific Ocean, 105
Pacific salmon, pteropod and, 166
paclitaxel, 255
Pakistan, population growth rates in, 213
paleoanthropologists, 260
paper
  dematerialization and, 387
  recycling of, 381
paradox of uneven development, 76
Pareto optimality, 356
Paris, France, 211*t*
pasture lands, degradation of, under conditions of open access, 52
pathogens, 138
paying attention, liberal education and, 426
Peace Corps, 269
Pearl Harbor, Japan launches attack on, 183

Pentagon, September 11, 2001 attacks and targeting meaning of, 43
permafrost, methane release and melting of, 160
permanent settlements, 135, 136
pesticides, 133, 138, 231, 232, 246, 294, 295, 296, 298, 331, 343
  bees and, 303
  cancer, radiation and, 297
  DDT, 7–8
  indirect pressure on ecosystems and, 248
petroleum, 175, 176, 178
  U.S. primary consumption of, 2010, *177*
pharmaceuticals, from natural products, 255
philanthropic foundations, 313, 339, 342–44
Philippines, 270
photosynthesis, 119, 122, 127, 128, 145, 146, 172
photovoltaic cells, *177,* 177–78, 419
physics, 33, 35*t*
phytoplankton, 122, 128, *129*
Pilgrims, landing at Plymouth, Massachusetts, 52
pioneers of the West, privatization and American myth of, 61
place, 38–41
  developmental concept of, 39
  human loyalty to, 415
  landscapes and idea of, 19
  subjective sense of, 39–40
  urban sense of, 207–8
  weakened loyalties to, 69
  world without edges and erosion of, 139
plague, 138, 297
planetary-scale limits, 149, 151–53
Planet Earth, role of humans on, 127
Planned Parenthood Federation, 200–201
plants, 111, 146, 235

climate and, 156
  in crown groups on tree of life, *117*
  cultivated, settlement and, 135–36
  domestication of, 133*t,* 134, 261
  in food web, 128, 129
  humans benefiting from half of sunbeams captured by, 151
  increase in number of species in different ecosystems, *238*
  invasive species, human dominance or alteration of, *152*
plastics
  petroleum and manufacture of, 176
  substituting biological products for, 386
pluralism, 329
  environmental movement and, 325
  struggle of groups and theory of, *321,* 321–22
plutonium, 179
Pogo cartoon, *391*
polar bears
  climate change and survival of, 11
  loss of sea ice and, 169
polar climates
  of land areas on Earth, *109*
  in Western Hemisphere, *106*
Polar Easterlies, *110*
polar front, *110*
polar high, *110*
polar ice
  extent of sea ice, at its minimum in 2009, *171*
  melting of, 11
polar regions, global warming and, 169
policy, adaptive management and, 423–24
policy decisions, understanding of "nature" and, 22
polis, 369, *370*
political power
  environmental problems and disparities in, 50

limited competence and, 330–33

political science, 33, 35*t*, 311, 341

political will, sustainability transition and, 414

politics, geology, economics, and, 182–83

Pollard, Elizabeth, 196

pollination services, by honeybees, 144

polluters
cap-and-trade policy and, 367, 368–69
environmental regulations and, 366–67

polluting firms, disproportionality of environmental impact across set of, 82–84, 98

pollution, 11, 59, 80, 82–84, 357, 362
air. *see* air pollution
American consumption and, 294
industry, environmental responsibility and, 58
mining and, 185
as negative externality, 363
progress and, 30, *30*
spatial mismatches of responsibility and, 70
tragedy of the commons and, 53
water. *see* water pollution

pollution-control policy, public trust doctrine and, 63

pollution-control programs, 429–30

poor communities, environmental injustice in, 79, 370–71

poor countries, institutions for commons and, 63–64

Popper, Frank, 305

population, 374
actual and projected changes in, 410*t*
of Egypt since sixth century B.C., *195*
of Maldives, 167

natural selection and, 113–14

permanent settlements and, 136

projections *vs.* predictions about, 262

stabilization of, 200, 224, 259, 282

sustainable development and, 15

world population since 1950, *199*

Population Bomb, The (Ehrlich), 140

population growth, 224, 262
biodiversity hotspots and, 244–45, 245, *246*
deceleration in, 204, 259, 262, 408
dematerialization and, 387
demographic transition, 193–95, *194*
falling birth rate and, 196–98
falling death rate and, 195–96
Hardin's analysis of the commons and, 74
projections of, 5, 198, 200–201, 204, 409
prosperity and, 264
urbanization and, 191–92
in urban *vs.* rural environments, 218*t*
world population estimates, 2010, 193*t*
world population projected to 2050, *201*

Poseidon, 377

poverty, 15, 270, 409
absolute, 279
disproportionality, pollution and, 306
ecosystem services and, 277
price spikes and, 410
urban, social and political conditions of, 223
urbanization and, 213, 215–17

power, 137, 139
liberal education and tempered use of, 428

nature, wealth, and, 271–72

powerless citizens, systematic exclusion of, 55

power plants, 62

practical politics, civic science and, 416–17

prairie, depletion of, 52

precautionary principle, 292

precipitation, 156
annual, measures of, 105
climates of Western Hemisphere and, 106–7
net, 107, *107*

predation, militarized, 138

pressure
direct, 243–45, 251
indirect, 246–50, 251
rising economic activity and, 277
sixth great extinction and, 240, 242–50

pressure-state-response, 230–31

pressure-state-response framework, sustainable development and, *231*

prevailing winds, 108

Price, David E., 297

prices and pricing, *354*, 367, 380
choices and, 349–51, 357, 359
mutually beneficial market outcomes and, 357
political unrest and spikes in, 410
price mechanism, magic of the market and, 354

primary consumers, *129*

primary metals sector, of U.S. economy, percentage of toxic emissions by, 85

Primary Nonferrous Metals sector, of U.S. economy, toxic releases by, *83*, 83–84

principals, 340

private ownership of land, landscape reshaped by, 60, 61, 134

private property, community-based governance and, 65–66

privatization, 59–61, 64
polluting plants and, 83–84
sustainable development and, 255

problem solving
integrative approach to, 13, 16–17, 416
liberal education and, 427

producers, *129*

production costs, 352
social cost and, 366

profits, *354*
maximizing, 352, *352*

progress, environmental awareness arising from, 29–31

Progressive Era, 289, 351

property, 47
American Indian concept of, 48–49, 384
environment and, 48–49

property law, 49

property rights, markets and, 59

Prospect Park, Brooklyn, NY, 95
Endale Arch in, *41*, 41–42

prosperity
energy and, 173–75
population growth relative to, 264

protected areas
percentage of land and marine areas protected by world governments, 2010, 254*t*
types of, 253–54

proteomics, 33

psychology, 33

pteropods, 166

public health
falling birth rate and, 197
rising survival rate and, 196

public policy
rent-seeking and, 54
responding to environmental problems and, 43
tradition and, 69–70

public resources, government management of, 62

public trust
in atmosphere, 63
regulation and, 61–44

public trust doctrine, pollution-control policy and, 63

Puritans, 21, 23, 132

Pynchon, Thomas, 328

Pynchon, William, 48

questions, thinking carefully and deliberately about, 16–17

Qur'an, 261

race, 293

radiation, toxic chemicals in pesticides and, 297

railroads
emergence of ranching economy and, 52, 53
public land going to, 61

rainfall, 106–7, 108, 119

rain forest, 101, 106, 107, 121, 135, 235, 240, 242, 255
diversity and tree height in, *119*
old-growth, in Sumba, Indonesia, *120*
protected areas in, 253

Rajastan Desert (India), 108

ranching economy, emergence of, after Civil War, 52

Ranch Life and the Hunting-Trail (Roosevelt), 52

rational choice, theories of, 52

rational herdsman, in Hardin's parable of the commons, 51–52

raw materials, American consumption, thirty-five pounds of waste and, 78–79, 98

reading, liberal education and, 426

Reagan, Roland, 417

recession of 2008, 265

recycling, 47, 50, 231, 289, 292, 301, 308, 318, 324, 380–82, 386, 387, 391, 398, 429

Red Cross, 342

redesign, 382–83, 386

red maple, 36

reductionism, 34

Rees, William, 145

reforestation, in North America, 387

regulated industries, 331

regulation, 64
conservative critics of, 61
of greenhouse gases, demands for, 64
protecting the commons and, 59
public trust and, 61–44
scientific discovery leading to, 14

reliable knowledge, science and pursuit of, 420

religion, as private matter in United States, 27

religious significance, beautiful landscapes imbued with, 24

Renaissance, 260, 261, 370, 415

renewable energy
sources of, 176
U.S. primary consumption of, 2010, *177*

rent-seeking, 54
science and, 58
suburban sprawl and, 73

reproduction, natural selection and, 113–14

Republican Party, 317, 321, 336

resilience
biodiversity and, 232
ecosystem function and, 233–34

resource degradation, industry, environmental responsibility and, 58

resource prices, economic growth and, 141–42

resources, finite, 15

responsibility

cap-and-trade policy and, 72

mismatch, environmental problems and, 70–73

responsible consumption, 396–99

retirement, 263

reuse, 381

revenues, 352

rhinoviruses, 138

Riboud, Marc, 30, *30*

rice, 230, 342

Richard Arkwright's Cotton Mill, Cromford (Day), *91*

rich nations, ecosystem services harvested by, 277, 279

Rift Valley, Africa, emergence of Homo sapiens in, 260

Rio de Janeiro, Brazil, 211*t*
Earth Summit (1992) in, 11, 12, 14

risk analysis, 302–3

river ecology, global warming and, 165

rivers, 105

Rockefeller, John D., 342, 343

Rockefeller Foundation, 273, 342

Rockefeller University, dematerialization study at, 387

Roman Empire, 139

Romans, religious observances, place and, 377

Romanticism, 21, 24, 29, 92

Romantic movement, 20

Roosevelt, Theodore, 52, 87

root crops, domestication of, 131

Rotary Club, 338

rotation of Earth
trade winds and, 108
weather and, 104, *104*
westerly tendency of the winds and, 109

rural population, global, 1950-2030, *209*

Russia
birth rate decline in, after end of cold war, 200
energy resources of, 182

Sahara Desert, 108

sanctions, graduated, community governance and, 67

Sand County Almanac, A (Leopold), 289, 293

San Francisco, 23

San Francisco Bay seawall, life-forms living on, 233, *234*

San Gabriel Mountains, 40

sanitation, in urban *vs.* rural environments, 218*t*

San Miguel Island
estimated number of Channel Island foxes on, 1993-99, *10*
rebounded number of foxes living on, 9

Santa Ana wind, Didion's description of, 39–40

São Paulo, Brazil, 210*t*

Saturn, bands on, 111

Saudi Arabia
oil production in, 182
well-being of people in, 273

savings and savings rate
reducing greenhouse gas emissions and, 386–87
in United States, 1959-2010, *392*

scale of our competence, 279–83
enlarging, 281, 310, 313, 344, 398, 431
understanding, 282

SCAM. *See* scientific uncertainty augmentation method (SCAM)

scarcity effect, 142

scenario, 193

scholarship, activism *vs.,* 17

Schubert, Franz, 20

science, 43, 57, 305, 345
citizen-based activism and, 301–3, 307–8
rent-seeking and, 58
two approaches to, 32–35
in White's analysis of Judeo-Christian tradition, 378

Science journal, 377

scientific knowledge, action linked with, 420–22
scientific philanthropy, 342
scientific theory, 418–19
scientific uncertainty augmentation method (SCAM), 303
seabirds, bycatch and, 243
seafood, sustainably caught, 396–98
Seafood Watch buying guide, *397,* 397–98
sea ice, polar bears and loss of, 169
sea-level rise, 250
  global warming and, 72, 165
  Maldives and, 168
Sea Round Us, The (Carson), 294, 296
sea walls, 165, 168
seawater, changing chemistry of, 250
secondary consumers, *129*
second law of thermodynamics, 390
Second Nature, 221, 222, 222*t,* 224, 271, 315
sedentary societies, 136
sediment budget, of Mississippi River, *56*
SEEDS. *See* Strategies for Ecology Education, Diversity, and Sustainability (SEEDS)
Selborne, England, 112, 261
  White's nature writing about, 89–92, 95
self-criticism, liberal education and, 427
self-government, 54, 139
self-interest, collective benefits and, 351
self-rule, 139, 349
self-sufficient society, 91
sellers
  ideal markets and, 357
  market equilibrium and, 354, *354,* 355
sense of place, 38–41, 87
  Gifford and, 94–95
  history behind, 88
  sensibilities and, 65

Thoreau and, 92–94
White and, 89–92
Seoul, Korea, 210*t*
September 11, 2001 attacks, 43, 78, 281
serfdom, 139
Sesame Street, 308
settlements
  disease organisms and, 138
  permanent, 135, 136
sewers, 222
SFI. *See* Sustainable Forestry Initiative (SFI)
Shakespeare, William, 100
shelled creatures, ocean acidification and, 166, 250
Shenzhen, China, 211*t*
shifting agriculture, 132
shifting baseline syndrome, 238–39
shrimp, bycatch and, 243
Sierra Club, 300, 307, 325, 331
Sierra Leone, income distribution and, 81
Sierra Nevada, 165, 247
Silent Spring (Carson), 8, 288, 303, 307, 317, 319, 320, 324, 328, 344
  classical environmentalism and, 296–300
  launching of modern environmental movement and, 294–96
  themes in, 297–98
Simon, Julian, 153
  betting on limits with Paul Ehrlich and, 140–43, 148, 185
Singapore, 209, 270
sixth great extinction, pressure and, 240, 242–50
sky islands, 166, 226
slavery, 88, 281, 290, 293
slums, rapid growth of, 217–19
smallpox, 49, 228, 230, 297
Smith, Adam, 59
smog, 294
snowfall, in Williamstown, Massachusetts, 1900-2002, 37, *37*
Snowy Egret (Audubon), *119*

Snyder, Gary, 39, 69, 121
  on Hardin's discussion of commons, 64–65
social capacity, sustainability transition and, 414
social capital, 336–37, 343, 425
social change, 425
  age structure and, 202–4
  classical environmentalism and, 289
  organizing for, 333–35
social cost
  cap-and-trade policy and, 367, 368–69
  externalities and, *363*
  market prices and, 362
  production cost and, 366
Social Darwinism, 302
social investing, shareholder campaigns and, 400
social justice, 306
  consumption tied to, 400
  sensible use of natural resources and, 70
social learning
  learning and, 424–25, 428–31
  methods of, 429
social pressure
  commons and, 64
  community-based governance and, 67
social psychology, 35*t*
social sciences, 13, 33, 75
Social Security, 197, 202, 263
social services, urban *vs.* rural environments and access to, 218*t*
social systems, resilience in, 234
social understanding, integration of science and, 418
social welfare, environmental quality and, 14
society, settlement, complexity and, 135–37
socioeconomic inequalities, sustainable development and, 5
sociology, 311, 341
soil
  erosion of, 133

human imprint and, 11
tropical, 119
solar energy, 127, *159, 177,* 178, 386
  net primary productivity and, 122
  planet's rotation and, 104
  weather and life powered by, 103
solar power, 387
solar radiation, *159*
solid-state physics, 419
Songo, Mali, *271,* 271–73, *272*
Sonoran Desert, 108
South Africa, 334
South America
  land suitable for conversion to agriculture in, 242
  oil reserves in, *183*
Southern California, conservation in, 323
South Korea, economic growth in, 270
South Pole, 261
Soviet Union, 349
  collapse of, 213, 316
  former, gross domestic product per capita in, 1700-2000, *267*
  oil reserves in, *183*
Spain, 176
spam e-mail, 336, 337
specialized roles in society, rise of, 135
species
  defined, 227
  DNA barcode for, 241
  endangered, 118
  equatorial, 116, 245
  human pressures and extinction of, 226–27
  interconnectedness of, 256
  irreplaceable genetic heritage of, 228
  natural selection and emergence of, 113–14
  spatial range of, 237
  world without edges and transfer of, 246–47
sports, 357

Tokyo, 206, 210*t*
tolerance, liberal education
and, 427
topography
defined, 101
of Jupiter, 111
tornadoes, 157
total energy consumption
continued climb in, 408
human-energy equivalent
of, based on average
human metabolic rate of
100 watts, 175*t*
total fertility rate, 193*t*
tourism, in Maldives, 167
toxic emissions, 79
from major U.S. industrial
sectors, 1993, *82*
small number of industries
responsible for, 80
Toxics Release Inventory
(TRI), 79, 80, 85, 306,
309, 399
toxic wastes, 105, 123
toxic waste sites, cleanup
rules for, 323
trade, 139, 347, 413
colonial landscapes, European economies and,
88, 261
trade winds, 108
traditions, 69
tragedy of dispossession, 54
tragedy of the commons, 52,
140, 194
Aral Sea ecosystem and,
150–51, 153
beaver pelt trade, New
England colonists and,
52–53, 57
cattle drives and, 52
community governance
institution and, 64
environmental problems
and, 52–53
forestalling, enforcement
of rules and, 386
in Mississippi Delta,
56–57
open access and, 70
overcoming, 311, 313
privatization and, 59–61,
64

regulation and, 61–64
"Tragedy of the Commons,
The" (Hardin), 50–51
tragedy of the uncommon,
American economy and,
86–87
transaction cost, carbon
offsets and, 358
Transcendentalism, 24, 25,
42, 377
appreciation of nature
in, 28
White's nature writing as
precursor to, 89
Transcendentalists, 288
prominent, 92, 94
vision of nature by, 29
transparency, responsibility
and, 71
transportation
social costs and, 366
U.S. primary consump-
tion of, 2010, *177*
world without edges and,
376, 379
trash
collecting in Accra,
Ghana, *217*
kinds of, 382
trawling, 239
treaty, 188
tree of life, 117, *117*
trees, 116, 146
diversity and height of, in
rain forests, *119*
seasonal fluctuations and,
112, 119
sustainably harvested,
396
TRI. *See* Toxics Release
Inventory (TRI)
trial and error, 423
tropical climates, in Western
Hemisphere, *106*
tropical diseases, changing
weather and movement
of, 166
tropical forests
biological diversity in, *236*
clearing of, 5
competition in, 237
species proliferation in,
118

tropical regions, temperate
regions *vs.*, in human
ecosystem, 278*t*
tropical soils, 119
tropical storms, 107
tropics
biodiversity in, 269
winds in, 108
trust
environmental awareness
and, 281
loss of social capital and
erosion of, 336
tsunamis, 168, 179
tuberculosis, 196
tubeworms, on San Fran-
cisco Bay seawall, 233, *234*
tuna, dolphin-safe, 243
Tunisia, 316
turbines, 176
Twain, Mark, 258
typhoid, 196

Ultimate Resource, The
(Simon), 141
UN. *See* United Nations
uncertainty, 55, 57–58
underground water, 105
understanding, liberal
education and, 426
unemployment, during
Great Depression, 265
uneven development, para-
dox of, 76
unions, 197
United Nations, 11, 15, 58,
146, 188, 214, 251, 254,
270
Conference on the
Human Environment,
Sweden, 288
Development Programme,
273
Environment Programme,
31, 253, 288
IPCC, policy choices and,
422
Millennium Ecosystem
Assessment, 228
Population Division, 200
World Commission on
Environment and Devel-
opment, 407

United States
age structure in, 202
coal reserves in, 185
consumption patterns
projected to whole
planet, 374, 375*t*
DDT banned in, 8
energy consumption in,
1969-2010, *389*
energy production in,
1775-2009, *172*
environmental conditions
exerted by, 87–88
falling birth rate in, 196,
197
fuel use in, *173*
GDP and energy use in,
174
human-energy equivalent
of total energy con-
sumption in, based on
average metabolic rate of
100 watts, 173–74, 175*t*
income distribution and,
81
income inequality in, 5,
274, 411
incomes in, 15
infant mortality in, *276*
international treaties on
climate change and, 188
nuclear plants in, 179
oil production in, 184,
185
personal savings rate in,
1959-2010, *392*
primary energy consump-
tion in, 176, *177*
raw materials required by
economy in, 78
religion as private matter
in, 27
responsibility and green-
house gas emissions
in, 72
sources of oil imported
into, 182
westward expansion in, 21
uranium, 179
urban environments, rural
environments *vs.*, in
accessing ecosystem and
social services, 218*t*

urban growth, 205, 208–9, 213–15

urban habitats, various species in, 219, 221

urbanization, 5, 11, 20, 21, 43, 92, 98, 229, 259, 262, 264, 283, 310, 410
  accelerating rate of, 262–64
  challenge of, 204–5
  falling birth rate and, 197
  golden eagles and impact of, 8
  grand challenge of, 222–24
  natural systems transformed by, 11
  population growth and, 191–92
  poverty, environment, and, 213, 215–17
  slums and, 217–19

urban penalty, 218–19

urban population
  global, 1950-2030, 209
  growth of, in high-, low-, and middle-income countries, 1950-2030, 214
  by region, 1950, 1990, and 2005, 212
  by size of settlement, 212

Uruk, site in Iraq, 136

U.S. Agency for International Development, 253, 270

U.S. Energy Information Administration, 408

U.S. Fish and Wildlife Service, 294

U.S. Forest Service, 290, 325

U.S. Green Buildings Council, LEED program, 397

U.S. Steel, 342

usufruct right, 48–49

vaccinations, mass, 196

Vaux, Calvert, 42

vegetarians, 129

Venezuela, oil production in, 182

Venus flytrap, 128

Vermont, Hurricane Irene in, 165

video monitors, 380, 381, 386

Viking raiders, 138

viruses, 229

Vitousek, Peter, 149

voting, 139

Wackernagel, Mathis, 145

wages, 354

Walden; or, Life in the Woods (Thoreau), 21, 92, 93, 94, 393, 395

Walden Pond, 266

Walden Pond State Reservation (Concord, Mass.), mugs sold at, 395

Wall Street Journal, 343, 408

Walmart, 282, 301, 386, 398

wants, needs *vs.*, 11–12

War of 1812, 22

waste, 362
  American consumption and thirty-five pounds of, 78–79, 98, 281, 394
  industrial economies and generation of, 78–79
  nuclear energy and disposal of, 179–80

water, 144
  agriculture, disproportionality and consumption of, 84, 85
  agriculture in American West and control of, 370
  biogeochemical cycle and, 161
  expanding demand for, 5
  human dominance or alteration of, 152
  planet's rotation and, 104
  sun, weather and, 105–7
  in urban *vs.* rural environments, 218t

water pollution, 98, 323
  human imprint and, 11
  IMC-Agrico company (Louisiana) and, 85–8

water rights, basis of, 72

watersheds, 157

water vapor, 108, 158

wealth

Gini coefficient and, 81, 81–82
  health and, over half a century, 275, 275–76, 276
  nature, power, and, 271–72

weather, 156
  climate *vs.*, 101
  evolving into climate, 107
  heat plus rotation and, 103–4
  imprint of human activity on, 44
  variations in, 157
  water, sun and, 105–7

weather forecasts, 103

well-being, of poor people, 273

Western Hemisphere, 108
  climates of, 106, 106
  satellite image of autumn day in, 107, 108

West Nile virus, 147, 148

westward expansion, *American Progress* and celebration of, 29, 29

wetlands, 243, 250
  coastal, mangroves in, 237
  global warming and, 165
  of Louisiana, erosion of, 56–57

wheat, 230, 342

White, Gilbert, 89–92, 93, 94, 95, 96, 101, 112, 197, 235, 237, 261, 315, 378

White, Lynn, 377, 378

whooping cough, 196

Wilbur, Richard, 404

wilderness, 360
  protection of, 255
  Snyder's use of word, 39

Wilderness Act of 1964, 319, 330

wilderness-protection policies, classical environmentalism and, 13

Wilderness Society, 290, 325

wildfires, 147, 157

wildlife, illegal imports of, 251

wild places, 39, 40, 40–41

William and Flora Hewlett Foundation, environment program at, 343–44

Williams College, 116, 268

Williamstown, Massachusetts, snowfall in, 1900-2002, 37, 37

Wilson, Edward O., 102, 112

windmills, 358, 379

wind power, 176, 179, 185, 387

winds
  Jupiter and patterns of, 111, 111
  prevailing, 108
  soil erosion and, 133
  trade winds, 108

winter, in New England, invisible present of, 37

wireless communications, auctioning of electromagnetic spectrum and, 60

wisdom, liberal education and, 427

Wkipedia.org, 337

wolves
  ancestors of, 117
  deer populations and, 226–27, 230, 290

women
  childbearing range for, 200
  falling birth rate and status of, 197, 204, 264
  wood gathering by, 268

women's movement, 289, 301

wood
  as fuel source, 387
  as renewable energy source, 176–77

wood gatherers, 268
  in Java, Indonesia, 270

wood products, population, affluence and consumption of, 387

Woods, Tiger, 139

Wordsworth, William, 20

"work that does no damage," finding, 280, 281, 335